Crystal Physics of Interaction Processes

PURE AND APPLIED PHYSICS

A SERIES OF MONOGRAPHS AND TEXTBOOKS

CONSULTING EDITORS

H. S. W. MASSEY

University College, London, England

KEITH A. BRUECKNER

University of California, San Diego
La Jolla, California

Crystal Physics of Interaction Processes

WARREN P. MASON

DEPARTMENT OF ENGINEERING AND APPLIED SCIENCE
COLUMBIA UNIVERSITY
NEW YORK, NEW YORK

1966

ACADEMIC PRESS   New York and London

ACADEMIC PRESS, INC.
111 Fifth Avenue, New York, New York 10003

United Kingdom Edition published by
ACADEMIC PRESS, INC. (LONDON) LTD.
Berkeley Square House, London W.1

LIBRARY OF CONGRESS CATALOG CARD NUMBER: 65-26402

Second Printing, 1969

PRINTED IN THE UNITED STATES OF AMERICA

Preface

The use of the crystalline properties of materials in research and industry is becoming increasingly widespread. The largest application is probably the use of piezoelectric crystals in selective wave filters, in frequency control, and in transducers for transforming electric energy into mechanical energy and vice versa; many other crystalline properties, however, are rapidly growing in importance through their application in light modulators, in photoelastic analysis of stresses, as thermoelectric devices, as piezoresistive strain gages, and as Hall-effect and magnetoresistance transducers.

When the physical constants of these crystalline properties are expressed in tensor form, the many similarities between them are brought out. This book aims at a unified presentation of the tensor description of these physical constants. Another fundamental building block of crystal physics is the thermodynamic expressions for various combinations of the electric, magnetic, elastic, and thermal energies, which allow one to determine all the equilibrium properties by simple tensor derivatives of a set of thermodynamic potentials. The terms present in a general tensor description are uniquely determined when the point-group symmetry is known. There are thirty-two such symmetry groups, which—as discussed in Chapter 5—are determined by the elements of symmetry specified with respect to a point in the center of the unit cell. Part I of this book describes these three fundamental building blocks of crystal physics.

Part II of this book applies the fundamental development of Part I to those properties that can be described by equilibrium thermodynamics. These include the properties of piezoelectric crystals used both as resonators and transducers, and the electrostrictive and magnetostrictive materials formed from sintered ceramics or polycrystalline metals. The latter are important transducer materials that can be given crystalline properties by a poling method that aligns crystalline domains in a direction parallel to the poling force. Crystal optics are included in this section because the ordinary and extraordinary rays present in crystalline materials are equilibrium properties determined by the symmetries of the crystal. The effect of electric fields—electro-optic effect—and mechanical stresses—piezo-optic effect—is to break up the light propagated along an optic axis into ordinary and extraordinary rays. These rays can be made to add or subtract and are applied in electro-optic modulators and as sources for determining stress patterns in

v

69633

complicated crystalline or isotropic materials. This technique is usually called photoelasticity, and it is widely used in industry. The formulation in terms of thermodynamic potentials can be taken to higher derivatives to describe the quadratic ("Kerr") electro-optic effect, which is more sensitive than the linear electro-optic effect and which is receiving considerable application. A rotation of the plane of polarization of light waves in the presence of a magnetic field can also be treated by equilibrium thermodynamics, as discussed in Chapter 8.

Transports of heat and electric charges are crystalline properties that cannot be described by reversible thermodynamics. Although steady state conditions can be realized, there is a continuous change in one direction in the sources and sinks of these quantities. To obtain expressions for these properties in crystalline media, it is necessary to use Onsager's principle, which is applicable to nonequilibrium thermodynamics. Part III of this book describes the flow of heat and electricity alone and coupled together through the thermoelectric effect. The flow of electricity is affected by stress—the piezoresistive effect—and by magnetic fields, which produce the Hall effect and the magnetoresistance effect. These effects are described in Chapters 10 and 11, and practical applications are described for various types of transducers associated with these effects. Since all useful devices based on these effects are made of semiconducting materials, a short description of the physical properties of semiconductors is given. When magnetic fields are applied to thermoconducting materials, there arise a number of new properties, which are discussed in Chapter 12. These include the Ettinghausen effect, the Nernst effect, the Righi-Leduc effect, and the magnetothermo-resistive effect, all of which can be expressed in terms of the symmetry properties of crystals.

This book grew out of a seminar presented at the invitation of the Institute for the Study of Fatigue and Reliability of the Department of Civil Engineering and Engineering Mechanics of the School of Engineering and Applied Science of Columbia University. The level of the treatment is well within the grasp of seniors or graduate students. The author has also had in mind the research worker in solid-state physics, metallurgy, or engineering mechanics who needs an introductory text on the uses of tensors, thermodynamic descriptions of properties, and the effects of crystal symmetries on these properties. The rationalized meter-kilogram-second system of units has been used throughout the book since this system is receiving growing acceptance from engineers and physicists. The system of notation used is that sponsored by the piezoelectric crystals committee of the Institute of Electrical and Electronic Engineers; because most users of piezoelectric crystals and transducers have adopted this notation, it appears to be the logical one for crystal properties. The principal differences from other notations is that the letters T_{ij} and S_{kl} denote stresses and strains, while the Greek capital letter Θ and the Greek letter σ denote absolute temperatures and entropies.

Most of the text has been written at Bell Telephone Laboratories where, for many years, the writer has been associated with piezoelectric research and research on the mechanical properties of crystals and other solids. I should like to express my gratitude to Drs. H. F. Tiersten and R. N. Thurston for many helpful discussions. At Columbia University I should like to express my gratitude to Professors A. M. Freudenthal and R. D. Mindlin for encouraging me to write the book and to present the course there. Thanks are due to Mr. Arthur Rosenburg for providing an index to the volume and for eliminating some errors in the proof.

Columbia University W. P. M.
New York
February, 1966

Contents

Part II—Application of Equilibrium Interaction Processes

CHAPTER 8.   **Optical Activity and the Faraday Effect**

Part III—Transport Crystalline Properties That Depend on Irreversible Thermodynamics

CHAPTER 9.   **Thermal and Electrical Conductivities**

CHAPTER 10.   **Piezoresistivity**

Crystal Physics of Interaction Processes

CHAPTER 1 • Introduction

1.1 Principal Interaction Phenomena

The interactions of mechanical waves with the electric and magnetic properties of solids have been used extensively in studying and utilizing the wave propagation in these solids. Examples are the use of piezoelectric crystals to detect the presence and to evaluate the stress-time characteristics of the stress waves propagated in the linear region[1] and the plastic region[2] of solids. Such crystals and polarized ferroelectric ceramics have been used as spark generators for gasoline engines and as detonators for hand grenades and for various types of bombs.

Conversely, mechanical waves have been generated in gases, liquids, and solids by piezoelectric crystals, polarized ferroelectric ceramics, and polarized ferromagnetic materials. Almost all underwater sound transducers for the detection of submarines make use of such interaction phenomena, and power levels up to the megawatt range have been maintained for short pulse times. Nonmilitary uses for transducer-generated waves in solids are dispersive and nondispersive delay lines[3] used for information storage. These lines find uses in radar systems and in certain types of computers. An even larger use is in very selective mechanical and electromechanical filters[4] that can separate one conversation or voice channel from as many as 2000 other conversations carried over a single cable or radio beam. This separation, which depends on a frequency difference between the various channels, results from the very low dissipation inherent in mechanically vibrating

---

[1] For properties of quartz under transient stresses, see Graham [1].
[2] For properties of tourmaline under hydrostatic pressure and uniaxial stress, see Hearst *et al.* [2].
[3] See May [3].
[4] See Mason [4].

1

systems, and the high temperature-frequency stability inherent in certain cuts of quartz crystals.

Other interaction effects are also of great importance in material mechanics. The photoelastic effect[5] has been used for many years to determine stresses and stress distributions in complicated sections for which they cannot be calculated. In this effect the stress causes a birefringence that results in a light wave being broken up into an ordinary ray and an extraordinary ray, which are propagated with slightly different velocities. An optic pattern results from the interference of such rays. The pattern depends on the stress distribution throughout the sample, and this stress distribution can be analyzed from the optical pattern[6] after making use of certain boundary conditions. Similarly, the application of an electric field can produce a birefringence[7] for certain directions in particular crystals. This effect has been used to produce a very fast modulation of light that is useful in studying the rapid motion of a body. On account of the very high frequency of this effect it is receiving attention for modulating laser beams.

Magnetic effects, which have been used as measuring tools, are the Hall effect[8] and the magnetoresistance effect in certain semiconductors. In the Hall effect a voltage is generated in a direction perpendicular to the current-flow direction. The magnitude is nearly proportional to the applied magnetic flux. In the magnetoresistance effect,[9] the resistance of a sample is changed by the presence of a magnetic flux. The Hall effect has been used practically in the measurement of magnetic fluxes. Both effects have been proposed for the measurement of small displacements and in various types of transducers.

Another resistance variation, which is now being used widely in material mechanics, is the piezoresistance[10] effect. In this effect a semiconducting sample, cut in a definite direction with respect to the crystal axes, increases or decreases its resistance value when a strain is applied to the section. This type of response is similar to that occurring in an ordinary strain gage, but it is about one hundred times as sensitive. Such strain gages find applications[11] in measuring very small strains and in eliminating amplifiers when large strains are to be measured.

Temperature-interaction effects are of considerable importance in material mechanics. One of the simplest is the change in dimensions of a specimen due to a change in temperature. These changes are determined by the

---

[5] See Coker and Filon [5].

[6] *Ibid.*, Chapters IV, V, and VI.

[7] See Mason [6].

[8] See Mason *et al.* [7].

[9] See Mason [8].

[10] See Dean [9].

[11] See Dean [9].

temperature-expansion coefficients relating strains to a change in the temperature. If these strains are prevented by external constraints, stresses that may become very large develop in the sample.

Thermoelectric effects are also of interest. In the Seebeck effect differences in temperatures between junctions between two metals or semiconductors can produce voltages between the junctions. This effect has been extensively used for temperature measurements. The efficiency of conversion of thermal to electrical energy can be rather high in certain semiconducting materials, and the thermoelectric effect has been considered for electric power generation. Conversely, the flow of electric current across a junction can abstract heat from the junction, and this effect (Peltier heat effect) has been used in refrigerating devices.

These interaction effects are the principal ones of use in material mechanics. A number of other effects will be mentioned, but because they have not been applied to any extent, only the equations for them are discussed.

1.2 Interaction Diagrams

The type of relations involved in interaction phenomena can be illustrated by the diagram of Fig. 1.1. This illustrates the relations between the mechanical variables—stresses T_{ij} and strains S_{kl}—and the electrical variables—the fields E_m and the dielectric displacements D_n. The vertical connections represent the relations between the stresses and the strains on the left side and between the fields and the electric displacements on the right-hand side. There are two types of relations, depending on which variables are considered dependent and which are independent. If the arrow points to the dependent variable—for example, the stress—the relation between the stresses and the strains can be written in the form of the tensor equation

$$T_{ij} = c_{ijkl}S_{kl} \tag{1.1}$$

Tensor equations, which are discussed in the next section, are necessary because the stresses are a second-rank tensor, as are also the strains. Their ratio, the elastic stiffness constants c_{ijkl}, is a fourth-rank tensor, as indicated by the four indices. This equation is general enough to fit any type of crystal symmetry. Such considerations are essential because many of the interaction phenomena—for example, piezoelectricity—appear only for symmetries different from the isotropic case. Furthermore, the particular type of constant existing depends on the crystal symmetry.

If the strains are regarded as the dependent variables, the relation between the strains and stresses can be written

$$S_{kl} = s_{ijkl}T_{ij} \tag{1.2}$$

where the s constants are known as the elastic compliances. There are relations between the compliance and elastic constants, as discussed in detail in Section 3.6.

These equations are general if they apply to an ordinary material or crystal without any piezoelectric effect. If the material has such an effect, the relations (1.1) and (1.2) are still valid, but one has to specify in addition the

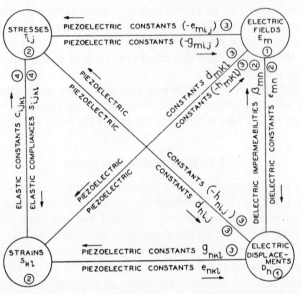

FIG. 1.1 Square showing relations between mechanical and electrical variables. Numbers indicate order of tensor representation.

electrical state of the crystal. The two extreme cases are given by the conditions that the crystal has no plating so that no flow of charge occurs anywhere in the crystal—constant electric displacement D—and by the condition that the crystal is fully plated on two sides and the two platings are shorted—constant E. Intermediate cases occur, which, however, can be related to the two extreme conditions. Equations (1.1) and (1.2) are still valid, but one has to specify the electrical condition of the sample. This is usually done by the use of a superscript, so that the two types of elastic constants are usually written

$$c_{ijkl}^{E} \quad \text{or} \quad c_{ijkl}^{D} \tag{1.3}$$

For complicated crystal structures it may be necessary to specify which fields are constant or which displacements are constant. Both the elastic constants c_{ijkl} and the elastic compliances s_{ijkl} can be called primary constants because they are relations between the mechanical variables, the stresses and strains. They are fourth-rank tensors, as indicated by the circled numbers of Fig. 1.1.

The other set of primary constants shown by Fig. 1.1 are the dielectric constants ε_{mn} or the dielectric impermeabilities β_{mn}. These relate the electric fields and the dielectric displacements by the equation

$$E_m = \beta_{mn}D_n ; \qquad D_n = \varepsilon_{mn}E_m \qquad (1.4)$$

Because the fields and dielectric displacements are vectors (first-rank tensors), the dielectric and impermeability constants are second-rank tensors. These relations hold for nonpiezoelectric crystals; but when there is a coupling to the mechanical variables through the piezoelectric constants, it is necessary to specify the states of the mechanical variables. Again the extreme cases are when the stresses T_{ij} or the strains S_{kl} are invariant. These conditions are represented by the superscripts

$$\beta_{mn}^T \quad \text{or} \quad \beta_{mn}^S \quad \text{and} \quad \varepsilon_{mn}^T \quad \text{or} \quad \varepsilon_{mn}^S \qquad (1.5)$$

In a piezoelectric crystal there are, in addition to the primary constants, a set of interaction or cross constants that relate the mechanical variables to the electrical variables. Again depending on which variables are considered dependent and which independent there are four sets of interaction parameters known as piezoelectric constants. As shown by the circled numbers, these are third-rank tensors, since they relate a second-rank tensor for the mechanical variable to a first-rank tensor for an electrical variable. For example, if the strains and electric displacements are regarded as the dependent mechanical and electrical variables and the stresses and electric fields are regarded as the independent variables, the relations can be written in the tensor form

$$S_{kl} = s_{ijkl}^E T_{ij} + d_{mij}E_m$$
$$D_n = \varepsilon_{mn}^T E_m + d_{nij}T_{ij} \qquad (1.6)$$

Writing the same set of piezoelectric constants in the relation between the strain and fields at constant stress and the relation between the electric displacements and the stresses measured at constant electric field implies a relation that is later proved. It turns out that the implied relationship is a result of the Maxwell relations in thermodynamics. In fact, as discussed in Chapter 4, thermodynamic derivations are the simplest means for determining the forms of the relations between the different variables. They are not able to determine any values for the resulting constants—this usually involves physical concepts that are beyond the scope of the present book—but are the most general and simplest method for determining relationships. There are three other sets of equations that involve the piezoelectric constants

e_{mij}, g_{mij}, and h_{nij}, all of which are obvious from Fig. 1.1. Sometimes the signs given are negative in order to agree with transformations from one set of equations to another.

Another set of variables of interest in piezoelectric crystals are the temperature variables—the temperature θ and the entropy σ. The symbols T and S have not been used for these because they conflict with the symbols for stress and strain. This follows the nomenclature standardized by the IEEE. Ordinarily it is not necessary to designate the thermal state of a vibrating piezoelectric crystal because the alternations of stress are so rapid that there is not time for the interchange of heat between the hot and cold parts and the conditions are adiabatic. This condition is designated by writing the elastic constants with a superscript σ to indicate an adiabatic condition—that is,

$$c_{ijkl}^{E,\sigma} \tag{1.7}$$

The superscript σ is usually understood and not written. If the elastic constants are measured from the resonant frequencies of plated crystals, the superscript E results.

Elastic constants, however, are often measured by applying static stresses and observing the resulting strains. The elastic constants are then isothermal. Since they are usually measured for unplated crystals, the constants measured are usually

$$c_{ijkl}^{D,\Theta} \tag{1.8}$$

where the superscript Θ designated constant temperature. It is then necessary to relate the two types of constants. The methods for doing this are discussed in Chapter 4.

The increased number of relations involving the three sets of variables is indicated by the interaction diagram of Fig. 1.2. In addition to the four forms of the piezoelectric relations, the elasticity, and the dielectric properties, there are thermal-expansion coefficients α_{ij}, thermal stresses, the piezocaloric effect, the heat of deformation, the heat capacity C, the heat of polarization, and the pyroelectric effect between the field and the entropy designated by the variable q_m. All of these relations are of some importance in the study of piezoelectric and ferroelectric crystals. They are most easily discussed by means of reversible thermodynamics.

If we replace the electric field and electric displacement by the magnetic field H and the magnetic flux B, a similar diagram can be obtained. The relations between the mechanical variables and the magnetic variables result in piezomagnetic relations that were first discussed by Voigt.[12] For single crystals no piezomagnetic constant has ever been shown to be large

---

[12] See Voigt [10].

enough to measure. However, if one polarizes a ferromagnetic substance such as nickel or a ferrite, measurable ratios exist between the strain and magnetic flux, for example, which follow the same rules as for the piezo-magnetic constants of Voigt. Following the standard adopted by the IEEE

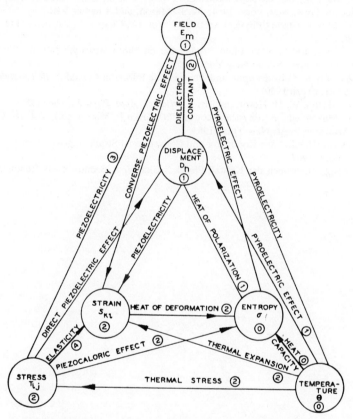

FIG. 1.2 Triangle showing relations between mechanical, electrical, and thermal variables.

committee on piezoelectric crystals,[13] such materials are called piezomagnetic; they are discussed in Chapter 4.

Interaction diagrams are useful only for processes that are thermo-dynamically reversible, since they imply a reciprocal action between the two types of variables. Many other interaction phenomena are also of interest, but they usually require nonequilibrium thermodynamics and the associated Onsager relations. Some of these are discussed in later chapters.

[13] Standard in course of adoption.

REFERENCES

1. R. A. Graham, *J. Appl. Phys.* **33,** 1755 (1962).
2. R. Hearst, L. B. Geesman, and D. V. Power, "Piezoelectric Sensitivity and Reproducibility of Z-cut Tourmaline," *J. Appl. Phys.*, to be published.
3. J. E. May, Jr., *in* "Physical Acoustics" (W. P. Mason, ed.), Vol. IA, Chapter VI. Academic Press, New York, 1964; W. P. Mason, *ibid*, Chapter VII.
4. W. P. Mason. ed., *in* "Physical Acoustics," Vol. IA, Chapter V. Academic Press, New York, 1964.
5. E. G. Coker and L. N. G. Filon, "A Treatise on Photoelasticity," 2nd ed. Cambridge Univ. Press, London and New York, 1955.
6. W. P. Mason, "Electro-optic and Photoelastic Effects in Crystals," *Bell System Tech. J.* **29,** 161, (April 1950).
7. W. P. Mason, W. H. Hewitt, and R. F. Wick, *J. Appl. Phys.* **24,** 166 (1953).
8. W. P. Mason, ed., *in* "Physical Acoustics" (Warren P. Mason, ed.), Vol. IB, Chapter X. Academic Press, New York, 1964.
9. M. Dean, III, ed., "Semiconductor and Conventional Strain Gages." Academic Press, New York, 1962.
10. W. Voigt, "Lehrbuch der Kristall Physik," Chapter 8, Section 8. B. Teubner, 1928.

Part I

General Principles

CHAPTER 2 • Introduction to Tensor Analysis

As discussed in the preceding chapter, most of the interaction phenomena occur only in nonisotropic materials. Although it is possible to deal with the resulting relations by writing down a number of simultaneous equations, a great simplification occurs when the equations are written in tensor form. Tensors also facilitate the transference of equations from one set of coordinates to a set rotated with respect to the first set. For the most general case of coordinate transformation, it is necessary to preserve the covariant or contravariant nature of the transformation; but when transformations are made from one orthogonal set to another orthogonal set, there is no difference and a simplified notation involving Cartesian tensors[1] is adequate. Because Cartesian tensors are adequate to transform from a set of coordinates Ox, Oy, Oz of Fig. 2.1 to a spherical or cylindrical set of coordinates, they are general enough for all the applications covered in this book, and discussion will be limited to this tensor form.

2.1 Definitions of Tensors

If we have two sets of rectangular axes Ox, Oy, Oz and Ox', Oy', Oz', as shown in Fig. 2.1, which have the same origin O, the coordinates of any point P with respect to the second set are given in terms of the first set of axes by the equations

$$x' = l_1 x + m_1 y + n_1 z$$
$$y' = l_2 x + m_2 y + n_2 z \qquad (2.1)$$
$$z' = l_3 x + m_3 y + n_3 z$$

---

[1] See Jeffreys [1] for an exposition of Cartesian tensors.

The quantities $(l_1, \ldots, n_3)$ are the cosines of the angles between the new and old axes, related according to the table

	x	y	z
x'	l_1	m_1	n_1
y'	l_2	m_2	n_2
z'	l_3	m_3	n_3

$$\text{(2.2)}$$

By solving Eqs. (2.1) simultaneously, the coordinates of the point P can be expressed in terms of the new variables by the equations

$$x = l_1 x' + l_2 y' + l_3 z'$$
$$y = m_1 x' + m_2 y' + m_3 z' \tag{2.3}$$
$$z = n_1 x' + n_2 y' + n_3 z'$$

The writing of Eqs. (2.1) and (2.3) can be shortened considerably by changing the notation. Instead of x, y, z let us write x_1, x_2, x_3 and in place

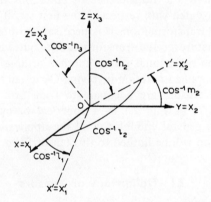

FIG. 2.1   Fixed and rotated axes showing definitions of direction cosines.

of x', y', z' we write x_1', x_2', x_3'. We can now say that the coordinates with respect to the first system are x_i, while those with respect to the second system are x_j' where i and j may take the values 1, 2, and 3. Then in (2.1) each coordinate x_j' is expressed as the sum of three terms involving the three values of x_i. Each x_i is multiplied by the cosine of the angle between the directions of x_i and x_j' both increasing. Let us denote this cosine by a_{ij}. Then for all values of j

$$x_j' = a_{1j} x_1 + a_{2j} x_2 + a_{3j} x_3 = \sum_{i=1}^{3} a_{ij} x_i \tag{2.4}$$

Conversely Eqs. (2.3) can be written

$$x_i = a_{1i}x_1' + a_{2i}x_2' + a_{3i}x_3' = \sum_{j=1}^{3} a_{ji}x_j \qquad (2.5)$$

where the a_{ij} have the same values as in (2.4) for the same values of i and j. Such a set of three quantities involving a relation between two coordinate systems is called a tensor of the first rank or a vector.

We note that each of Eqs. (2.4) and (2.5) is a set of three equations in which a summation of three terms occurs on the right-hand side. When such a summation occurs, a suffix j is repeated in the expression for the general term $a_{ij}x_j'$. Making use of the Einstein summation convention, which states that when a letter suffix occurs twice in the same term, summation with respect to this suffix is to be automatically understood, then (2.4) and (2.5) can be written simply as

$$x_j' = a_{ij}x_i \qquad (2.4a)$$
$$x_i = a_{ij}x_j' \qquad (2.5a)$$

Tensors of rank up to the sixth occur in interaction processes. There are single quantities such as mass and distance, which are the same for all systems of coordinates. These are called tensors of the zero rank or scalars. A second-rank tensor is a set of nine quantities referred to one set of axes, which can be transformed to another set of axes by the rule

$$w_{jl}' = a_{ij}a_{kl}w_{ik} \qquad (2.6)$$

Tensors of the second rank can arise also as the product of two vectors. Suppose that the components of one vector are multiplied by each component of the other; then we obtain a set of nine quantities expressed by u_iv_k, where i and k are independently given the values 1, 2, and 3. These components can be transformed to another set of axes by applying the transformation equations for vectors (2.4a). The components $u_j'v_l'$ in the new coordinates will be

$$u_j'v_l' = w_{jl}' = (a_{ij}u_i)(a_{kl}v_k) = a_{ij}a_{kl}(u_iv_k) = a_{ij}a_{kl}w_{ik} \qquad (2.7)$$

Higher-order tensors can be obtained by taking the product of more vectors. Thus a set of 3^n quantities that transforms like the vector product of n vectors is called a tensor of rank n, where n is the number of factors.

2.2 Symmetry Properties of Tensors

In Eq. (2.7) the right-hand side has the "dummy" suffixes, which are given the numbers 1 to 3 and summed. It therefore makes no difference which we call i and which k, so that

$$w_{jl}' = a_{ij}a_{kl}w_{ik} = a_{kj}a_{il}w_{ki} \qquad (2.8)$$

Hence w_{ki} transforms by the same rules as w_{ik} and hence is a tensor of the second rank. The importance of this is that if we have a set of parameters given by (2.9a), which we know to be a tensor of the second rank, the set of quantities in (2.9b) is also a tensor of the second rank

$$\begin{vmatrix} w_{11} & w_{12} & w_{13} \\ w_{21} & w_{22} & w_{23} \\ w_{31} & w_{32} & w_{33} \end{vmatrix} \tag{2.9a}$$

$$\begin{vmatrix} w_{11} & w_{21} & w_{31} \\ w_{12} & w_{22} & w_{32} \\ w_{13} & w_{23} & w_{33} \end{vmatrix} \tag{2.9b}$$

Hence the sum $(w_{ik} + w_{ki})$ and the difference $(w_{ik} - w_{ki})$ are also tensors of the second rank. The first of these has the property that it is unaltered by interchanging i and k, and therefore it is called a symmetric tensor. The second tensor has its components reversed in sign when i and k are interchanged and is therefore an antisymmetric tensor. In the antisymmetric tensor it is obvious that the leading diagonal terms will all be zero, whereas the terms on the top side of this diagonal will have the reverse sign from those on the bottom. Now since

$$w_{ik} = \tfrac{1}{2}(w_{ik} + w_{ki}) + \tfrac{1}{2}(w_{ik} - w_{ki}) \tag{2.10}$$

any tensor of the second rank can be considered as the sum of a symmetric tensor and an antisymmetric tensor. Most tensors expressing interaction properties are symmetrical. Exceptions are in such phenomena as thermoelectricity, magneto- and electrostriction, and so on.

2.3 Contraction of Tensors

Another operation of interest is the one of putting two suffixes in the tensor equal and summing the terms. This is known as the contraction of a tensor; it results in a tensor two ranks lower than the original one. For example, if we contract the second-rank tensor $E_i E_j$ by putting $i = j$, this results in a scalar

$$E_1{}^2 + E_2{}^2 + E_3{}^2 = |E|^2 \tag{2.11}$$

which gives the square of the absolute value of the vector E_i. This is a tensor of zero rank, which is invariant to the coordinate system.

2.4 Transformation of Tensors from One Set of Axes to Another

One of the great advantages of writing relations in the form of tensor equations is that it is a relatively simple matter to transform the quantities into another set of coordinates. For a tensor of the first rank—a vector— the form is given by Eqs. (2.1) and the reverse form by Eqs. (2.3). The direction cosines can be expressed in the form of a table.

	x_1	x_2	x_3
x_1'	$l_1 = \dfrac{\partial x_1'}{\partial x_1} = \dfrac{\partial x_1}{\partial x_1'} = a_{11}$;	$m_1 = \dfrac{\partial x_1'}{\partial x_2} = \dfrac{\partial x_2}{\partial x_1'} = a_{21}$;	$n_1 = \dfrac{\partial x_1'}{\partial x_3} = \dfrac{\partial x_3}{\partial x_1'} = a_{31}$
x_2'	$l_2 = \dfrac{\partial x_2'}{\partial x_1} = \dfrac{\partial x_1}{\partial x_2'} = a_{12}$;	$m_2 = \dfrac{\partial x_2'}{\partial x_2} = \dfrac{\partial x_2}{\partial x_2'} = a_{22}$;	$n_2 = \dfrac{\partial x_2'}{\partial x_3} = \dfrac{\partial x_3}{\partial x_2'} = a_{32}$
x_3'	$l_3 = \dfrac{\partial x_3'}{\partial x_1} = \dfrac{\partial x_1}{\partial x_3'} = a_{13}$;	$m_3 = \dfrac{\partial x_3'}{\partial x_2} = \dfrac{\partial x_2}{\partial x_3'} = a_{23}$;	$n_3 = \dfrac{\partial x_3'}{\partial x_3} = \dfrac{\partial x_3}{\partial x_3'} = a_{33}$

$$(2.12)$$

These equations are obvious from Fig. 2.1. For example the direction cosine n_2 is either the projection of the unit vector along the x_3 axis on the x_2' axis, or the projection of a unit vector along the x_2' axis on the x_3 axis. These projections are given respectively by

$$\frac{\partial x_2'}{\partial x_3} \quad \text{and} \quad \frac{\partial x_3}{\partial x_2'} \tag{2.13}$$

as in Eqs. (2.12).

Using these forms, Eqs. (2.5a) and (2.5b) can be written

$$x_j' = \frac{\partial x_j'}{\partial x_i} x_i = a_{ij} x_i ; \qquad x_i = \frac{\partial x_i}{\partial x_j'} x_j' = a_{ij} x_j' \tag{2.14}$$

Similarly, because a tensor of the second rank can be regarded as the product of two vectors, it can be transformed according to the equation

$$x_j' x_l' = \left(\frac{\partial x_j'}{\partial x_i} x_i \right) \left(\frac{\partial x_l'}{\partial x_k} x_k \right) = \frac{\partial x_j'}{\partial x_i} \frac{\partial x_l'}{\partial x_k} x_i x_k \tag{2.15}$$

which can also be expressed in the generalized form

$$w_{jl}' = \frac{\partial x_j'}{\partial x_i} \frac{\partial x_l'}{\partial x_k} w_{ik} \tag{2.16}$$

In general the transformation equation of a tensor of the nth rank can be written

$$x_{k1}' \cdots x_{kn}' = \frac{\partial x_{k_1}'}{\partial x_{j_1}} \frac{\partial x_{k_2}'}{\partial x_{j_2}} \cdots \frac{\partial x_{k_n}'}{\partial x_{j_n}} x_{j_1} x_{j_2} \cdots x_{j_n} \tag{2.17}$$

2.5   Vector and Scalar Products—Gradient, Divergence, and Curl of Vectors

The scalar and vector products of two vectors are quantities of interest in vector analysis. They can be represented by the reduced tensor product (2.18a) and the twice-reduced fifth-rank tensor product (2.18b)

$$p_i q_i \qquad \qquad (2.18a)$$

$$\varepsilon_{ijk} p_j q_k \qquad \qquad (2.18b)$$

where ε_{ijk} is the rotation tensor, which is zero if $i = j$, $i = k$, or $j = k$ but has the values ± 1 if all the suffixes are different. If they are in rotation— that is, 123, 231, or 312—the value is $+1$; if not, the value is -1.

The reduced tensor of (2.18a) has the scalar value

$$A = p_1 q_1 + p_2 q_2 + p_3 q_3 \qquad \qquad (2.18c)$$

Since A will have the same value in any set of coordinates, we can pick one for which the vector p_i lies along the $Ox_1{}'$ axis. The components of p_i will be $p_1{}'$, whereas those of q_i will be $q_1{}', q_2{}', q_3{}'$. The reduced tensor has the value

$$A = p_1{}' q_1{}' = |p|\, |q| \cos \alpha \qquad \qquad (2.19)$$

since $p_1{}'$ is the absolute value of p and $q_1{}'$ is the absolute value of q times the direction cosine of α, the angle between $p_1{}'$ and $q_1{}'$.

Making use of the fact that $\varepsilon_{ijk} = 0$ unless all the suffixes are different and equals 1 if the suffixes are in rotation and -1 if they are out of rotation, Eq. (2.18b) defines a vector r_i having the components

$$r_1 = p_2 q_3 - q_2 p_3 \; ; \qquad r_2 = p_3 q_1 - p_1 q_3 \; ; \qquad r_3 = p_1 q_2 - p_2 q_1 \quad (2.20)$$

If we pick a coordinate system for which q lies along the x_3 axis while p makes an angle α in a plane making an angle φ with the x_1 axis—as shown by Fig. 2.2—the components of p will be

$$p_1 = |p| \sin \alpha \cos \varphi; \qquad p_2 = |p| \sin \alpha \sin \varphi; \qquad p_3 = |p| \cos \alpha \qquad (2.21)$$

Since $q_1 = q_2 = 0$, the components of the vector r are, from (2.20),

$$r_1 = [|p|\, |q| \sin \alpha] \sin \varphi; \qquad r_2 = -[|p|\, |q| \sin \alpha] \cos \varphi; \qquad r_3 = 0 \quad (2.22)$$

The absolute value of r is

$$|r| = \sqrt{r_1{}^2 + r_2{}^2} = |p|\, |q| \sin \alpha \qquad \qquad (2.23)$$

and the direction of r_i is perpendicular to the plane containing q_k and p_j, in agreement with the definition of the vector product of two vectors. The perpendicularity follows from the fact that the direction cosines between r_1 and x_1 and x_2 are

$$\cos \psi = \cos (90° - \varphi) = \sin \varphi; \qquad -\sin \psi = -\sin (90° - \varphi) = -\cos \varphi$$

$$(2.24)$$

in agreement with the values in Eq. (2.22). As will be discussed later, the axes x_1', x_2', $x_3 = x_3'$ form a right-handed coordinate system.

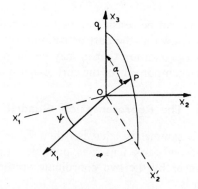

FIG. 2.2 Coordinate system for defining the vectors p and q for a vector product.

Differential operators can also be expressed in tensor form. The simplest one is the gradient of a scalar. If U is the scalar quantity, its gradient will be $\partial U/\partial x_i$ or $\partial U/\partial x_j'$, depending on the set of axes used. But since

$$\frac{\partial U}{\partial x_j'} = \frac{\partial x_i}{\partial x_j'} \frac{\partial U}{\partial x_i} = a_{ij} \frac{\partial U}{\partial x_i} \qquad (2.25)$$

the gradient transforms according to the vector rule.

Similarly, the gradient of a vector is a tensor of rank two. For if u_i and u_j' are the components of a vector with respect to two different sets of axes having the components x_i' and x_k, we have

$$\frac{\partial u_j'}{\partial x_i'} = \frac{\partial x_k}{\partial x_i'} \frac{\partial u_j'}{\partial x_k} = a_{ki} \frac{\partial}{\partial x_k} (a_{ij} u_i) = a_{ij} a_{ki} \frac{\partial u_i}{\partial x_k} \qquad (2.26)$$

and hence the gradient of a vector transforms like a second-rank tensor.

The tensor $\partial u_i/\partial x_k$ gives a scalar on contraction that is known as the divergence of u_i. This has the value

$$\frac{\partial u_i}{\partial x_i} = \frac{\partial u_1}{\partial x_1} + \frac{\partial u_2}{\partial x_2} + \frac{\partial u_3}{\partial x_3} = \text{divergence of } u_i \qquad (2.27)$$

The tensor $\partial u_j / \partial x_k$ can have symmetric and antisymmetric parts, which are

$$\frac{1}{2}\left(\frac{\partial u_j}{\partial x_k} + \frac{\partial u_k}{\partial x_j}\right) \quad \text{and} \quad \frac{1}{2}\left(\frac{\partial u_j}{\partial x_k} - \frac{\partial u_k}{\partial x_j}\right) \tag{2.28}$$

The first form is of use in defining strains, as will be discussed in Chapter 3. Since the components of the antisymmetric form of (2.28) can be obtained from the equation

$$V_i = \varepsilon_{ijk}\frac{\partial u_k}{\partial x_j} \tag{2.29}$$

the antisymmetric tensor can be considered to be associated with the vector V_i. The operation is known as the curl or rotation of u_i. The vanishing of the curl is a necessary condition that u_i may be the gradient of a scalar. The values of V_i are the components of the curl of u_i around each coordinate axis.

2.6 The Representation Quadric

When a tensor T_{ij} is symmetric—that is, when $T_{ij} = T_{ji}$—a geometrical representation of a second-rank tensor is possible. Let us multiply the components of the tensor by the two coordinate vectors $x_i x_j$ and set the result equal to 1—that is,

$$T_{ij}x_i x_j = 1 \tag{2.30}$$

Performing the summations with $T_{ij} = T_{ji}$ and collecting terms, we have

$$T_{11}x_1^2 + T_{22}x_2^2 + T_{33}x_3^2 + 2T_{23}x_2 x_3 + 2T_{31}x_3 x_1 + 2T_{12}x_1 x_2 = 1 \tag{2.31}$$

This is the general equation of a second-degree surface (a quadric) referred to its center as an origin. It may in general be an ellipsoid or a hyperboloid. Furthermore, the quadric surface can be transformed from one set of axes to another by the same type of transformations that apply for a tensor. To show this we transform the vectors x_i and x_j to a new set of coordinates by employing Eqs. (2.5a). The resulting equations are

$$x_i = a_{ki}x_k' \quad \text{and} \quad x_j = a_{lj}x_l' \tag{2.32}$$

Introducing these values into Eq. (2.30) we have

$$T_{ij}\alpha_{ki}\alpha_{lj}x_k' x_l' = 1 \tag{2.33}$$

If we compare this with the tensor transformation equation (2.6) it is evident that

$$T_{ij}\alpha_{ki}\alpha_{lj} = T_{kl}' \tag{2.34}$$

and hence the quadric surface in the new coordinate system has the equation

$$T_{kl}'x_k' x_l' = 1 \tag{2.35}$$

An important property of a quadric surface is that it possesses three directions at right angles, such that when the quadric surface is referred to them its equation takes the simpler form

$$T_1 x_1{}^2 + T_2 x_2{}^2 + T_3 x_3{}^2 = 1 \qquad (2.36)$$

These directions in the quadric surface are known as the principal axes, and the components T_1, T_2, and T_3 are known as the principal components of the tensor T_{ij}. By comparing (2.36) with the standard equation for a quadric

$$\frac{x^2}{a^2} + \frac{y^2}{b^2} + \frac{z^2}{c^2} = 1 \qquad (2.37)$$

it is seen that the semiaxes of the representation quadric have the lengths $1/\sqrt{T_1}$, $1/\sqrt{T_2}$, $1/\sqrt{T_3}$. If T_1, T_2, and T_3 are all positive, the surface is an

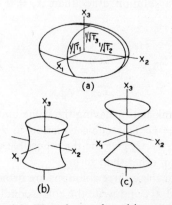

FIG. 2.3   Three forms of quadric systems.

ellipsoid, as shown by Fig. 2.3a. If two coefficients are positive and one negative the quadric surface is a hyperboloid of one sheet, Fig. 2.3b, whereas if two coefficients are negative, the surface is a hyperboloid of two sheets as shown by Fig. 2.3c. If all three coefficients are negative the surface is an imaginary ellipsoid.

2.61   Determination of the Principal Axes

From the theory of quadric surfaces it is known that the determination of the principal axes requires the solution of a cubic equation. For convenience in defining the principal axes, we make use of the relation that a vector defined by the operation

$$X_j' = T_{ij} X_i \qquad (2.38)$$

must be proportional to X_i when X_i lies along a principal axis. In this equation X_i is a vector from the origin to a point P on the quadric. This relation is obvious because, for a set of coordinates coinciding with the principal axes, Eq. (2.38) becomes

$$X_1' = T_1 X_1 ; \qquad X_2' = T_2 X_2 ; \qquad X_3' = T_3 X_3 \qquad (2.39)$$

and hence

$$\lambda X_j = T_{ij} X_i \qquad (2.40)$$

is an equation representing a solution for the three principal axes. This equation represents three homogeneous linear equations in the variable X_i, which can be written

$$0 = (T_{11} - \lambda)X_1 + T_{12}X_2 + T_{13}X_3$$
$$0 = T_{12}X_1 + (T_{22} - \lambda)X_2 + T_{23}X_3 \qquad (2.41)$$
$$0 = T_{13}X_1 + T_{23}X_2 + (T_{33} - \lambda)X_3$$

These equations have a solution other than $X_i = 0$ if the determinant of (2.41) is zero—that is,

$$\begin{vmatrix} T_{11} - \lambda & T_{12} & T_{13} \\ T_{12} & T_{22} - \lambda & T_{23} \\ T_{13} & T_{23} & T_{33} - \lambda \end{vmatrix} = |T_{ij} - \lambda \delta_{ij}| = 0 \qquad (2.42)$$

where δ_{ij} is a second-rank tensor having the value unity for any component if $i = j$ and zero if $i \neq j$. This cubic equation in λ is called the *secular equation*. The roots λ_1, λ_2, λ_3 give the three possible values of λ that insure that (2.41) has a solution other than zero.

It is easy to show that these three solutions are orthogonal. Consider any two of them, say λ_1 and λ_2, and write the corresponding vectors

$$T_{ij}X_{i_1} = \lambda_1 X_{j_1} ; \qquad T_{ij}X_{i_2} = \lambda_2 X_{j_2}. \qquad (2.43)$$

Multiplying the first equation by X_{j_2} and the second by X_{j_1} and subtracting, we have

$$T_{ij}[X_{i_1}X_{j_2} - X_{i_2}X_{j_1}] = (\lambda_1 - \lambda_2)X_{j_1}X_{j_2} \qquad (2.44)$$

Since T_{ij} is symmetrical the left-hand side is equal to zero. Hence, since $X_{j_1}X_{j_2}$ represents the scalar product of the two vectors X_{j_1} and X_{j_2}, they must be at right angles as long as $\lambda_1 \neq \lambda_2$.

The directions of the principal axes can be determined from Eq. (2.40) after the values of λ are found. If we let X_i be a unit vector, the components will be given by the direction cosines α, β, and γ. Hence (2.40) can be written

$$T_{11}\alpha + T_{12}\beta + T_{13}\gamma = \alpha\lambda$$
$$T_{12}\alpha + T_{22}\beta + T_{23}\gamma = \beta\lambda \qquad (2.45)$$
$$T_{13}\alpha + T_{23}\beta + T_{33}\gamma = \gamma\lambda$$

By taking the ratio of the first equation to the second and the first to the third it is found that

$$\alpha = \left[\frac{T_{13}(\lambda - T_{22}) + T_{12}T_{23}}{(\lambda - T_{22})(\lambda - T_{11}) - T_{12}^2}\right]\gamma; \quad \beta = \left[\frac{T_{23}(\lambda - T_{11}) + T_{12}T_{13}}{(\lambda - T_{22})(\lambda - T_{11}) - T_{12}^2}\right]\gamma$$

(2.46)

Making use of the relations

$$\alpha^2 + \beta^2 + \gamma^2 = 1 \tag{2.47}$$

all three direction cosines can be found. This is a complicated procedure and is usually not necessary for any crystals up to the triclinic type because the symmetry usually produces one unique axis that is a principal axis.

2.62 The Mohr Circle Diagram

The circular construction of Fig. 2.4, due to Otto Mohr, is used by engineers in the analysis of stress and strain. It is equally applicable for obtaining

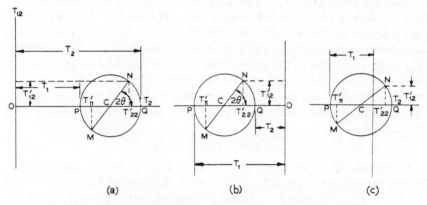

FIG. 2.4   Methods for specifying stresses using Mohr Circle diagram.

components of any other second-rank tensor. If one of the principal axes is known, it is often desirable to transform the components of a symmetrical tensor for one set of coordinates to a new set, which are obtained by rotating about the principal axes. If in Fig. 2.2 we take the x_3 axis as the principal axis and rotate the other two axes in a counterclockwise direction—that is, $\theta = \varphi; \varphi + \psi = 90°$—the direction cosines of Eq. (2.12) become

	x_1	x_2	x_3
x_1'	$\cos\theta$	$-\sin\theta$	0
x_2'	$+\sin\theta$	$\cos\theta$	0
x_3'	0	0	1

(2.48)

where θ is the counterclockwise angle of rotation. Inserting these values in the tensor transformation formula (2.16) the components of the new tensor in terms of the old axes—which were assumed to be principal axes—are

$$T'_{11} = T_1 \cos^2 \theta + T_2 \sin^2 \theta$$
$$T'_{22} = T_1 \sin^2 \theta + T_2 \cos^2 \theta \tag{2.49}$$
$$T'_{12} = (T_1 - T_2) \sin \theta \cos \theta$$

This equation can be written in the form

$$T'_{11} = \tfrac{1}{2}(T_1 + T_2) - \tfrac{1}{2}(T_2 - T_1)(\cos^2 \theta - \sin^2 \theta)$$
$$= \tfrac{1}{2}(T_1 + T_2) - \tfrac{1}{2}(T_2 - T_1) \cos 2\theta$$
$$T'_{22} = \tfrac{1}{2}(T_1 + T_2) + \tfrac{1}{2}(T_2 - T_1) \cos 2\theta \tag{2.50}$$
$$T'_{12} = \tfrac{1}{2}(T_1 - T_2) \sin 2\theta$$

This result can be expressed by using the plane diagram of Fig. 2.4a. We suppose that $T_1 < T_2$. On the horizontal axis the point P corresponds to T_1 and the point Q to T_2. The midpoint C is $\tfrac{1}{2}(T_1 + T_2)$. Now suppose that we draw a circle about C with a radius $\tfrac{1}{2}(T_2 - T_1)$. Then a diameter MCN drawn so that it makes an angle 2θ with the PQ axis, measured in a counterclockwise direction, will determine the values of T'_{11}, T'_{22}, and T'_{12}. T'_{11} is the projection of the point M on the base line while T'_{22} is the projection of the point N. Similarly, the component T'_{12} is the height of N above the base line. It will have its maximum value when $2\theta = \pi/2$ or $90°$, for which case the values of T'_{11} and T'_{22} are both equal to $\tfrac{1}{2}(T_1 + T_2)$. In fact it is seen that

$$T'_{11} + T'_{22} = T_1 + T_2 \tag{2.51}$$

Conversely, if we are given the values of T'_{11}, T'_{22}, and T'_{12} it is possible to find the values of T_1 and T_2. The point C is determined by half the value of $T'_{11} + T'_{22}$. The point N can be plotted on the diagram from the coordinates T'_{12}, T'_{22}, while M is at the position $-T'_{12}$, T'_{11}. The value of the radius of the circle is given by

$$r = CM = \sqrt{\frac{(T'_{22} - T'_{11})^2}{4} + (T'_{12})^2} \tag{2.52}$$

The principal values T_1 and T_2 are given by

$$T_1 = \tfrac{1}{2}(T'_{11} + T'_{22}) - r; \qquad T_2 = \tfrac{1}{2}(T'_{11} + T'_{22}) + r$$

and the angle θ is given by the equation

$$\tan 2\theta = \left(\frac{2T'_{12}}{T'_{22} - T'_{11}} \right) \tag{2.53}$$

Figures 2.4b and 2.4c show that the same construction works when T_1 and T_2 are both negative or when one is negative and the other positive.

2.7   Tensors Involving the Direction Cosines a_{ij}

The direction cosines a_{ij} of Eqs. (2.12) are a set of nine quantities, but there are six relations between the nine quantities. This is readily seen by considering the number of operations required to obtain the most general rotated system. If the axes Ox_1, Ox_2, Ox_3 are given, as shown by Fig. 2.5, one new axis, here labeled x_3', requires the specification of two angles, here shown as θ and φ. The other two axes are at right angles to Ox_3' and it

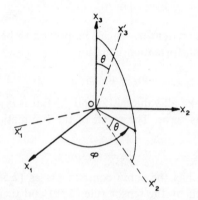

FIG. 2.5   Number of axes angles to specify an orientation.

requires only one other angle to specify the rotation of the other two axes about x_3'.

Since each row of the array of Eqs. (2.12) is a set of direction cosines of one straight line with respect to the three right-angle axes Ox_1, Ox_2, Ox_3, then the squares of the three elements in each row equal unity or

$$a_{11}^2 + a_{21}^2 + a_{31}^2 = 1; \qquad a_{12}^2 + a_{22}^2 + a_{32}^2 = 1; \qquad a_{13}^2 + a_{23}^2 + a_{33}^2 = 1$$

$$(2.54)$$

These three relations can be represented by the general relations

$$a_{ik}a_{jk} = 1 \qquad \text{if} \qquad i = j \tag{2.55}$$

Again since each new axis is perpendicular to the other two, their scalar product has to be zero. Since each axis is represented by the direction cosines of its row, the resulting equations are

$$a_{11}a_{12} + a_{21}a_{22} + a_{31}a_{23} = 0$$

$$a_{11}a_{13} + a_{21}a_{23} + a_{31}a_{33} = 0 \tag{2.56}$$

$$a_{12}a_{13} + a_{22}a_{23} + a_{32}a_{33} = 0$$

Equations (2.54) and (2.56), which are called the *orthogonality relations*, can be combined into a single equation

$$a_{ik}a_{jk} = \delta_{ij} \quad \text{where } \delta_{ij} = 1 \text{ if } i = j \text{ and } 0 \text{ if } i \neq j. \quad (2.57)$$

The components of the δ_{ij} tensor are

$$\begin{vmatrix} 1 & 0 & 0 \\ 0 & 1 & 0 \\ 0 & 0 & 1 \end{vmatrix} \quad (2.58)$$

and it is called a unitary tensor. It can be proved to be a tensor because it satisfies the tensor transformation equation

$$\delta'_{jl} = a_{ij}a_{kl}\,\delta_{ik} \quad (2.59)$$

The suffix k has to take all values 1, 2, and 3. But if $k \neq i$, $\delta_{ik} = 0$ and the corresponding term is zero. The result of the summation with regard to k is

$$\delta'_{jl} = a_{ij}a_{il} \quad (2.60)$$

But from (2.57) this has the same components as (2.58), and hence δ_{ij} is transformed into itself by the tensor rule (2.59) and therefore is a tensor of the second rank.

The tensor δ_{ij} operating on the vector p_i can be written

$$\delta_{ij}p_i = p_j \quad (2.61)$$

since δ_{ij} is zero unless $i = j$. A similar transformation occurs to a second-rank tensor:

$$\delta_{ij}T_{jl} = T_{il} \quad \text{and} \quad \delta_{il}T_{jl} = T_{jl} \quad (2.62)$$

Because of this convenient property δ_{ij} is sometimes called the subsitution tensor.

2.71   The Value of $|a_{ij}|$

We have so far considered only right-handed systems of coordinates, such as shown by Fig. 2.6a. Some crystal operations, such as a reflection in a plane or an inversion through a center, change the handedness of the axes from the right-handed systems of Fig. 2.6a to the left-handed systems of Fig. 2.6b. We wish now to show that the determinant of the transformation tensor a_{ij}, which we write as $|a_{ij}|$, is equal to $+1$ if the handedness of the system does not change, but is equal to -1 if the handedness of the axes changes.

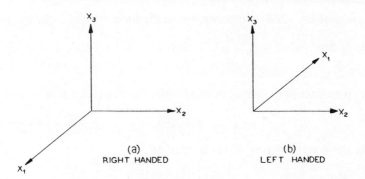

FIG. 2.6   Right- and left-handed systems of axes.

The determinant of the tensor a_{ij} is

$$|a_{ij}| = \begin{vmatrix} a_{11} & a_{12} & a_{13} \\ a_{21} & a_{22} & a_{23} \\ a_{31} & a_{32} & a_{33} \end{vmatrix} \tag{2.63}$$

$$= a_{11}(a_{22}a_{33} - a_{23}a_{32}) - a_{21}(a_{12}a_{33} - a_{32}a_{13}) + a_{31}[a_{12}a_{23} - a_{22}a_{13}]$$

We wish now to discuss the value of the determinant, following the work of Jeffries.[2]

We can solve the first two equations of (2.56) for the ratios of a_{11}, a_{21}, a_{31}, obtaining

$$\frac{a_{11}}{a_{32}a_{23} - a_{33}a_{22}} = \frac{a_{21}}{a_{33}a_{12} - a_{13}a_{32}} = \frac{a_{31}}{a_{13}a_{22} - a_{23}a_{12}} = -k \tag{2.64}$$

Substituting these values back in the first equation of (2.54),

$$1 = -k[a_{11}(a_{32}a_{23} - a_{33}a_{22}) + a_{21}(a_{33}a_{12} - a_{13}a_{32}) + a_{31}(a_{13}a_{22} - a_{23}a_{12})$$

$$= +k \begin{vmatrix} a_{11} & a_{12} & a_{13} \\ a_{21} & a_{22} & a_{23} \\ a_{31} & a_{32} & a_{33} \end{vmatrix} \tag{2.65}$$

We have also by squaring the relations in (2.64) and substituting in the first of (2.54)

$$k^2[(a_{32}a_{23} - a_{33}a_{22})^2 + (a_{33}a_{12} - a_{13}a_{32})^2 + (a_{13}a_{22} - a_{23}a_{12})^2] = 1 \tag{2.66}$$

We now make use of the identity

$$(a^2 + b^2 + c^2)(a'^2 + b'^2 + c'^2) - (aa' + bb' + cc')^2$$
$$= (bc' - cb')^2 + (ca' - ac')^2 + (ab' - ba')^2 \tag{2.67}$$

[2] See Jeffreys [1], p. 9.

to show that Eq. (2.66) can be written in the form

$$k^2[(a_{13}^2 + a_{23}^2 + a_{33}^2)(a_{12}^2 + a_{22}^2 + a_{32}^2) - (a_{13}a_{12} + a_{23}a_{22} + a_{33}a_{32})^2] = 1 \tag{2.68}$$

The first two terms are equal to unity while the third term is zero, and hence

$$k = \pm 1 \tag{2.69}$$

Hence the determinant of (2.65) is equal to

$$|a_{ij}| = \pm 1 \tag{2.70}$$

If two rows of $|a_{ij}|$ are interchanged, the sign reverses. Hence the sign of $|a_{ij}|$ depends on how the axes are numbered. If we apply the identical transformation

$$\delta_{ij} = \begin{vmatrix} 1 & 0 & 0 \\ 0 & 1 & 0 \\ 0 & 0 & 1 \end{vmatrix} \tag{2.71}$$

to a set of axes, this results in the same set of axes as the one originally specified. It was shown by Eq. (2.60) that a transformation to a rotated set results in the same identity tensor, so that the sign of a set of axes is not changed by this transformation. Hence the value of $|a_{ij}| = 1$ for a transformation of either left- or right-handed axes from the original set. If, however, we apply the tensor

$$\delta_{ij} = \begin{vmatrix} -1 & 0 & 0 \\ 0 & 1 & 0 \\ 0 & 0 & 1 \end{vmatrix} \tag{2.72}$$

to a set of axes, we obtain

$$x_1' = -x_1 ; \quad x_2' = x_2 ; \quad x_3' = x_3 \tag{2.73}$$

and $|a_{ij}| = -1$. An inspection of Figs. 2.6a and 2.6b shows that the right-handed system has been changed to a left-handed system or vice versa. Hence any transformation that changes the handedness of the axis system has a value $|a_{ij}| = -1$.

Another useful relation is obtained by putting $k = \pm 1$ in Eq. (2.64). This results in

$$a_{11} = \pm(a_{22}a_{33} - a_{32}a_{23}); \quad a_{21} \equiv \pm(a_{13}a_{32} - a_{33}a_{12});$$

$$a_{31} = \pm(a_{23}a_{12} - a_{13}a_{22}) \tag{2.74}$$

These equations are identical with those expressed in the usual notation of solid geometry by

$$l_1 = (m_2 n_3 - n_2 m_3); \qquad m_1 = (l_3 n_2 - l_2 n_3); \qquad n_1 = (m_3 l_2 - l_3 m_2) \quad (2.75)$$

which are valid as long as the handedness of the axes does not change.

2.8   Polar and Axial Vectors

A vector that transforms according to the usual vector transformation equation (2.4) is called a polar vector. Examples of such vectors are forces, displacements, velocities, electric currents, electric fields, electric displacements, thermal currents, and so on. For these vectors, when the handedness of the axes changes, one or more of the components may change sign.

There are, however, a series of vectors that transform according to the law

$$r_1' = \pm a_{11} r_1 \pm a_{21} r_2 \pm a_{31} r_3 \quad (2.76)$$

where the plus sign refers to transformations in the same handed coordinate system and the negative sign refers to a change from one handed system to the opposite handed system. Such vectors are called axial vectors.

The simplest example is the vector product of two polar vectors, which is a vector at right angles to the other two. This process was discussed in Section 2.5. It was there shown that the vector product results in an antisymmetric tensor having the components

$$\varepsilon_{ijk} p_j q_k = \begin{vmatrix} 0 & -(p_1 q_2 - p_2 q_1) & (p_3 q_1 - p_1 q_3) \\ (p_1 q_2 - p_2 q_1) & 0 & -(p_2 q_3 - p_3 q_2) \\ -(p_3 q_1 - p_1 q_3) & (p_2 q_3 - p_3 q_2) & 0 \end{vmatrix} \quad (2.77)$$

which can be written with the vector components

$$\begin{vmatrix} 0 & -r_3 & r_2 \\ r_3 & 0 & -r_1 \\ -r_2 & r_1 & 0 \end{vmatrix} \quad (2.78)$$

where the r's are related to the tensor components by Eqs. (2.20).

When the axes are changed from x_i to x_i', the components of p_i and q_k will change to p_i' and q_k'. If the definition of (2.20) is to continue to hold for the new axis systems, then

$$r_1' = p_2' q_3' - p_3' q_2' = a_{i2} p_i a_{j3} q_j - a_{j3} p_j a_{i2} q_i \quad (2.79)$$

from Eq. (2.4a), since p_i and q_j are assumed to be polar vectors. Hence

$$r_1' = a_{i2} a_{j3} (p_i q_j - p_j q_i) \quad (2.80)$$

When this equation is expanded, we obtain

$$r_1' = (a_{22}a_{33} - a_{23}a_{32})(p_2q_3 - p_3q_2) + (a_{32}a_{13} - a_{12}a_{33})(p_3q_1 - p_1q_3)$$
$$+ (a_{12}a_{23} - a_{22}a_{31})(p_1q_2 - p_2q_1) \qquad (2.81)$$

From Eq. (2.74) this expansion equation can be written

$$r_1' = \pm a_{11}r_1 \pm a_{21}r_2 \pm a_{31}r_3 \qquad (2.82)$$

where the plus sign refers to a transformation of axes that does not involve a change of handedness and the negative sign holds when the handedness

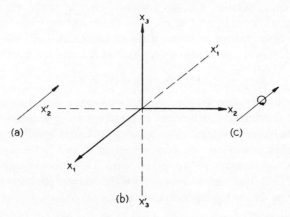

FIG. 2.7   (a) Polar vector; (b) left- and right-handed axes determined by mirror reflections; (c) axial vector.

of the axes changes. Hence this type of vector does not transform like an ordinary polar vector, and it is called an axial vector.

A polar vector can be represented unambiguously by a directed arrow, as shown by Fig. 2.7a. If, for example, we change to a left-handed set of axes by reflecting from three planes parallel to the basal planes, as shown by the dashed lines of Fig. 2.7b, the sign of the polar vector changes. Axial vectors have associated with them the idea of a right-hand screw motion, and axial vectors are taken as positive in the direction of advance of a right-hand screw, as indicated by Fig. 2.7c. For example, the vector product of two vectors is a third vector at right angles to the other two in a right-handed system of coordinates. If the two vectors of the vector product lie along the x_1 and x_2 axes of Fig. 2.7b, the third vector given by the product

$$r_3 = p_1q_2 \qquad (2.83)$$

lies along the axis x_3. By introducing the mirror reflections of Fig. 2.7b, both p_1 and q_2 will change sign, but their product r_3 still lies along the x_3

axis. A number of axial vectors are angular velocity, angular momentum, mechanical couples, the curl of a polar vector, and hence the magnetic field and the magnetic flux.

As long as one is considering the relation between two axial vectors, as for example the permeability of a solid, the operations are the same as for polar vectors. If one considers the relation between axial and polar vectors, however, as in the piezomagnetic transducer discussed in Chapter 6, one has to introduce such concepts as the "gyrator," which inverts the fields and currents or the forces and velocities in the equivalent circuits of such devices. Another difference is in the number and kinds of the piezomagnetic constants that relate[3] the strains to the magnetic flux in a crystal of definite symmetry.

2.9 Some Integral Theorems of Interest

In addition to the differential operations discussed in Section 2.5, some integral relations are of interest in electrical and mechanical theory. These are rather simply expressed in tensor form.

2.91 Gauss' Theorem

Gauss' theorem involves a relation between the divergence of a vector throughout a volume and the integral of the vector times the surface normal taken over the surface of the volume. It can be expressed in the form

$$\iiint \frac{\partial V_i}{\partial x_i} \, d\tau = \iint V_i n_i \, dS \tag{2.84}$$

where the left-hand integral is taken over the volume and the right-hand integral over the surface S. $d\tau$ is an element of volume, dS an element of the surface, and n_i is the normal to the surface.

A simple illustration that makes this theorem obvious is obtained by considering the infinitesimal parallelopiped shown by Fig. 2.8. At $x_1 = x_{1_0}$

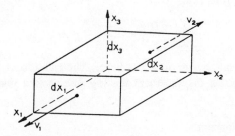

FIG. 2.8   Infinitesimal parallelopiped showing flow vectors.

[3] See Voigt [2].

an amout of fluid $V_{1_1} dx_2 dx_3$ passes through the face $dx_2 dx_3$. Only the normal component of V contributes to the value of V passing through the surface. This amount enters the volume, since the normal is directed along the negative x_1 direction. At the face $x_{1_0} + dx_1$, an amount $dx_2 dx_3 V_{1_2}$ leaves the second surface. But

$$V_{1_2} = V_{1_1} + \frac{\partial V_1}{\partial x_1} dx_1 \tag{2.85}$$

and hence

$$\frac{\partial V_1}{\partial x_1} dx_1 dx_2 dx_3 = [V_{1_2} - V_{1_1}] dx_2 dx_3 \tag{2.86}$$

Similar results hold for the other two directions so that we can write

$$\frac{\partial V_i}{\partial x_i} dx_1 dx_2 dx_3 = [V_i n_i] dS \tag{2.87}$$

Integrating this over the volume and surfaces

$$\iiint \frac{\partial V_i}{\partial x_i} d\tau = \iint V_i n_i \, dS \tag{2.88}$$

By dividing any finite volume into a large number of elementary cubes it is seen that the expression is general for any shaped surface.

If as much "material" in the sense of the absolute value of $V_i n_i$ flows into the surface as leaves it, the value of the surface integral is zero and the divergence is zero at every point. If this is not true, there must be sources or sinks in the interior. An example from Maxwell's equations is

$$\frac{\partial D_i}{\partial x_i} = \rho \tag{2.89}$$

where ρ is the charge density at the point of the divergence. If the charge density is different from zero only over a small spherical volume of radius r_0, this equation integrates to

$$\iiint \frac{\partial D_i}{\partial x_i} d\tau = \frac{4\pi r_0^3}{3} \rho = e \tag{2.90}$$

where e is the total charge on the sphere. The relation of (2.88) shows that there is more electric displacement leaving one surface than enters through the other.

2.92 Stokes' Theorem

Another theorem of interest is Stokes' theorem, which states that the line integral of the vector u_i, taken over a closed curve C, is equal to the surface integral of the curl of u_i, taken over any surface having C as a boundary. Expressed in tensor form, this equation can be written

$$\int_C u_i \, ds_i = \iint_S l_i \varepsilon_{ijk} \frac{\partial u_k}{\partial x_j} \, dS \tag{2.91}$$

where ds_i is an element of the curve C and l_i are the direction cosines of the normal to the elementary area dS. The sense of the normal is such that if the contour is described in a positive sense about an axis x_i, l_i is taken positive when the normal is in the direction of x_i increasing.

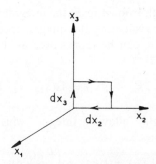

FIG. 2.9   Illustration of Stokes' theorem.

A simple illustration of Stokes' theorem is shown by Fig. 2.9. The area is the infinitesimal area $dx_2 \, dx_3$ normal to the x_1 axis. If we start at the origin and traverse the contour in a clockwise direction, the integral on the left of (2.91) takes the form

$$u_{3_0} \, dx_3 + u_{2_1} \, dx_2 - u_{3_1} \, dx_3 - u_{2_0} \, dx_2 \tag{2.92}$$

where u_{3_0} is the x_3 component of u_i along the first leg, u_{2_1} the x_2 component of u_i along the top line, and so on. But

$$u_{3_1} = u_{3_0} + \frac{\partial u_3}{\partial x_2} dx_2 \; ; \qquad u_{2_1} = u_{2_0} + \frac{\partial u_2}{\partial x_3} dx_3 \tag{2.93}$$

Hence

$$\int u_i \, ds_i = \left(\frac{\partial u_3}{\partial x_2} - \frac{\partial u_2}{\partial x_3} \right) dx_2 \, dx_3 \tag{2.94}$$

The expression in brackets is from (2.29), the x_1 component of the vector $\varepsilon_{ijk}(\partial u_k/\partial x_j)$. Because the normal to the surface lies along the positive x_1 direction, (2.94) is seen to be a special case of the general formula (2.91).

By taking an infinitesimal area whose normal has components along all three axes, it is readily shown[4] that (2.91) holds generally for such an area.

[4] See, for example, Joos [3].

Finally, since an area can be divided into elementary areas of this type, and all the terms on common boundaries cancel out, the remaining value will be the integral of (2.91) around the boundary.

An important result of Eq. (2.91) is that if the curl of the vector is zero, the vector can be represented as the gradient of a scalar function U. Since the left side of (2.91) is zero it follows that the integral

$$\int_A^B u_i \, dx_i = -\int_B^A u_i \, dx_i \qquad (2.95)$$

or if one traverses any path from A to B, the integral is equal to the negative of the value obtained by evaluating the integral from B to A. This is the necessary condition that u_i can be represented as the gradient of a scalar, since

$$U(x_{1_0} + dx_1, x_{2_0} + dx_2, x_{3_0} + dx_3) = U(x_{1_0}, x_{2_0}, x_{3_0}) + \frac{\partial U}{\partial x_i} \, dx_i \qquad (2.96)$$

Hence $U(B)$ is given by the following integral, which holds independently of what path is taken:

$$U(B) = U(A) + \int_A^B \frac{\partial U}{\partial x_i} \, dx_i \qquad (2.97)$$

RESUME CHAPTER 2

I. Tensors

Tensors of rank n are defined as any set of 3^n quantities that transforms from one set of coordinates to another rectangular set by means of the tensor transformation equation

$$x'_{k_1} x'_{k_2} \cdots x'_{k_n} = \frac{\partial x'_{k_1}}{\partial x_{j_1}} \times \frac{\partial x'_{k_2}}{\partial x_{j_2}} \cdots \frac{\partial x'_{k_n}}{\partial x_{j_n}} \, x_{j_1} x_{j_2} x_{j_n}$$

where the partial derivatives are specified by the direction cosines variously written as

	x_1		x_2		x_3	
x_1'	$l_1 = \dfrac{\partial x_1'}{\partial x_1} = \dfrac{\partial x_1}{\partial x_1'} = a_{11}$;		$m_1 = \dfrac{\partial x_1'}{\partial x_2} = \dfrac{\partial x_2}{\partial x_1'} = a_{21}$;		$n_1 = \dfrac{\partial x_1'}{\partial x_3} = \dfrac{\partial x_3}{\partial x_1'} = a_{31}$	
x_2'	$l_2 = \dfrac{\partial x_2'}{\partial x_1} = \dfrac{\partial x_1}{\partial x_2'} = a_{12}$;		$m_2 = \dfrac{\partial x_2'}{\partial x_2} = \dfrac{\partial x_2}{\partial x_2'} = a_{22}$;		$n_2 = \dfrac{\partial x_2'}{\partial x_3} = \dfrac{\partial x_3}{\partial x_2'} = a_{32}$	
x_3'	$l_3 = \dfrac{\partial x_3'}{\partial x_1} = \dfrac{\partial x_1}{\partial x_3'} = a_{13}$;		$m_3 = \dfrac{\partial x_3'}{\partial x_2} = \dfrac{\partial x_2}{\partial x_3'} = a_{23}$;		$n_3 = \dfrac{\partial x_3'}{\partial x_3} = \dfrac{\partial x_3}{\partial x_3'} = a_{33}$	

A reduced tensor is obtained by putting two suffixes in the tensor equal and summing the terms. It results in a tensor two ranks lower.

Any second-rank tensor can be considered as the sum of a symmetric tensor and an antisymmetric tensor having the terms

$$\begin{vmatrix} w_{11} & \frac{1}{2}(w_{12}+w_u) & \frac{1}{2}(w_{13}+w_{31}) \\ \frac{1}{2}(w_{12}+w_u) & w_u & \frac{1}{2}(w_{23}+w_{32}) \\ \frac{1}{2}(w_{13}+w_{31}) & \frac{1}{2}(w_{23}+w_{32}) & w_{32} \end{vmatrix} \text{ and } \begin{vmatrix} 0 & -\frac{1}{2}(w_{12}-w_u) & -\frac{1}{2}(w_{13}-w_{31}) \\ -\frac{1}{2}(w_{12}-w_u) & 0 & \frac{1}{2}(w_{23}-w_{32}) \\ -\frac{1}{2}(w_{13}-w_{31}) & -\frac{1}{2}(w_{23}-w_{32}) & 0 \end{vmatrix}$$

2. Standard Operations in Tensor Form

The scalar and vector products of two vectors can be represented by the equations

$$p_i q_i \text{ (scalar product)}; \qquad r_i = \varepsilon_{ijk} p_j q_k \text{ (vector product)}$$

The gradient of a scalar is a vector—that is,

$$\frac{\partial U}{\partial x_j} = y_j$$

The gradient of a vector is a second-rank tensor:

$$\frac{\partial u_j{}'}{\partial x_i} = a_{ij} a_{kl} \frac{\partial u_l}{\partial x_k}$$

The divergence of a tensor is

$$\frac{\partial u_i}{\partial x_i} = \frac{\partial u_1}{\partial x_1} + \frac{\partial u_2}{\partial x_2} + \frac{\partial u_3}{\partial x_3}$$

The curl of a vector can be expressed in the form

$$V_i = \varepsilon_{ijk} \frac{\partial u_k}{\partial x_j}$$

The vanishing of the curl is a necessary condition that u_i may be the gradient of a scalar.

3. Representation Quadric

A geometrical representation of a second-rank tensor can be obtained by multiplying the tensor by two coordinate vectors and setting the product equal to unity—that is,

$$T_{ij} x_i x_j = 1$$

This equation can be transformed into a set of axes, called the principal axes, for which

$$T_1 x_1{}^2 + T_2 x_2{}^2 + T_3 x_3{}^2 = 1$$

The surface is respectively an ellipsoid, a hyperboloid of one sheet, a hyperboloid of two sheets, or an imaginary ellipsoid if the coefficients are all positive, two positive and one negative, one positive and two negative, or all negative.

For five out of the seven crystal systems the principal axes coincide with the crystallographic axes. If one axis coincides with a principal axis, the remaining two can be determined by a Mohr circle diagram.

4. Tensors Involving the Direction Cosines

There are six sets of relations between the nine direction cosines of the table in 1. The value of the determinant of $|a_{ij}| = \pm 1$. The reduced tensor product

$$a_{ij} a_{il} = \delta_{jl}$$

is a unitary tensor having the matrix

$$\begin{vmatrix} 1 & 0 & 0 \\ 0 & 1 & 0 \\ 0 & 0 & 1 \end{vmatrix}$$

The product of a tensor w_{ij} by a unitary tensor reproduces w_{ij}. An axial vector transforms according to the equation

$$r_1' = \pm a_{11}r_1 \pm a_{21}r_2 \pm a_{31}r_3$$

where the plus sign refers to a transformation between axes of the same handedness whereas the negative sign refers to a change in handedness. A polar vector transforms according to the relation

$$r_1' = a_{11}r_1 + a_{21}r_2 + a_{31}r_3$$

for either the same or different axes. Some polar vectors are force, displacement, velocity, electric currents, fields and displacements, thermal currents, and so on. Some axial vectors are angular velocity, angular momentum, magnetic fields and fluxes, and so on.

5. Integral Theorems

Gauss' theorem and Stokes' theorem can be expressed in the forms

$$\iiint \frac{\partial V_i}{\partial x_i}\, d\tau = \iint V_i n_i\, dS \qquad \text{(Gauss' theorem)}$$

where V_i is a vector and n_i is the outward-drawn normal from the surface.

Stokes' theorem states that the line integral of the vector u_i taken over a closed curve C is equal to the surface integral of the curl of u_i taken over any surface having C as a boundary. It takes the form

$$\int_C u_i\, ds_i = \iint l_i \varepsilon_{ijk} \frac{\partial u_k}{\partial x_j}\, dS \qquad \text{(Stokes' theorem)}$$

where ds_i is an element of length on the curve C, dS an element of area on the surface, and l_i are the direction cosines of the normal to the elementary area dS.

PROBLEMS CHAPTER 2

1. Using the direction cosine matrix

	x_1	x_2	x_3
x_1'	$\dfrac{\partial x_1'}{\partial x_1} = \cos\theta;$	$\dfrac{\partial x_1'}{\partial x_2} = \sin\theta;$	$\dfrac{\partial x_1'}{\partial x_3} = 0$
x_2'	$\dfrac{\partial x_2'}{\partial x_1} = -\sin\theta;$	$\dfrac{\partial x_2'}{\partial x_2} = \cos\theta;$	$\dfrac{\partial x_2'}{\partial x_3} = 0$
x_3'	$\dfrac{\partial x_3'}{\partial x_1} = 0;$	$\dfrac{\partial x_3'}{\partial x_2} = 0;$	$\dfrac{\partial x_3'}{\partial x_3} = 1$

transform the tensor

$$\begin{vmatrix} w_{11} & w_{12} & 0 \\ w_{12} & w_{22} & 0 \\ 0 & 0 & w_{33} \end{vmatrix}$$

to the x_1', x_2', x_3' axes.

2. Prove that the vector product of an axial vector and a polar vector is a polar vector.

3. Transform the following tensors to their principal axes, using the Mohr circle construction.

$$\begin{vmatrix} 3 & 3 & 0 \\ 3 & 3 & 0 \\ 0 & 0 & 7 \end{vmatrix} \qquad \begin{vmatrix} 5 & 0 & -3 \\ 0 & 15 & 0 \\ -3 & 0 & 2 \end{vmatrix}$$

REFERENCES CHAPTER 2

1. H. Jeffreys, "Cartesian Tensors," 2nd ed. Cambridge Univ. Press, London and New York, 1952.
2. W. Voigt, "Lehrbuch der KristallPhysik,," Chapter 8, Section 8. B. Teubner, 1928.
3. G. Joos, "Theoretical Physics," p. 26. G. E. Stechert (Hafner), London 1934.

CHAPTER 3 • Definition of Fundamental Variables Involved in Interaction Effects

As discussed in Chapter 1 the principal interaction phenomena of interest have to do with the transference of electrical into mechanical energy or vice versa, the transference of magnetic energy into electrical energy, and the transference of thermal energy into electrical energy. Hence the principal variables of interest are stress, strain, electric fields and displacements, electric currents, magnetic fields and fluxes, temperature, entropy, quantity of heat, and thermal currents. It is the purpose of this chapter to discuss their definitions and to derive some elementary relationships between them.

3.1 Electric Fields and Displacements

3.11 Dielectric Permittivity of a Crystal Condenser

If we have a parallel-plate condenser, as shown by Fig. 3.1, across which an electrical potential V is placed, the electric field is defined as the ratio of the potential to the separation x_3 or

$$E = \frac{V}{x_3} \tag{3.1}$$

If for any reason the variation with distance is not uniform, the field is defined as

$$E = \frac{\partial V}{\partial x_3} \tag{3.2}$$

E is a vector quantity since it is the space derivative of a scalar quantity. By shifting to another set of coordinates, having the direction cosines a_{11}, a_{21}, a_{31} with respect to the direction of the field, x_1', E_i' will have the components

$$E_i' = \alpha_{ji}E_j \tag{3.3}$$

When the battery of potential V is applied to the plates, it will be found that a charge per unit area has collected on the plates whose value depends on separation of the plates—that is, the field E—and on the permittivity of the medium, or

$$D_0 = \varepsilon_0 E \tag{3.4}$$

In the MKS system of units, for which E is measured in volts per meter and the electric displacement in coulombs per square meter, the value of ε_0 for a vacuum is 8.85×10^{-12} farads/meter.

If, now, an isotropic material is inserted between the plates, in such a way that the plates are integral with the material, there occurs an additional

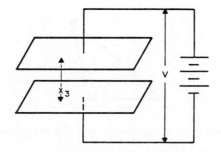

FIG. 3.1 Parallel-plate condenser with air dielectric.

separation of charge, which is called the polarization P of the sample. This polarization adds to the electric displacement in the vacuum so that the total electric displacement is equal to

$$D = D_0 + P = \varepsilon_0 E + P \tag{3.5}$$

But the polarization—except at very high field strengths—is proportional to the field, so that we can write

$$D = \varepsilon_0[1 + \eta]E = \varepsilon E \tag{3.6}$$

where ε is called the permittivity and η the dielectric susceptibility. The ratio of ε to ε_0

$$\frac{\varepsilon}{\varepsilon_0} = K \tag{3.7}$$

is called the relative dielectric constant.

For crystals, however, the polarization does not occur in the same direction as the field, and as shown by Fig. 3.2, the electric displacement does not lie in the same direction as the electric field. In general the relation between the electric displacement D_n and the electric field has to be written in the form of a tensor equation

$$D_n = \varepsilon_{mn} E_m \tag{3.8}$$

It will be shown shortly that ε_{mn} is a symmetric second-rank tensor, and hence from the discussion of Section 2.6 there are three principal axes. A measurement of three dielectric constants along these directions suffices to determine the values in any other direction. The effects of crystal symmetries on the properties of a second-rank tensor are discussed in Chapter 5, and it is shown that for all classes except triclinic and monoclinic crystals, the crystallographic axes coincide with the principal axes. For the monoclinic system, the b crystallographic axis is a principal axis and the other two can be determined from a Mohr diagram. Only in the case of the most unsymmetric crystal, the triclinic crystal, is it necessary to solve the cubic

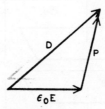

FIG. 3.2   Relative directions of the electric field, the electric displacement, and the polarization.

equations of Section 2.6 to obtain the principal axes. Table 1 gives the relative dielectric constants of a number of commonly used crystals. These are the static or low-frequency values, which are usually valid up to frequencies of 10^{10} cycles or higher. In the case of parallel-faced crystals between the two parallel plates of a condenser of the type shown by Fig. 3.1, it is not at once obvious what component of the permittivity ε_{mn} is measured. Let us consider the parallel-plate condenser with a crystalline medium in between and suppose that the air gaps between the plates and crystal are very small compared to the crystal thickness. Because the plates are assumed to have large lateral dimensions and are themselves equipotential surfaces, the equipotential surfaces through the air gap and crystal must run parallel to the surfaces. Hence the fields outside and inside must be perpendicular to the surfaces and hence lie along the z axis. Furthermore, since the potential V is equal to

$$V = 2E_A l_A + E_c l_c \qquad (3.9)$$

where E_A is the field in the air gaps, E_c the field in the crystal, l_A the air gap thickness and l_c the crystal thickness, it follows that since l_A is very small compared to l_c

$$E_c = \frac{V}{l_c} \qquad (3.10)$$

The charges on the plates with the crystal present (σ') and with the crystal absent (σ) are

$$\sigma = \varepsilon_0 E_A \; ; \qquad \sigma' = \varepsilon_{33}\left(\frac{V}{2l_A + l_c}\right) \doteq \varepsilon_{33}\frac{V}{l_c} \qquad (3.11)$$

These relations satisfy one requirement for a solution, namely that the tangential components of E across the boundary be equal; in this case they are both zero.

Table 1

RELATIVE CONSTANTS OF SOME COMMONLY USED CRYSTALS[a]

Crystal	Crystal Class	K_1	K_2	K_3
LiF	cubic	9.27	—	—
NaCl	cubic	5.9	—	—
KCl	cubic	4.68	—	—
ZnS	cubic	8.2	—	—
MgO	cubic	9.05	—	—
AgCl	cubic	12.3	—	—
Al_2O_3	hexagonal	9.34	9.34	11.54
$NH_4H_2PO_4$ (ADP)	tetragonal	56.0	56.0	14.0
KH_2PO_4 (KDP)	tetragonal	44.5	44.5	22.0
TiO_3 (Rutile)	tetragonal	87.5	87.5	180
SiO_2 (Quartz)	trigonal	4.5	4.5	4.6
$CaCO_3$ (Calcite)	trigonal	8.5	8.5	8.0
CdS	hexagonal	9.35	9.35	10.33
ZnO	hexagonal	8.1	8.1	8.5
HIO_3	orthorhombic	7.2	8.0	6.9
Tourmaline	trigonal	8.2	8.2	7.5

[a] Values from "Amer. Inst. of Phys. Handbook," pp. 9–93 to 9–95 (McGraw-Hill, New York, 1963); D. Berlincourt, H. Jaffe, and L. R. Shiozawa, *Phys. Rev.* **129**, 1009 (1963); Von Hippel, "Dielectric Materials and Applications," Wiley, New York, 1954. When the crystals are piezo-electric, the relative dielectric constants are the free dielectric constants ε_{ij}^T.

In the gap the electric displacement D_A is of course parallel to E, but in the crystal D is not parallel to E. It follows directly from Gauss' theorem (Section 2.9) that, since there are no free charges on the crystal surface, the normal component of D equals the electric displacement in the gap. Therefore,

$$D_3 = D_A = \sigma \qquad (3.12)$$

where σ is the charge per unit area on the electrodes, and hence the ratio of the capacitance with the crystal in to that with the crystal out is

$$\frac{C}{C'} = \frac{\sigma}{\sigma'} = \frac{D_3}{D_A} = \frac{\varepsilon_{33}}{\varepsilon_0} = K_3 \qquad (3.13)$$

As a special case it follows that when one of the principal permittivity directions is normal to the plates, the ratio of capacitances gives the principal dielectric constants directly.

3.12 The Energy of a Polarized Crystal

When the polarization of a crystal is changed, then provided the field is entirely confined to the crystal, it is proved in standard books on electromagnetic theory[1] that the work done is

$$dW = vE_i \, dD_i \tag{3.14}$$

where v is the volume of the crystal. As a special case we may consider the effect of a change of potential V for the system of Fig. 3.3. The effect of a

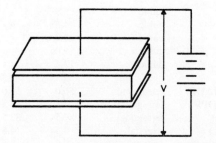

FIG. 3.3   Parallel-plate condenser with crystalline dielectric.

change of polarization of the crystal is to produce new surface charges $d\sigma$ and $-d\sigma$ per unit area. The amount of work done is

$$dW = AV \, d\sigma \tag{3.15}$$

where A is the area of the plates. This energy comes from the battery. Since it is equal to the normal component of D in the crystal and $E_3 l_c = V$, where l_c is the thickness of the crystal, the work is

$$dW = A l_c E_i \, dD_i = v E_i \, dD_i \tag{3.16}$$

Using Eq. (3.8) this can be written

$$dW = v\varepsilon_{ij} E_i \, dE_j \tag{3.17}$$

If we write this out in full, the equation becomes

$$dW = v[\varepsilon_{11}E_1 \, dE_1 + \varepsilon_{12}E_1 \, dE_2 + \varepsilon_{13}E_1 \, dE_3 + \varepsilon_{21}E_2 \, dE_1 + \varepsilon_{22}E_2 \, dE_2$$
$$+ \varepsilon_{23}E_2 \, dE_3 + \varepsilon_{31}E_3 \, dE_1 + \varepsilon_{32}E_3 \, dE_2 + \varepsilon_{33}E_3 \, dE_3] \tag{3.18}$$

[1] See Schelkunoff [1].

Hence

$$\frac{1}{v}\frac{\partial W}{\partial E_1} = \varepsilon_{11}E_1 + \varepsilon_{21}E_2 + \varepsilon_{31}E_3$$

and   (3.19)

$$\frac{1}{v}\frac{\partial W}{\partial E_2} = \varepsilon_{12}E_1 + \varepsilon_{22}E_2 + \varepsilon_{32}E_3$$

W is a function only of the independent variables E_1, E_2, and E_3. It is shown in Chapter 4 that for such functions the order of differentiation makes no difference so that

$$\frac{1}{v}\frac{\partial}{\partial E_1}\left(\frac{\partial W}{\partial E_2}\right) = \frac{1}{v}\frac{\partial}{\partial E_2}\left(\frac{\partial W}{\partial E_1}\right)$$ (3.20)

and hence

$$\varepsilon_{12} = \varepsilon_{21}$$

Similarly we find that

$$\varepsilon_{13} = \varepsilon_{31} \;; \qquad \varepsilon_{23} = \varepsilon_{32} \qquad \text{and in general} \qquad \varepsilon_{ij} = \varepsilon_{ji} \quad (3.21)$$

Equation (3.18) can be obtained by differentiating the function

$$W = v[\tfrac{1}{2}\varepsilon_{11}E_1^{2} + \varepsilon_{12}E_1E_2 + \varepsilon_{13}E_1E_3 + \tfrac{1}{2}\varepsilon_{22}E_2^{2} + \varepsilon_{23}E_2E_3 + \tfrac{1}{2}\varepsilon_{33}E_3^{2}] \quad (3.22)$$

or in tensor notation

$$W = \tfrac{1}{2}v\varepsilon_{ij}E_iE_j \quad (3.23)$$

3.2   Magnetic Variables

3.21   Magnetic Fields, Magnetic Fluxes, and Permeabilities

The electric fields and electric displacements of the last section are the fundamental electric variables for piezoelectric and electrostrictive transducers, which are discussed in Chapter 6. For another set of transducers, the piezomagnetic or magnetostrictive types, magnetic fields and magnetic fluxes are the fundamental variables.

Magnetic fields are usually produced by the flow of electric currents in conductors. Three configurations of special interest are shown in Fig. 3.4. The first structure is formed by two concentric tubes extending in the direction normal to the plane of the paper. The inner tube carries current flowing toward the reader while the other tube carries an equal and opposite current away from the reader. This arrangement gives a magnetic field whose value is independent of the separation of the cylinders but varies inversely as the radial distance. The curved arrow indicates the direction of force on the north-seeking positive end of a compass needle.

The second arrangement, shown by Fig. 3.4b, consists of two parallel flat plates running normal to the plane of the paper. $+i$ is flowing toward the reader, $-i$ away from the reader, and the direction of the magnetic field is shown by the arrow labeled H. Between the two plane sheets the magnetic field is uniform; it depends on the current density in the sheets but not on the distance between them. This current density can be taken as

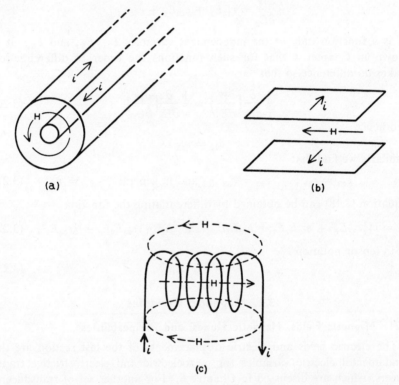

(a)

(b)

(c)

FIG. 3.4 Configurations that give magnetic fields that are independent of the dimensions of the structures: (a) two concentric conductors; (b) two infinite parallel plates; (c) an elongated helix.

a measure of H, and in the MKS system of units H has the dimensions of amperes per meter. One ampere per meter is equal to $(4\pi/10^3)$ oersteds, the cgs units of field strengths. However, a more usual definition is the one obtained from Fig. 3.4c.

The arrangement shown by Fig. 3.4c is a cylindrically wound coil; the magnetic field is independent of the shape and size of the cross section of the cylinder, is parallel to the axis of the coil, and is equal to the circulating current per unit length of the coil. In this case H is equal to the number of ampere turns—that is, the product of the current in the coil in amperes times the number of turns per meter. It is more usual to define H in terms of ampere turns per meter than in terms of amperes per meter.

If the coil has no magnetic material in it, this field produces a flux density B equal to

$$B_i = \mu_0 H_i \tag{3.24}$$

where μ_0 is the permeability of a vacuum, which in rationalized MKS units is $4\pi/10^7 = 1.257 \times 10^{-6}$. The flux density is measured in webers per square meter. This unit is 10^4 times as large as the cgs unit, the gauss.

If an isotropic magnetic material is placed in the coil, a magnetic intensity I per unit volume is generated in the material and B is given by

$$B_i = \mu_0 H_i + I_i \tag{3.25}$$

In many isotropic solids the magnetic intensity I_i is proportional to the magnetic field, and hence we can write

$$I_i = \mu_0 \psi H_i \tag{3.26}$$

where ψ is a constant called the *magnetic susceptibility*. I_i and $\mu_0 H_i$ have the same dimensions, and hence ψ is a dimensionless ratio.

Combining (3.25) and (3.26)

$$B = (1 + \psi)\mu_0 H = \mu H \qquad \text{where} \qquad \mu = \mu_0(1 + \psi) \tag{3.27}$$

is called the permeability of the substance.

In a crystal I_i is in general not parallel to H_j but can be written

$$I_i = \mu_0 \psi_{ji} H_j \tag{3.28}$$

This relation holds for all paramagnetic crystals and for ferromagnetic crystals at low field strengths. For these conditions

$$B_i = \mu_{ji} H_j \qquad \text{where} \qquad \mu_{ji} = \mu_0(\delta_{ji} + \psi_{ji}) \tag{3.29}$$

The number of components in this form is nine, but it can be proved from the form of the energy that $\mu_{ij} = \mu_{ji}$ and hence six components suffice to express the relation (3.29). We have seen also that a symmetrical tensor can be expressed in terms of three principal axes, and it turns out that these usually lie along crystallographic axes. A crystal is said to be paramagnetic along one of the principal axes if ψ is positive for this particular axis and diamagnetic if ψ is negative along this axis. The principal susceptibilities of such crystals are in the order of $+10^{-5}$ and -10^{-5} respectively. Nye[2] has given a table for a few crystals, but since they are not of interest in transducer applications, they are not given here.

[2] See Nye [2].

All of the ferromagnetic materials of interest here are polycrystalline materials such as iron, nickel, cobalt—and alloys of the three metals—or sintered ferrite materials. As transducers they are magnetically polarized along an axis. This process has the effect of changing an isotropic material to one that has a unique axis along the direction of polarization. The

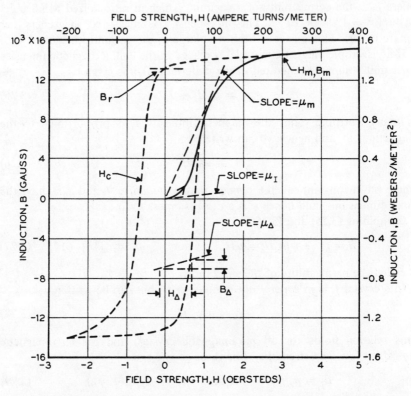

FIG. 3.5 Magnetization curve (solid) and hysteresis loop (dotted). Some important magnetic quantities are illustrated (after Bozorth).

symmetry for this type of axis—that is, ∞/m as discussed in Chapter 5—results in an intensity of magnetization that lies along the unique axis for a field applied along this axis. Hence it is not necessary to consider magnetic flux components that do not lie along the direction of the applied field.

The magnetic intensity for these types of materials is, however, a nonlinear function of the applied field. Figure 3.5 shows a typical B-H curve for the magnetic material iron. The curve starts out with an initial slope μ_I, increases its slope to a value μ_m and then ends up with a slope proportional to $\mu_0 H$. If we subtract $\mu_0 H$ from B, it is found that there is a saturation flux due to the material equal to B_S. As the field is taken off, the flux follows the dashed line and has the remnant value B_r when $H = 0$. By making H negative, the flux is reversed in sign. The coercive field H_c occurs when the value of flux is reduced to zero.

It is necessary to distinguish between different types of permeabilities as follows:

1. The initial permeability μ_I is the limit approached by the normal permeability as B and H approach zero. This ratio is a constant when H is sufficiently weak.

2. The *maximum permeability* μ_m is the largest slope of the B-H curve.

3. *The incremental permeability* μ_Δ refers to the permeability measured when there is a superposed field H_S that is applied and held constant. Then, as shown by Fig. 3.5, when a superposed field H_Δ is applied and alternated cyclically, it will cause B_Δ to traverse a small hysteresis loop. μ_Δ is then the average ratio B_Δ/H_Δ. When H_Δ approaches zero, μ_Δ approaches a limiting value known as the reversible permeability μ_r. Both μ_Δ and μ_r are dependent on the value of H_S; μ_Δ is also dependent on the magnitude of H_Δ. Since transducers are sometimes operated at remanence—$H_c = 0$—the reversible permeability at remnance, μ_R, is also of interest. It usually turns out that the electromechanical coupling factor is higher at some other value of bias, and most transducers are operated at a definite bias usually provided by a permanent magnet. In tables the permeabilities are usually listed as the ratio of μ to μ_0, the permeability of a vacuum. Table XII of Chapter 4 lists the properties of a number of substances of interest for transducers.

3.22   Demagnetizing Effects

When a crystal is placed in a magnetic field and develops an intensity of magnetization I_i, it sets up a magnetic field of its own that depends on the susceptibility, the shape, and the size of the crystal. The actual field acting in the crystal is therefore different from the applied field. If we denote the field in the crystal by H_{c_i} and the applied field by H_{a_i}, the actual or total field H_{t_i} will be the sum of the two or

$$H_{t_i} = H_{a_i} + H_{c_i} \qquad (3.30)$$

The field that enters Eqs. (3.29) is then the total field H_{t_i}.

For paramagnetic materials the field set up by the intensity of magnetization is usually only one part in 10^5 as large as the applied field and can usually be neglected. For a magnetized polycrystalline material or a polarized ceramic ferrite the demagnetizing field lies in the same direction as the applied field and subtracts from it. Demagnetizing factors have been calculated for various shapes. If the material closes on itself, as in Fig. 3.7, or if a bar is

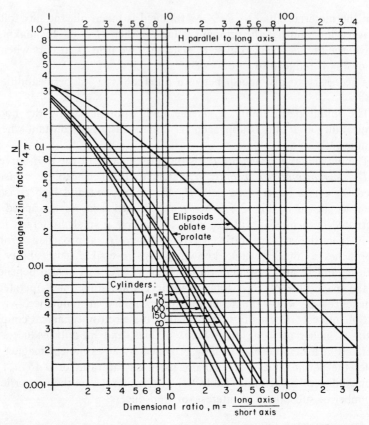

FIG. 3.6   Demagnetizing factors of ellipsoids and cylinders (after Bozorth).

used whose length is much larger than its diameter, the demagnetizing field is zero. For rods and ellipsoids it has been shown[3] that the demagnetizing field ΔH is proportional to the magnetic intensity $I = (B - \mu_0 H)$ with a factor of proportionality $N/4\pi\mu_0$, or

$$\Delta H = \frac{NI}{4\pi\mu_0} = \frac{N}{4\pi}\left(\frac{B}{\mu_0} - H\right) \tag{3.31}$$

Figure 3.6 shows calculated values of N as a function of the ratio of the long-axis length to the short-axis length.

3.23   Energy Associated with a Magnetized Crystal

The energy stored in a magnetic body is found by calculating how much work is done in magnetizing the body. This is done by setting up a differential

[3] See Bozorth [3].

equation by considering how much work is done in producing a small change in magnetism. The simplest system to consider is the closed-circuit arrangement of Fig. 3.7, in which it is assumed that similar legs are joined together at the corners and that this joining volume is small compared to the volume in the material. The solenoid has n turns per unit length and hence has a field ni produced in it. This solenoid has a resistance R, and the current i is maintained by a battery with an emf equal to V. We seek an expression for the work done when there is a small change in the magnetization of the crystal or polarized ferromagnetic material. Let B_n be the

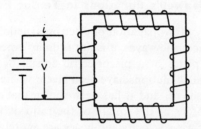

FIG. 3.7 Crystal square for finding the energy associated with a magnetized crystal.

component of B along the axis. Then as B changes there is an induced emf in the circuit equal to

$$-Anl\frac{dB_n}{dt} \tag{3.32}$$

(see next section) where A is the area of the crystal. Therefore, adding the emf's around the circuit, we have

$$V - Anl\frac{dB_n}{dt} = iR \tag{3.33}$$

The work done by the battery in time dt is

$$Vi\,dt = i^2R\,dt + Alni\,dB_n = i^2R\,dt + AlH_n\,dB_n \tag{3.34}$$

The term $i^2R\,dt$ is the Joule heat loss, whereas the second term represents the energy stored in the crystal. This can be written as

$$U = vH_i\,dB_i = v\mu_{ij}H_i\,dH_j \tag{3.35}$$

since Al is the volume of the crystal. By an argument similar to that for Eq. (3.31) it is evident that

$$\mu_{ij} = \mu_{ji} \tag{3.36}$$

Written out in complete form, the internal energy for a paramagnetic material or a ferromagnetic material at low field strengths becomes

$$U = v[\tfrac{1}{2}\mu_{11}H_1{}^2 + \mu_{12}H_1H_2 + \mu_{13}H_1H_3 + \tfrac{1}{2}\mu_{22}H_2{}^2 + \mu_{23}H_2H_3 + \tfrac{1}{2}\mu_{33}H_3{}^2]$$
$$= \tfrac{1}{2}v\mu_{ij}H_iH_j \tag{3.37}$$

For a ferromagnetic material at higher field strengths, B_i is a nonlinear function of H_i and one has to know the functional relationship to obtain U. For a polarized crystal with a small excursion in H_i around the superposed field H_S, the only constant entering is the last term of Eq. (3.37). For this case the energy term is

$$U = U_0 + v[\tfrac{1}{2}\mu_r(H_3 - H_s)^2]$$ (3.38)

where μ_r is the reversible permeability.

3.3 Maxwell's Equations in Tensor Form

Because the velocity of propagation of electromagnetic waves is about 10^5 times the velocity of acoustic waves, it is not ordinarily necessary to consider wave motion in the plates of a piezoelectric crystal or the solenoid of a piezomagnetic material. Solutions have been given[4] where such wave motion has been taken account of, and it has been shown that correction terms to velocities are in the order of the square of the ratio of the sound velocity to the electromagnetic wave velocity and hence are entirely negligible.

When we consider the piezo-optic effect and the electro-optic effect, however, it is necessary to consider electromagnetic wave propagation in the crystal and the effects of strain and electric fields on this wave propagation. Hence it is necessary to consider Maxwell's equations. Because the propagation takes place in dielectric material it is not necessary to consider anisotropic magnetic quantities.

In tensor form Maxwell's equations can be written in the form

$$\varepsilon_{ijk}\frac{\partial H_k}{\partial x_j} = \frac{\partial D_i}{\partial t} + I_i ; \qquad \varepsilon_{ijk}\frac{\partial E_k}{\partial x_j} = -\frac{\partial B_i}{\partial t} ; \qquad \frac{\partial D_i}{\partial x_i} = \rho; \qquad \frac{\partial H_j}{\partial x_j} = 0$$

(3.39)

where I_i is the current density in the ith direction and ρ is the charge density. For a dielectric crystal I_i and ρ are zero and $B_i = \mu_0 H_i$. Hence Eqs. (3.39) take the form

$$\varepsilon_{ijk}\frac{\partial H_k}{\partial x_j} = \frac{\partial D_i}{\partial t} ; \qquad \varepsilon_{ijk}\frac{\partial E_k}{\partial x_j} = -\mu_0\frac{\partial H_i}{\partial t} ; \qquad \frac{\partial D_i}{\partial x_i} = 0; \qquad \frac{\partial H_j}{\partial x_j} = 0 \quad (3.40)$$

A derivation of the propagation of an electromagnetic wave in a crystal is given in Appendix A.

It is shown in books on electromagnetic theory[5] that the first equation represents the magnetic field generated around a surface through which the

---

[4] See Kyame [4] and Hutson and White [5].
[5] See, for example, Schelkunoff [6].

current density I_i is flowing—that is, Amperes law. The total term contains the displacement current $\partial D_i/\partial t$ as well as the actual current flowing. If we consider the component along the x_1 axis and perform the integration

$$\iint_S \varepsilon_{1jk} \frac{\partial H_k}{\partial x_j} \, dS = \iint \left(\frac{\partial D_1}{\partial t} + I_1 \right) dS = \int_c H_i \, ds_i \qquad (3.41)$$

we can set this integral equal to the right-hand side of (3.41) by using Stokes's theorem. If we take the area dS as being a circle of radius r_0, then the surface integral becomes

$$\frac{\pi r_0^2}{2} \left[I_1 + \frac{\partial D_1}{\partial t} \right] = i_1 \qquad (3.42)$$

where i_1 is the total current—convective as well as displacement—through the area. From symmetry it is obvious that the component along a circle of radius r_1 around the cylinder is going to be the same at all points of the circle and equal to H_r. Hence the field at a distance r_1 from the center of the current carrying area is

$$H_r = \frac{i}{2\pi r_1} \qquad (3.43)$$

The integral of $H_i \, ds_i$ along any path is called the magnetomotive force in analogy to the similar quantity, the electromotive force (emf), which is the integral of the electric field along any path.

The second equation of (3.39) is an expression of Faraday's observation that the electromotive force around a closed path is equal to the time rate of change of the magnetic flux through it. By taking the area as a circle perpendicular to the x_1 axis as before and employing Stokes's theorem we find

$$\int_c E_i \, dx_i = -\frac{d\Phi}{dt} = -\frac{\pi^2 r_0^2 \dot{B}}{2} \qquad (3.44)$$

assuming that the flux is uniform through the circular area. The integral on the left of (3.44) is the integral of the electric field in the direction of the path displacement dx_i and hence is the electromotive force (emf). If the emf is expressed in volts, Φ is expressed in webers. The dimension of the weber is the volt second. The magnetic flux density B is the number of webers per square meter. This unit is too large for practical magnetic fields, and the gauss, equal to 10^{-4} weber/meter^2 is often used. The weber is equal to 10^8 maxwells, the cgs unit of flux.

The third equation expresses the fact that the divergence of the electric displacement is zero except in the presence of a charge distribution having the density ρ. The charge distribution causes the electric displacement leaving a volume to be larger than the electric displacement entering the volume. The fourth equation reflects the fact that there are no magnetic charges and the magnetic field is a solenoidal vector with no divergence.

A fifth equation is usually given, which expresses the force on a moving or stationary charged particle in an electric and a magnetic field. This equation takes the form

$$F_i = qE_i + q\varepsilon_{ijk}v_jB_k \tag{3.45}$$

where q is the charge on the particle and v_j its velocity The product of the charge times its velocity is equivalent to an electric current This equation is of interest for moving-coil or moving-armature electromagnetic transducers, but is not considered in this book.

Finally, some boundary conditions are of interest when one goes from one medium to another. These follow from the Gauss and Stokes theorems. These relations[6] are that the tangential components of the electric and magnetic field strengths are continuous across the boundary, and the normal components of the electric current density and the magnetic flux density are continuous across the boundary. The equations can be written

$$E_{2_{\tan}} = E_{1_{\tan}} \; ; \quad H_{2_{\tan}} = H_{1_{\tan}} \; ; \quad I_{2_n} + \frac{\partial}{\partial t}D_{2_n} = I_{1_n} + \frac{\partial}{\partial t}D_{1_n} \; ; \quad B_{2_n} = B_{1_n}$$

$$\tag{3.46}$$

where *tan* refers to the tangential components and n to the normal components. From Gauss' equation, we find that if there is a surface density of charge equal to q_S, the normal components of the electric displacement differ by

$$D_{1_n} - D_{2_n} = q_S \tag{3.47}$$

3.4 Stress Tensors

The effect of an applied voltage or current on a piezoelectric or a piezomagnetic transducer is to cause a stress or strain to be generated. It is the purpose of this section to discuss the definitions of these terms and describe their interaction.

3.41 Definition of Stresses

The stresses exerted on any elementary cube of material with its edges along the three rectangular axes x_1, x_2, x_3 can be specified by considering the stresses acting on each face of the cube illustrated in Fig. 3.8. The total force acting on the face $ABCD$ can be represented by a resultant force R_D with its point of application at the center of the face, plus a couple that takes account of the variation of the stress across the face. The force R is directed

[6] See Schelkunoff [6], p. 257.

outward, since a stress is considered positive if it exerts a tension. As the face is shrunk in size, the force R will be proportional to the area of the face, while the couple will vary as the cube of the edge dimension. Hence in the limit the couple can be neglected with respect to R. The stress (force per unit area) due to R can be resolved into three components along the three axes, which are given the designation

$$T_{11_2}, \quad T_{21_2}, \quad T_{31_2} \tag{3.48}$$

Here the first number designates the direction of the stress component and the second number 1_2 denotes the second face of the cube normal to the x_1

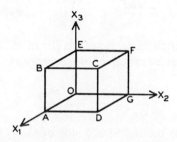

FIG. 3.8 Elementary cube for determining stresses.

axis. The remaining stress components on the other five faces have the designations

$$
\begin{array}{llll}
\text{Face } OEFG & T_{11_1}, & T_{21_1}, & T_{31_1} \\
OABE & T_{12_1}, & T_{22_1}, & T_{32_1} \\
CFGD & T_{12_2}, & T_{22_2}, & T_{32_2} \\
OADG & T_{13_1}, & T_{23_1}, & T_{33_1} \\
BCFE & T_{13_2}, & T_{23_2}, & T_{33_2}
\end{array}
\tag{3.49}
$$

The resultant force in the x_1 direction is obtained by summing all the forces with components in the x_1 direction, or

$$T_1 = (T_{11_2} + T_{11_1})\, dx_2\, dx_3 + (T_{12_2} + T_{12_1})\, dx_1\, dx_3 + (T_{13_2} + T_{13_1})\, dx_1\, dx_2 \tag{3.50}$$

But

$$T_{11_2} = -T_{11_1} + \frac{\partial T_{11}}{\partial x_1}\, dx_1 \,; \qquad T_{12_2} = -T_{12_1} + \frac{\partial T_{12}}{\partial x_2}\, dx_2 \,;$$

$$T_{13_2} = -T_{13_1} + \frac{\partial T_{13}}{\partial x_3}\, dx_3$$

Hence F_1 can be written in the form

$$F_1 = \left(\frac{\partial T_{11}}{\partial x_1} + \frac{\partial T_{12}}{\partial x_2} + \frac{\partial T_{13}}{\partial x_3}\right) dx_1\, dx_2\, dx_3 \qquad (3.51)$$

If in addition there is a body force such as gravity acting on the element, the total force along the x_1 axis is

$$F_1 = \left(\frac{\partial T_{1i}}{\partial x_i} + \rho g_1\right) dx_1\, dx_2\, dx_3 \qquad (3.52)$$

Similar expressions hold for the other two directions, and hence we can write generally

$$F_i = \left(\frac{\partial T_{ij}}{\partial x_j} + \rho g_i\right) dx_1\, dx_2\, dx_3 \qquad (3.53)$$

Because force is equal to mass times acceleration—by Newton's laws of motion—a fundamental equation of wave propagation can be written in the form

$$\frac{\partial T_{ij}}{\partial x_j} + \rho g_i = \rho\, \frac{\partial^2 x_i}{\partial t^2} \qquad (3.54)$$

Under static conditions the right-hand side of (3.54) is zero. The resulting equation is known as the equation of equilibrium and is much used in the theory of elasticity. It will be noted that the space derivative of a second-rank tensor is a third-rank tensor, but because it is contracted by the common use of j in numerator and denominator, it reduces to a first-rank tensor—that is, a vector.

The other set of equations of interest in the determination of the stresses on a body is the set involving the rotation of the elementary volume about an axis. If, for example, we consider the rotation of the volume about the x_3 axis, the stresses of interest are those shown by Fig. 3.9. The stress couples T_{21_2} and T_{21_1} tend to rotate the cube in a clockwise direction by producing the couples

$$\left[T_{21_2}\frac{dx_1}{2} + T_{21_1}\left(-\frac{dx_1}{2}\right)\right] dx_2\, dx_3 \qquad (3.55)$$

Since

$$T_{21_2} = -T_{21_1} + \frac{\partial T_{21}}{\partial x_1} dx_1$$

this couple equals

$$T_{21_0}\, dx_1\, dx_2\, dx_3 + \frac{\partial T_{21}}{\partial x_1} dx_1{}^2\, dx_2\, dx_3 \qquad (3.56)$$

where T_{21_0} is the value of T_{21} at the center of coordinates. The variation term is an order of magnitude smaller than the term involving T_{21_0} and hence can be neglected as the elementary volume becomes very small. The term involving T_{12} produces a couple in the opposite direction. Hence the surface forces produce a couple equal to

$$[T_{21} - T_{12}] \, dx_1 \, dx_2 \, dx_3 \tag{3.57}$$

If there is a body torque G_3, the total torque can be set equal to the angular acceleration times the moment of inertia I_3, or

$$[T_{21} - T_{12} + G_3] \, dx_1 \, dx_2 \, dx_3 = -I_3 \frac{\partial^2 \theta}{\partial t^2} \tag{3.58}$$

Because the moment of inertia has the dimensions $\rho[dx]^5$, then as the volume becomes very small, the right-hand side approaches zero faster than the left-hand side, and hence

$$T_{21} - T_{12} + G_3 = 0 \tag{3.59}$$

with similar equations for the other two axes.

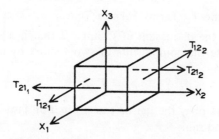

Fig. 3.9   Shearing stresses about the x_3 axis.

If a body has a permanent magnetic or electric moment, such as can occur for a ferromagnetic or ferroelectric material, a body torque can exist in the presence of a magnetic or an electric field. An example for which this effect is important is the third-order elastic constants of the ferromagnetic crystal nickel. If it is assumed that $T_{ij} = T_{ji}$, a series of third-order elastic constants result, which are proportional to the squares of the magnetic intensity.[7] These have been given the name morphic constants because they result from the change in the symmetry of the body caused by the magnetostrictive strains. In addition it has been found[8] that other deviations from the normal elastic constants occur, which are determined by the magnetic moment of the nickel in the presence of a magnetic field. These types of deviations have been called intrinsic constants. Both types of deviations are important in the theory of the coupling of magnetic spin waves to elastic waves.[9]

[7] See Mason [7].
[8] See Alers et al. [8].
[9] See Auld, et al. [9] and LeCraw and Comstock [10].

Presumably a similar effect should occur for a ferroelectric material such as barium titanate, but it has not yet been measured.

At this point it may be mentioned that a theory of elasticity has been evolved[10] that depends on the presence of body torques, which are assumed to exist in any type of material. This system results in 45 elastic moduli rather than the 21 elastic moduli normally assumed to be the maximum for the most unsymmetrical crystal. On the theoretical side it has been shown by Mindlin that such a system violates the conservation of energy and momentum principles,[11] and on the experimental side the best measurements[12] do not show the presence of these additional elastic constants. Hence in all the following work we shall assume that

$$T_{ij} = T_{ji} \tag{3.60}$$

except for ferromagnetic or ferroelectric crystals in the presence of magnetic or electric fields respectively.

3.42 Proof that the T_{ij} Terms Form a Tensor

We know from the discussion of Chapter 2 that if a set of symmetrical quantities T_{ij} relate the components of two vectors by an equation of the form

$$p_i = T_{ij}q_j \tag{2.38}$$

the T_{ij} components obey the tensor transformation laws and hence form a tensor of the second rank.

To obtain such an equation let us consider the area ABC of Fig. 3.10, formed by a plane cutting the x_1, x_2, x_3 axes. $p_1 l_1$, $p_2 l_2$, $p_3 l_3$ times the area ABC represent the components of force along the three axes exerted by the material on the positive side of the area on the tetrahedron; l_1, l_2, l_3 are the direction cosines of the normal with respect to the axes x_1, x_2, x_3. If this tetrahedron is to be in equilibrium, equal and opposite forces must be exerted on the other three faces. Resolving forces parallel to Ox_1, we have

$$p_1(ABC) = T_{11}(BOC) + T_{12}(AOC) + T_{13}(AOB) \tag{3.61}$$

But since BOC is the projection of ABC on the x_2, x_3 plane, we have

$$l_1 = \frac{BOC}{ABC} \; ; \qquad l_2 = \frac{AOC}{ABC} \; ; \qquad l_3 = \frac{AOB}{ABC} \; .$$

and hence

$$p_1 = T_{11}l_1 + T_{12}l_2 + T_{13}l_3 \tag{3.62}$$

---

[10] See Laval [11], Congr. Solvay [12], and Raman and Viswanathan [13].
[11] See Mindlin [14a] also Lax [14b].
[12] See Jaffe and Smith [15] and Beckman et al. [16].

Similarly for the other axes and in general

$$p_i = T_{ij}l_j \tag{3.63}$$

where l_j represents the direction cosines of the normal n. As the size of the tetrahedron becomes smaller these relations still hold, even if the stresses T_{ij} are not homogeneous—that is, if the stress is independent of the position of the element it acts on. If body forces are present they disappear at a more rapid rate than surface stresses, and hence (3.63) holds for this case also.

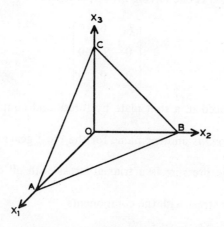

FIG. 3.10   Tetrahedron for proving the tensor nature of stresses.

Because the stress is a symmetric second-rank tensor there exist three \directions for which the stresses are normal to the surfaces and can be represented by the components

$$\begin{vmatrix} T_1 & 0 & 0 \\ 0 & T_2 & 0 \\ 0 & 0 & T_3 \end{vmatrix} \tag{3.64}$$

These are the principal stresses. We note that the stress tensor does not depend on crystal properties and in fact can have all six components in an isotropic material. Such tensors are called field tensors, whereas such tensors as those expressing dielectric permittivity or magnetic permeability are called matter tensors because they depend on the symmetry of the crystals in which they exist.

3.43   Special Forms of the Stress Tensor

Special forms of the stress tensor, all referred to the principal axes, are of interest.

1. *A uniaxial stress* is a stress produced in a long rod by a weight hanging on the end. This stress has the form

$$\begin{vmatrix} T_1 & 0 & 0 \\ 0 & 0 & 0 \\ 0 & 0 & 0 \end{vmatrix} \tag{3.65}$$

2. *A biaxial stress* is one having the stress tensor

$$\begin{vmatrix} T_1 & 0 & 0 \\ 0 & T_2 & 0 \\ 0 & 0 & 0 \end{vmatrix} \tag{3.66}$$

It can be produced in a thin plate by forces and couples applied to the edges.

3. *A triaxial stress* is another name for the most general stress system of Eq. (3.64).

4. A hydrostatic pressure is a triaxial stress with all components equal and negative.

5. A pure shear stress with the components

$$\begin{vmatrix} -T_1 & 0 & 0 \\ 0 & T_1 & 0 \\ 0 & 0 & 0 \end{vmatrix} \tag{3.67}$$

is a special case of a biaxial stress. It is seen from the Mohr circle construction of Fig. 2.4c, with the origin at the point C, that a change of axes

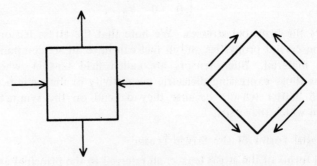

FIG. 3.11   Two equivalent forms for a shearing stress.

of 45° reduces the stress to the form

$$\begin{vmatrix} 0 & T_1 & 0 \\ T_1 & 0 & 0 \\ 0 & 0 & 0 \end{vmatrix} \tag{3.68}$$

Hence the name pure shear. This effect is illustrated by Fig. 3.11. The Ox_3 axis is the axis of shear.

3.5 Strain Tensors

The effect of applying a stress or a set of stresses to a solid body is to produce a set of displacements that are related to the strains in the body. It is the purpose of this section to discuss the definitions of the strain terms.

3.51 Material and Spatial Description of Strains

If we consider a fixed rectangular coordinate system Ox_1, Ox_2, Ox_3, any particular point can be specified at a time t by the vector r_i. A point that always moves with the material is called a material point. Lines and surfaces composed of particles or material points are called material lines or surfaces. We can also identify a particle by its coordinates at some reference time t_0 by means of a vector a_i having the coordinate values (a_1, a_2, a_3).

The vectors r_i and a_i both specify positions in a fixed Cartesian frame of reference. At any time t, r_i is the position vector of the particle initially at a_i. This connection can be written symbolically as

$$r_i = r_i(t, a_i) \qquad \text{or} \qquad x_i = x_i(t, a_1, a_2, a_3)$$

where $$\tag{3.69}$$

$$a_i = r(t_0, a_i) \qquad \text{or} \qquad a_i = x_i(t_0, a_1, a_2, a_3)$$

The coordinates a_i that identify particles are called material or Lagrangian coordinates. The inverse of Eqs. (3.69) can be written in the form

$$a_i = a_i(t, r_i) \qquad \text{or} \qquad a_i = a_i(t, x_1, x_2, x_3)$$

where $$\tag{3.70}$$

$$r_i = a_i(t_0, r_i) \qquad \text{or} \qquad x_i = a_i(t_0, x_1, x_2, x_3)$$

A spatial or Eulerian description uses the independent variables (t, x_1, x_2, x_3), the x_i's being called spatial coordinates. When used as independent variables, the x_i's merely specify a point in space.

The displacement vector from the reference position of a particle to its new position has the components

$$u_i = x_i - a_i \tag{3.71}$$

The term *strain* always refers to a change in the relative positions of the material points in a body. Consider some final configuration described at a time $t = $ a constant. Then t need no longer appear as a variable and we can write

$$x_i = x_i(a_1, a_2, a_3); \qquad a_i = a_i(x_1, x_2, x_3) \qquad (3.72)$$

In order to make use of the symmetries inherent in the unstrained crystal, material or Lagrangian strain terms are the most useful. This system regards the a_i components as the fundamental variables. The da_i are regarded as the fundamental increments, and the differentials of x_i are defined by the formulas

$$dx_i = \frac{\partial x_i}{\partial a_j} da_j \qquad (3.73)$$

Changes in distance between particles distinguish strains from rigid body rotations. Let the particle initially at the point (a_1, a_2, a_3) move to the point (x_1, x_2, x_3). The squares of the initial distance to a neighboring particle having the coordinates $a_i + da_i$ is

$$ds_0^2 = da_i \, da_i$$

In the Lagrangian or material description, the strains η_{jk} are defined by the equation

$$dx_i \, dx_i - da_i \, da_i = 2\eta_{jk} \, da_j \, da_k \qquad (3.74)$$

in which the a_i are regarded as the independent variables.

Expressions for η_{jk} are obtained by substituting from Eq. (3.73) into (3.71) and (3.74). Performing this substitution, we have

$$\eta_{jk} = \frac{1}{2}\left(\frac{\partial x_i}{\partial a_j}\frac{\partial x_i}{\partial a_k} - \delta_{jk}\right) = \frac{1}{2}\left[\left(\delta_{ij} + \frac{\partial u_i}{\partial a_j}\right)\left(\delta_{ik} + \frac{\partial u_i}{\partial a_k}\right) - \delta_{jk}\right]$$
$$= \frac{1}{2}\left(\frac{\partial u_j}{\partial a_k} + \frac{\partial u_k}{\partial a_j} + \frac{\partial u_i}{\partial a_j}\frac{\partial u_i}{\partial a_k}\right) \qquad (3.75)$$

These are the definitions of finite strains in a material or Lagrangian system. By direct inspection it is seen that they form a symmetrical tensor

$$\eta_{jk} = \eta_{kj}$$

For most applications of crystalline materials as transducers, the displacements are less than 10^{-3} times the initial length. This requirement is necessary in order that the material of the transducer does not fracture by brittle fracture or fatigue. Then for accuracies to 0.1 per cent, the products or square terms of (3.75) can be neglected. Furthermore, the ratio of the extension to the length does not vary much if either the initial or final lengths are employed. Hence infinitesimal strains, defined by the equations

$$S_{ij} = \frac{1}{2}\left(\frac{\partial u_j}{\partial x_i} + \frac{\partial u_i}{\partial x_j}\right) \tag{3.76}$$

are usually sufficient. An exception is the case of determinations of third-order elastic moduli by means of ultrasonic waves in crystals stressed by superposed stresses.[13] In this case finite strains are required. In this book finite strains are designated by the symbols η_{ij}, whereas infinitesimal strains are designated by S_{ij}. Because a complete discussion for finite strains requires extended ideas of stresses as well as strains, and because finite strains are not required for any of the present applications, the reader is referred to other texts[14] for finite-strain effects.

3.52 Engineering Strains, Rotations, Compatibility Equations

The type of strains present in a body and the rigid rotations of the body can be specified by considering two points P and Q of the body and considering their separation in the strained state. Let P be at the origin and Q have the coordinates (x_1, x_2, x_3) as shown by Fig. 3.12. If the crystal is strained homogeneously or rotated, the point P may go to P' and Q to Q'. In order to specify the strains we have to calculate the difference in length after straining as in Eq. (3.74). After the material has stretched, the point P' will have the coordinates $(u_{1_1}, u_{2_1}, u_{3_1})$ and Q' will have the coordinates

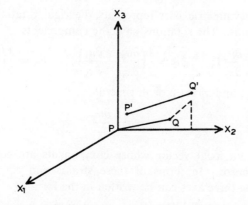

FIG. 3.12   Effect of strain on the position of the line PQ.

[13] See Bateman et al. [17].
[14] See Thurston [18].

$(x_1 + u_{1_2}, x_2 + u_{2_2}, x_3 + u_{3_2})$. But the displacement is a continuous function of the coordinates so that

$$u_{1_2} = u_{1_1} + \frac{\partial u_1}{\partial x_1} x_1 + \frac{\partial u_1}{\partial x_2} x_2 + \frac{\partial u_1}{\partial x_3} x_3$$

also

$$u_{2_2} = u_{2_1} + \frac{\partial u_2}{\partial x_1} x_1 + \frac{\partial u_2}{\partial x_2} x_2 + \frac{\partial u_2}{\partial x_3} x_3 \quad \text{or} \quad u_{i_2} = u_{i_1} + \frac{\partial u_i}{\partial x_j} x_j \qquad (3.77)$$

$$u_{3_2} = u_{3_1} + \frac{\partial u_3}{\partial x_1} x_1 + \frac{\partial u_3}{\partial x_2} x_2 + \frac{\partial u_3}{\partial x_3} x_3$$

Hence the separation of the two ends has the components

$$\delta u_i = \frac{\partial u_i}{\partial x_j} x_j \qquad (3.78)$$

The net elongations of the line in the three coordinate directions are

$$\frac{\partial u_1}{\partial x_1} x_1 ; \qquad \frac{\partial u_2}{\partial x_2} x_2 ; \qquad \frac{\partial u_3}{\partial x_3} x_3 \qquad (3.79)$$

and the elongations per unit length are defined as the *linear* strain components. The tensor of (3.78) has symmetric and antisymmetric parts, which are

$$\frac{\partial u_i}{\partial x_j} = \frac{1}{2}\left(\frac{\partial u_i}{\partial x_j} + \frac{\partial u_j}{\partial x_i}\right) + \frac{1}{2}\left(\frac{\partial u_i}{\partial x_j} - \frac{\partial u_j}{\partial x_i}\right) \qquad (3.80)$$

The symmetric part consists of the strain components

$$S_{ij} = \frac{1}{2}\left(\frac{\partial u_i}{\partial x_j} + \frac{\partial u_j}{\partial x_i}\right) \qquad (3.81)$$

whereas the antisymmetric part represents the rigid rotation of the body about the same axis. The rotations have the components

$$\omega_1 = \frac{1}{2}\left(\frac{\partial u_3}{\partial x_2} - \frac{\partial u_2}{\partial x_3}\right); \qquad \omega_2 = \frac{1}{2}\left(\frac{\partial u_3}{\partial x_1} - \frac{\partial u_1}{\partial x_3}\right); \qquad \omega_3 = \frac{1}{2}\left(\frac{\partial u_2}{\partial x_1} - \frac{\partial u_1}{\partial x_2}\right)$$

about the x_1, x_2, and x_3 axes, or in general

$$\omega_i = \varepsilon_{ijk} \frac{\partial u_k}{\partial x_j} \qquad (3.82)$$

The rotation is an axial vector whose components are equivalent to an antisymmetric tensor. In terms of these strains and rotations, the displacements of the three axes can be written in the form

$$\delta u_1 = x_1 S_{11} + x_2(S_{12} - \omega_3) + x_3(S_{13} + \omega_2)$$
$$\delta u_2 = x_1(S_{12} + \omega_3) + x_2 S_{22} + x_3(S_{23} - \omega_1) \qquad (3.83)$$
$$\delta u_3 = x_1(S_{13} - \omega_2) + x_2(S_{23} + \omega_1) + x_3 S_{33}$$

When the rotation components are zero, the form of (3.81) shows that the strain components form a symmetrical second-rank tensor. Hence the tensor has principal axes of strain where the only components of the strain are the principal strains S_1, S_2, and S_3. A shearing strain, say S_{12}, has a displacement δu_1 that increases in proportion to the dimension x_2, as shown by Fig. 3.13. Similarly, the displacement δu_2 is proportional to x_1. The

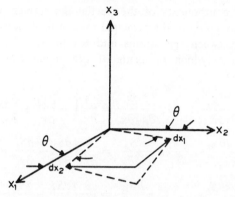

FIG. 3.13 Distortion of a square caused by a shearing strain.

resulting displacement turns a square—shown by the solid lines—into the rhombus shown by the dashed lines, with an angle θ between the legs of the rhombus and the coordinate axes.

$$\tan \theta = \frac{\delta u_2}{x_1 + \delta u_1} = \frac{\delta u_1}{x_2 + \delta u_2} \doteq S_{12} \doteq \theta \tag{3.84}$$

since δu_1 and δu_2 are very small compared to the dimensions x_1 and x_2 and θ is a very small angle. The cosine of the angle φ between the two legs is

$$\cos \varphi = \cos (90° - 2\theta) = \cos 90° \cos 2\theta + \sin 90° \sin 2\theta \doteq 2S_{12} \tag{3.85}$$

Just as in the case of stresses, shown in Fig. 3.11, a rotation of 45° of the defining axes turns a pure shear strain into a tensile stretch and a compression.

Unfortunately, engineering usage has defined the shearing strains as twice the values given by Eq. (3.81). At the same time, since there are six independent strains in the strain matrix, it has been customary to designate them by the subscripts 1 to 6. We make it a practice to refer to the engineering strains by the single subscript and to the tensor strains by the double subscript. With this convention the relations between the engineering strains and the tensor strains are

$$
\begin{aligned}
&S_1 = S_{11}; &&S_2 = S_{22}; &&S_3 = S_{33}; \\
&S_4 = 2S_{23} = 2S_{32}; &&S_5 = 2S_{13} = 2S_{31}; &&S_6 = 2S_{12} = 2S_{21}
\end{aligned}
\tag{3.86}
$$

A similar convention is used with respect to the stresses. In this case, however, there is no difference between the tensor stresses and the engineering stresses, so that we have

$$T_1 = T_{11} ; \qquad T_2 = T_{22} ; \qquad T_3 = T_{33} ;$$
$$T_4 = T_{23} = T_{32} ; \qquad T_5 = T_{13} = T_{31} ; \qquad T_6 = T_{12} = T_{21} \tag{3.87}$$

Since the nine components of the partial derivatives of (3.80) can be reduced to three by going to the principal axes, it is obvious that there must be six relations between the strains and rotations. These are the compatibility relations,[15] which in terms of the engineering strains can be written in the form

$$\frac{\partial^2 S_2}{\partial x_3{}^2} + \frac{\partial^2 S_3}{\partial x_2{}^2} = \frac{\partial^2 S_4}{\partial x_2 \, \partial x_3} ; \qquad 2\frac{\partial^2 S_1}{\partial x_2 \, \partial x_3} = \frac{\partial}{\partial x_1}\left(-\frac{\partial S_4}{\partial x_1} + \frac{\partial S_5}{\partial x_2} + \frac{\partial S_6}{\partial x_3}\right)$$

$$\frac{\partial^2 S_3}{\partial x_1{}^2} + \frac{\partial^2 S_1}{\partial x_3{}^2} = \frac{\partial^2 S_5}{\partial x_3 \, \partial x_1} ; \qquad 2\frac{\partial^2 S_2}{\partial x_3 \, \partial x_1} = \frac{\partial}{\partial x_2}\left(\frac{\partial S_4}{\partial x_1} - \frac{\partial S_5}{\partial x_2} + \frac{\partial S_6}{\partial x_3}\right) \tag{3.88}$$

$$\frac{\partial^2 S_1}{\partial x_2{}^2} + \frac{\partial^2 S_2}{\partial x_1{}^2} = \frac{\partial^2 S_6}{\partial x_1 \, \partial x_2} ; \qquad 2\frac{\partial^2 S_3}{\partial x_1 \, \partial x_2} = \frac{\partial}{\partial x_3}\left(\frac{\partial S_4}{\partial x_1} + \frac{\partial S_5}{\partial x_2} - \frac{\partial S_6}{\partial x_3}\right)$$

Most of the elastic constants, which turn out to be ratics of stress to strain under particular conditions, have been measured in terms of the engineering strains. It should be emphasized that they do not form a tensor and hence cannot be transformed to other axes by tensor formulas. It is usually simpler to work with tensor strains and then convert to engineering strains by multiplying the shear components by the factor 2.

3.53 Energy Stored in a Body Due to Small Strains

It is shown by Love[16] that the potential energy of deformation of a body under equilibrium is given by the expression

$$W = \frac{1}{2} \iiint \left(S_1 \frac{\partial W}{\partial S_1} + S_2 \frac{\partial W}{\partial S_2} + S_3 \frac{\partial W}{\partial S_3} + S_4 \frac{\partial W}{\partial S_4} + S_5 \frac{\partial W}{\partial S_5} + S_6 \frac{\partial W}{\partial S_6}\right) d\tau$$
$$= \frac{1}{2} \iiint S_m \frac{\partial W}{\partial S_m} d\tau \tag{3.89}$$

where m takes the values 1 to 6. The partial derivatives of the energy by the strains are the stresses—that is,

$$T_m = \frac{\partial W}{\partial S_m} \qquad m = 1 \text{ to } 6 \tag{3.90}$$

[15] See Love [19].
[16] See Love [19], p. 173.

Hence the increment of energy produced by an increment of strain dS_m can be written in the form

$$T_m \, dS_m \, dx_1 \, dx_2 \, dx_3 \tag{3.91}$$

This can be illustrated by means of the elementary cube of Fig. 3.8. If this is given a strain $S_{11} + dS_{11}$, the elongation in the x direction due to the infinitesimal strain dS_{11} is $dx_1 \, dS_{11}$. This works against a surface force $T_{11} \, dx_2 \, dx_3$. Hence the internal energy stored by this strain is

$$T_{11} \, dS_{11} \, dx_1 \, dx_2 \, dx_3 \tag{3.92}$$

Similar expressions hold for the other tensional components. For the shear components, it is evident from Eq. (3.83) that the displacements

$$\delta u_1 = x_2 \, dS_{12} \quad \text{and} \quad \delta u_2 = x_1 \, dS_{12} \tag{3.93}$$

occur when a change dS_{12} occurs in the shearing strain of the element of Fig. 3.9. The work done on the surfaces perpendicular to x_1 is then

$$\tfrac{1}{2} dx_1 \, T_{21_2} \, dS_{12} \, dx_2 \, dx_3 + \tfrac{1}{2} dx_1 \, T_{21} \, dS_{21} \, dx_2 \, dx_3 \doteq T_{21} \, dS_{12} \, dx_1 \, dx_2 \, dx_3 \tag{3.94}$$

Similarly, the work on the faces perpendicular to x_2 is given by

$$T_{12} \, dS_{12} \, dx_1 \, dx_2 \, dx_3$$

and the total work is

$$2T_{12} \, dS_{12} \, dx_1 \, dx_2 \, dx_3 \tag{3.95}$$

since $T_{21} = T_{12}$. Similar results hold for the other two shearing strains, and the total work can be written in the tensor form

$$T_{ij} \, dS_{ij} \, dx_1 \, dx_2 \, dx_3 = T_{11} \, dS_{11} + 2T_{12} \, dS_{12}$$
$$+ 2T_{13} \, dS_{13} + T_{22} \, dS_{22} + 2T_{23} \, dS_{23} + T_{33} \, dS_{33} \tag{3.96}$$

Since

$$T_{12} = T_6 \quad \text{and} \quad 2 \, dS_{12} = dS_6$$

this equation is the same as that given in (3.91).

3.6 Elasticity

It is a consequence of Hooke's law, valid for strains below elastic limits, that the stresses are proportional to the strains or vice versa. If we write this relation in terms of the engineering stresses and strains—as is usually done—the most general relationship for the stresses and the strains is

$$
\begin{aligned}
T_1 &= c_{11}S_1 + c_{12}S_2 + c_{13}S_3 + c_{14}S_4 + c_{15}S_5 + c_{16}S_6 \\
T_2 &= c_{21}S_1 + c_{22}S_2 + c_{23}S_3 + c_{24}S_4 + c_{25}S_5 + c_{26}S_6 \\
T_3 &= c_{31}S_1 + c_{32}S_2 + c_{33}S_3 + c_{34}S_4 + c_{35}S_5 + c_{36}S_6 \\
T_4 &= c_{41}S_1 + c_{42}S_2 + c_{43}S_3 + c_{44}S_4 + c_{45}S_5 + c_{46}S_6 \\
T_5 &= c_{51}S_1 + c_{52}S_2 + c_{53}S_3 + c_{54}S_4 + c_{55}S_5 + c_{56}S_6 \\
T_6 &= c_{61}S_1 + c_{62}S_2 + c_{63}S_3 + c_{64}S_4 + c_{65}S_5 + c_{66}S_6
\end{aligned} \tag{3.97a}
$$

or

$$T_m = c_{mn}S_n \qquad (3.97b)$$

In this equation c_{11}, for example, is an elastic constant expressing the proportionality of stress T_1 to the strain S_1 when all the other strains are zero. The elastic constants c_{mn} really form a fourth-rank tensor, as can be seen from the relation between stress and strain written in tensor form

$$T_{ij} = c_{ijkl}S_{kl} \qquad (3.98)$$

In this form the four-index elastic constants are related to the two-index constants in the same way as the two-index stress and strain tensors are related to the one-index stress and strain terms—that is, for any ij components, replace ij by m where

$$1 = 11, \quad 2 = 22, \quad 3 = 33, \quad 4 = 23 = 32, \quad 5 = 13 = 31, \quad 6 = 12 = 21$$
$$(3.99)$$

Similar relations hold for the kl terms that are replaced by the n terms. This tensor summation indicated by (3.98) automatically takes account of the difference between the definitions of the engineering strains and the tensor strains. For example, the fourth term of the first equation of (3.97) will be, in tensor form,

$$c_{1123}S_{23} + c_{1132}S_{32} = 2c_{1123}S_{23} = c_{14}S_4$$

In this form there are 36 elastic constants, but the existence of a strain energy function W can be used to show that

$$c_{mn} = c_{nm} \qquad \text{or} \qquad c_{ijkl} = c_{klij} \qquad (3.100)$$

Introducing the relations of (3.97) in (3.91) we have

$$dW = c_{mn}S_n\, dS_m \qquad (3.101)$$

Hence the first derivative is

$$\frac{\partial W}{\partial S_m} = c_{mn}S_n = T_n \qquad (3.102)$$

Differentiating both sides of this equation by S_n we have

$$\frac{\partial^2 W}{\partial S_n\, \partial S_m} = c_{mn} \qquad (3.103)$$

But since W is a function only of the state of the body, specified by the strain components, we can interchange the order of differentiation, and hence

$$c_{mn} = c_{nm} \qquad (3.104)$$

This symmetry reduces the number of elastic constants to 21 for the most unsymmetric crystals. As discussed in Chapter 5, the symmetries inherent in different types of crystals reduce the number of independent constants.

Table B.IV, Appendix B, shows the number of constants for the different types of symmetries. In particular, for an isotropic material only two constants are left, the λ and μ Lamé constants, which are related to the c_{ij} constants of Eq. (3.97) by the relations

$$\lambda + 2\mu = c_{11} = c_{22} = c_{33} \ ; \qquad \lambda = c_{12} = c_{13} = c_{23} \ ; \qquad \mu = c_{44} = c_{55} = c_{66}$$

(3.105)

All the other c constants are zero. Hence for an isotropic material the stress-strain relations are

$$
\begin{aligned}
T_1 &= (\lambda + 2\mu)S_1 + \lambda(S_2 + S_3); & T_4 &= \mu S_4 \\
T_2 &= (\lambda + 2\mu)S_2 + \lambda(S_1 + S_3); & T_5 &= \mu S_5 \\
T_3 &= (\lambda + 2\mu)S_3 + \lambda(S_1 + S_2); & T_6 &= \mu S_6
\end{aligned}
$$

(3.106)

It is obvious from Eq. (3.97) that the strains can be expressed in terms of the stresses by solving the six equations simultaneously. The result can be expressed in the form

$$
\begin{aligned}
S_1 &= s_{11}T_1 + s_{12}T_2 + s_{13}T_3 + s_{14}T_4 + s_{15}T_5 + s_{16}T_6 \\
S_2 &= s_{12}T_1 + s_{22}T_2 + s_{23}T_3 + s_{24}T_4 + s_{25}T_5 + s_{26}T_6 \\
S_3 &= s_{13}T_1 + s_{23}T_2 + s_{33}T_3 + s_{34}T_4 + s_{35}T_5 + s_{36}T_6 \\
S_4 &= s_{14}T_1 + s_{24}T_2 + s_{34}T_3 + s_{44}T_4 + s_{45}T_5 + s_{46}T_6 \\
S_5 &= s_{15}T_1 + s_{25}T_2 + s_{35}T_3 + s_{45}T_4 + s_{55}T_5 + s_{56}T_6 \\
S_6 &= s_{16}T_1 + s_{26}T_2 + s_{36}T_3 + s_{46}T_4 + s_{56}T_5 + s_{66}T_6
\end{aligned}
$$

or

$$S_i = s_{ij}T_j \ , \qquad i,j = 1 \text{ to } 6$$

(3.107)

where the compliance constants s_{ij} can be expressed in terms of the elastic constants c_{ij} by the determinantal relation

$$s_{mn} = \frac{(-1)^{m+n}\Delta_{mn}^c}{\Delta^c}$$

(3.108)

where Δ^c is the determinant of Eq. (3.97)—that is,

$$
\begin{vmatrix}
c_{11} & c_{12} & c_{13} & c_{14} & c_{15} & c_{16} \\
c_{12} & c_{22} & c_{23} & c_{24} & c_{25} & c_{26} \\
c_{13} & c_{23} & c_{33} & c_{34} & c_{35} & c_{36} \\
c_{14} & c_{24} & c_{34} & c_{44} & c_{45} & c_{46} \\
c_{15} & c_{25} & c_{35} & c_{45} & c_{55} & c_{56} \\
c_{16} & c_{26} & c_{36} & c_{46} & c_{56} & c_{66}
\end{vmatrix}
$$

(3.109)

and Δ_{mn}^c is the minor obtained by suppressing the mth row and nth column. From this it is obvious that $s_{mn} = s_{nm}$.

Equations (3.107) can also be written in the tensor form

$$S_{ij} = s_{ijkl} T_{kl} \qquad (3.110)$$

For this case the relation between the s_{mn} constants and the s_{ijkl} constants are determined by the replacements of Eqs. (3.99). However, for any number 4, 5, or 6 the s_{mn} compliance has to be divided by 2 to equal the corresponding s_{ijkl} compliance. If the numbers 4, 5, or 6 occur twice, the divisor has to be 4.

For an isotropic material the six equations can be written in the form

$$S_1 = \frac{T_1}{Y_0} - \frac{\sigma}{Y_0}(T_2 + T_3); \qquad S_4 = \frac{T_4}{\mu}$$

$$S_2 = -\frac{\sigma}{Y_0}(T_1 + T_3) + \frac{T_2}{Y_0}; \qquad S_5 = \frac{T_5}{\mu} \qquad (3.111)$$

$$S_3 = -\frac{\sigma}{Y_0}(T_1 + T_2) + \frac{T_3}{Y_0}; \qquad S_6 = \frac{T_6}{\mu}$$

where

$$s_{11} = s_{22} = s_{33} = \frac{1}{Y_0}; \qquad s_{12} = s_{13} = s_{23} = \frac{\sigma}{Y_0}; \qquad s_{44} = s_{55} = s_{66} = \frac{1}{\mu}$$

with all other s_{mn} constants equal to zero. Here Y_0 is Young's modulus and σ is Poisson's ratio. By solving (3.108) we find

$$s_{11} = \frac{1}{Y_0} = \frac{\lambda + \mu}{\mu(3\lambda + 2\mu)}; \qquad s_{12} = -\frac{\lambda}{2(\lambda + \mu)}\left[\frac{\lambda + \mu}{\mu(3\lambda + 2\mu)}\right] = -\frac{\sigma}{Y_0}$$

where

$$\sigma = \frac{\lambda}{2(\lambda + \mu)}$$

$$(3.112)$$

Another elastic modulus of some interest is the bulk modulus B, which is defined as the ratio of a hydrostatic pressure p to the change in volume Δ of the material. Since $\Delta = S_1 + S_2 + S_3$, we find from (3.106) by setting $T_1 = T_2 = T_3 = p$ that

$$\Delta = -\frac{3p}{(3\lambda + 2\mu)}; \qquad \text{hence,} \qquad B = \frac{3\lambda + 2\mu}{3} = \lambda + \frac{2}{3}\mu \qquad (3.113)$$

Table 2 shows the values of the elastic constants for the crystals whose

Table 2

ELASTIC CONSTANTS OF SOME COMMONLY USED CRYSTALS

Crystal	Crystal Symmetry[a]	c_{11}[b]	c_{12}	c_{13}	c_{33}(17)	c_{14}	c_{44}	c_{66}	Density ρ in kg/m³ × 10³ or g/cm³
LiF	$m3m$	11.35	4.8	—	—	—	6.35	—	2.646
NaCl	$m3m$	4.93	1.31	—	—	—	1.275	—	2.164
KCl	$m3m$	4.08	0.69	—	—	—	0.635	—	2.038
ZnS	$\bar{4}3m$	9.98	6.48	—	—	—	3.41	—	4.087
MgO	$m3m$	28.6	8.7	—	—	—	14.8	—	3.576
AgCl	$m3m$	6.01	3.62	—	—	—	0.625	—	5.56
Al₂O₃	$\bar{3}m$	49.6	16.4	11.1	49.8	-2.35	14.7	—	4.00
NH₄H₂PO₄ (ADP)	$\bar{4}2m$	6.76	0.59	2.0	3.38	—	0.867	0.688	1.804
KH₂PO₄ (KDP)	$\bar{4}2m$	7.4	1.8	2.7	6.8	—	1.35	0.63	2.338
TiO₂ (Rutile)	$4/mmm$	28.0	18.0	14.0	46.0	—	12.0	16.0	4.264
SiO₂ (α quartz)	32	8.68	0.709	1.19	10.59	+1.80	5.82	—	2.65
CaCO₃ Calcite	$\bar{3}m$	13.8	4.6	5.68	8.11	-2.0	3.5	—	2.712
CdS	$6mm$	8.42	5.21	4.64	9.39	—	1.489	—	4.825
ZnO	$6mm$	20.9	12.1	10.5	21.1	—	4.25	—	5.676
Tourmaline	$3m$	27.0	4.0	3.5	16.5	-0.68	6.5	—	3.1

[a] The significance of the symmetry symbols is explained in Chapter 5.

[b] All the elastic moduli c_{mn} are expressed in newtons per square meter × 10¹⁰. To get the cgs units of dynes/cm² multiply by 10.

[17] These values are obtained from Anderson [20]. Large tables of elastic constants are given there.

Table 3

LAMÉ ELASTIC CONSTANTS FOR A NUMBER OF METALS

Material	Density ρ kg/m³ ×10³ or g/cm³	$\lambda + 2\mu$	μ	E	Poisson's ratio σ	$V_{(\lambda+2\mu)}$	V_E	V_μ
		Units 10¹⁰ N/m² = 10¹¹ d/cm²				Meters per second		
Aluminum	2.711	11.01	2.62	7.06	0.345	6380	5100	3100
Duraluminum	2.70	11.05	2.63	1.08	0.345	6398	5120	3122
Beryllium	1.87	31.00	14.7	30.8	0.05	12,890	12,870	8,880
Brass[a]	8.45	16.15	3.73	10.06	0.35	4370	3450	2100
Chromium	7.16	31.39	11.53	27.9	0.21	6608	6230	4000
Constantan	8.88	23.8	6.12	16.24	0.327	5180	4280	2625
Copper	8.93	20.22	4.83	12.98	0.343	4760	3810	2325
Gold (hard drawn)	19.1	20.7	2.85	8.12	0.42	3240	2030	1200
Invar[b]	8.10	17.6	5.72	14.4	0.259	4660	4215	2660
Iron (soft)	7.85	27.86	8.16	21.14	0.293	5960	5190	3225
Iron (cast)	7.6	18.95	6.0	15.23	0.27	4995	4480	2810
Lead (rolled)	11.4	4.38	0.54	1.6	0.43	1960	1210	690
Magnesium	1.74	5.87	1.73	4.47	0.291	5820	5080	3165
Molybdenum	10.22	42.85	12.56	32.48	0.293	6475	5640	3500
Monel metal[c]	8.90	25.4	6.5	17.3	0.327	5360	4410	2710
Nickel (soft)	8.86	27.9	7.60	19.95	0.312	5610	4790	2930
Nickel (hard)	8.86	29.95	8.39	21.92	0.306	5814	4974	3078
Nickel silver[d]	8.62	19.80	4.97	13.25	0.333	4800	3920	2400
Niobium	8.58	22.04	3.75	10.49	0.397	5068	3497	2092
Ni Span C	8.05						4831	2799
Platinum	21.4	22.7	6.4	16.7	0.303	3260	2800	1730
Silver	10.5	14.4	3.03	8.27	0.367	3690	2810	1710
Steel (mild)	7.85	27.88	8.22	21.19	0.291	5960	5200	3235
Steel (tool)	7.83	27.5	8.22	21.16	0.287	5921	5200	3233
Steel (stainless)[e]	7.72	27.8	8.4	21.5	0.283	5980	5282	3300
Tantalum	16.68	28.85	6.92	18.6	0.342	4160	3340	2040
Tin	17.24	8.27	1.84	5.00	0.357	3390	2630	1600
Tungsten	19.20	58.1	13.4	36.2	0.35	5410	4320	2640
Vanadium	6.07	22.03	4.67	12.76	0.365	6025	4590	2775
Zinc	7.14	12.52	4.19	10.45	0.241	4190	3825	2420

[a] (70 Zn, 30 Cu)
[b] (36 Ni, 63.8 Fe, 0.2 C)
[c] (71 Ni, 27 Cu, 2 Fe)
[d] (55 Cu, 18.1 Ni, 27 Zn)
[e] (0.2 C, 0.5 Si, 0.7 Mn, 2 Ni, 18 Cr, 78.6 Fe)

relative dielectric constants are given in Table 1. The units used are newtons/meter^2 $\times$ 10^{10}. Table 3 gives the Lamé constants for a number of metals.

3.7 Thermal Variables

3.71 Temperature, Quantity of Heat, and Entropy

The final set of variables of interest in interaction processes are the thermal variables, temperature and entropy. In the interaction diagram of Fig. 1.2 it is seen that an increase in temperature produces strains through the thermal-expansion coefficient and produces electric displacements through the pyroelectric effect. Similarly strains and displacements react on the entropy of a substance through the heat of deformation and the heat of polarization. Hence, because of the interaction of thermal effects with other effects, such thermal variables are of fundamental importance in interaction processes.

From the first law of thermodynamics we have, "When work is transformed into heat, or heat into work, the amount of work is always equivalent to the heat." The unit of heat is usually taken to be the calorie, which is defined as the amount of heat that will raise the temperature of 1 gram of water from 14.5° C to 15.5° C. The mechanical equivalent of heat has been shown to be

$$1 \text{ calorie} = 4.186 \text{ joules} = 4.186 \times 10^7 \text{ ergs} \qquad (3.114)$$

It is found that it requires much less thermal energy to increase the temperature of a gram of copper by 1° C than it does for a gram of water. This is related to the smaller heat capacity of the copper, and a measure of the heat capacity is the proportionality factor between the heat added and the resulting temperature difference

$$C = \frac{Q}{\Theta_2 - \Theta_1} \qquad (3.115)$$

The third thermodynamic variable, the entropy σ, is usually defined by the equation

$$d\sigma = \frac{C}{\Theta} d\Theta \qquad (3.116)$$

From (3.115) if we let the temperature difference $\Theta_2 - \Theta_1$ becomes small so that $\Theta_2 - \Theta_1 = d\Theta$, then Eq. (3.116) can be written in the form

$$\Theta \, d\sigma = dQ \qquad (3.117)$$

In Eq. (3.116), Θ is the temperature in degrees Kelvin, measured from absolute zero—that is, $-273°$ C. According to the second law of thermodynamics, in any closed system the entropy always remains the same for any reversible process but increases for any irreversible process, or

$$d\sigma \geqq 0 \tag{3.118}$$

The entropy also has some significance in the statistical mechanics of an ensemble of particles—for example, a gas. The free energy of such an ensemble is defined as

$$A = -k\Theta \log Z \tag{3.119}$$

where Z is the partition function determining the states of the individual particles and k is Boltzmann's constant. The entropy σ is

$$\sigma = -\frac{\partial A}{\partial \Theta} = k \log Z \tag{3.120}$$

It can be shown that its value is a minimum when the distribution f is given by the Maxwell-Boltzmann configuration.[18] However, we shall not have occasion to consider such relations in the phenomenological treatment discussed in this book.

3.72 Thermal Expansions, Specific Heats, and Adiabatic and Isothermal Elastic Moduli

Since the specific heat of Eq. (3.115) depends on whether it is measured at constant volume—that is, zero strains—or at constant pressure—that is, zero or constant stress on the body—some results of the next chapter are assumed and relations are considered between the mechanical variables and the thermal variables.

As shown in Chapter 4, the strains can be represented in terms of the stresses and an increment in temperature $\delta\Theta$ by the equations below. A relation between the entropy, stresses, and temperature is also given

$$S_{ij} = s_{ijkl}^{\Theta} T_{kl} + \alpha_{ij} \delta\Theta$$
$$\delta\sigma = +\alpha_{kl} T_{kl} + \frac{\rho C_p}{\Theta} \delta\Theta \tag{3.121}$$

The elastic compliances have the superscript Θ to indicate they are measured under isothermal conditions, and the specific heat per gram has the subscript p to indicate that it is measured at constant stress. ρC_p is the specific heat per unit volume.

There are in general six temperature-expansion coefficients that relate the strains to the increment in temperature $\delta\Theta$. As shown by Table B. II in Appendix B, the number varies depending on the symmetry, and crystallographic axes are principal axes except for the triclinic and monoclinic

[18] See Joos [21].

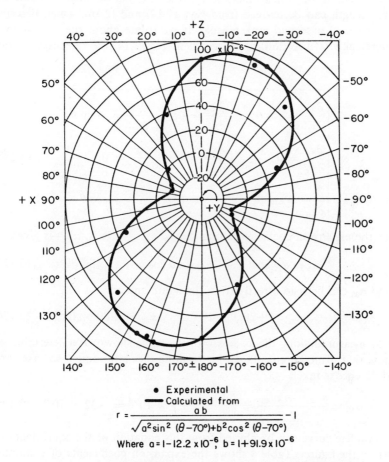

FIG. 3.14 Temperature expansion coefficients for an EDT crystal.

systems. The method of measurement usually consists of an optical system that measures the number of wavelengths change between two mirrors caused by the change in separations produced by the thermal expansion of the crystal. Very sensitive measurements have been obtained by employing special systems.[19] X-ray measurements of lattice spacings as a function of temperature are also employed. All methods measure the expansion normal to the length of the sample. Figure 3.14 shows the thermal expansion for

[19] See McCammon and White [22] and White [23].

the monoclinic crystal ethylene diamine tartrate (EDT)[20] as a function of orientation in the x_1, x_3 plane—here designated the x, z plane.

The equation of transformation for the thermal-expansion coefficient along a length can be derived from Eqs. (2.12) and (2.16), which describe the rotation of a second-rank tensor. If we rotate about the x_2 axis in a counterclockwise direction, as indicated in Fig. 3.14, the direction cosines are

$$
\begin{array}{c|ccc}
 & x_1 & x_2 & x_3 \\
\hline
x_1' & \dfrac{\partial x_1}{\partial x_1'} = \cos\theta; & \dfrac{\partial x_2}{\partial x_1'} = 0; & \dfrac{\partial x_3}{\partial x_1'} = -\sin\theta \\[2ex]
x_2' & \dfrac{\partial x_1}{\partial x_2'} = 0; & \dfrac{\partial x_2}{\partial x_2'} = 1; & \dfrac{\partial x_3}{\partial x_2'} = 0 \\[2ex]
x_3' & \dfrac{\partial x_2}{\partial x_3'} = \sin\theta; & \dfrac{\partial x_2}{\partial x_3'} = 0; & \dfrac{\partial x_3}{\partial x_3'} = \cos\theta
\end{array}
\tag{3.122}
$$

Hence the expansion coefficient along the x_3' axis—that is, α_{33}'—is given by

$$
\alpha_{33}' = \alpha_{11}\cos^2\theta + 2\alpha_{13}\sin\theta\cos\theta + \alpha_{33}\sin^2\theta
\tag{3.123}
$$

α_{11} and α_{33} can be read directly from the figure and are

$$
\alpha_{11} = 0; \qquad \alpha_{22} = +20.3 \times 10^{-6}; \qquad \alpha_{33} = +80 \times 10^{-6}
\tag{3.124}
$$

α_{22} was measured along the x_2 axis with the value shown. To determine α_{13} we note that if measurements are made along the $+45°$ and $-45°$ axes and the difference is taken

$$
\alpha_{13} = \frac{\alpha_{+45°} - \alpha_{-45°}}{2} = \left(\frac{+8-72}{2}\right) \times 10^{-6} = -32 \times 10^{-6}
\tag{3.125}
$$

The complete curve can be plotted from this equation or the equivalent one shown in the figure. Table 4 shows the expansion coefficients of a number of crystals and materials of interest. All coefficients are given in parts per million per degree C.

The relations of Eq. (3.121) can be used directly to determine the difference between the adiabatic and the isothermal elastic compliances. For an adiabatic condition there is no heat flow from one part of the crystal to the other and hence $d\sigma = 0$. Solving the last of Eqs. (3.121) for $d\Theta$ and inserting in the first equation we find

$$
S_{ij} = \left[s_{ijkl}^{\Theta} - \frac{\alpha_{ij}\alpha_{kl}\Theta}{\rho C_p} \right] T_{kl}
\tag{3.126}
$$

[20] See Mason [24].

Table 4

THERMAL EXPANSION COEFFICIENTS OF SOME MATERIALS OF INTEREST

Material	Crystal system	α_1	α_2	α_3	$\alpha_5 = \alpha_{13/2}$
Diamond	cubic	0.87	0.87	0.87	—
LiF	cubic	33.17	33.17	33.17	—
KCl	cubic	36.0	36.0	36.0	—
MgO	cubic	10.98	10.98	10.98	—
NaCl	cubic	39.2	39.2	39.2	—
AgCl	cubic	31.36	31.36	31.36	—
ZnS	cubic	6.7	6.7	6.7	—
Al_2O_3	trigonal	5.42	5.42	6.58	—
$NH_4H_2PO_4$ (ADP)	tetragonal	34.0	34.0	2.5	—
KH_2PO_4 (KDP)	tetragonal	48.4	48.4	26.6	—
TiO_2 (Rutile)	tetragonal	7.14	7.14	9.19	—
SiO_2 (α quartz)	trigonal	14.3	14.3	7.8	—
$CaCO_3$	trigonal	26.6	26.6	−5.6	—
CdS	hexagonal	6.5	6.5	4.0	—
Tourmaline	trigonal	3.7	3.7	9.0	—
Rochelle salt	orthorhombic	58.3	37.0	43.0	—
Ethylene diamine tartrate	monoclinic	0	+20.3	+80.0	−16.0

The term in brackets is the adiabatic compliance s^{σ}_{ijkl} and hence

$$s^{\sigma}_{ijkl} = s^{\Theta}_{ijkl} - \frac{\alpha_{ij}\alpha_{kl}\Theta}{\rho C_p} \qquad (3.127)$$

The stresses can be expressed in terms of the strains by solving the six components of the first equation of (3.121) simultaneously. The result is

$$T_{kl} = c^{\Theta}_{ijkl}S_{ij} - \lambda_{kl}\, d\Theta \qquad (3.128)$$

where

$$\lambda_{kl} = \alpha_{ij}c^{\Theta}_{ijkl}$$

In terms of the reduced notation

$$\lambda_1 = \alpha_1 c^{\Theta}_{11} + \alpha_2 c^{\Theta}_{12} + \alpha_3 c^{\Theta}_{13} + \alpha_4 c^{\Theta}_{14} + \alpha_5 c^{\Theta}_{15} + \alpha_6 c^{\Theta}_{16}$$
$$\cdots\cdots\cdots\cdots\cdots\cdots\cdots\cdots\cdots\cdots\cdots\cdots\cdots\cdots\cdots\cdots\cdots\cdots\cdots \qquad (3.129)$$
$$\lambda_6 = \alpha_1 c^{\Theta}_{16} + \alpha_2 c^{\Theta}_{26} + \alpha_3 c^{\Theta}_{36} + \alpha_4 c^{\Theta}_{46} + \alpha_5 c^{\Theta}_{56} + \alpha_6 c^{\Theta}_{66}$$

As can be seen from (3.128) the values of λ represent the ratios of stresses to the increment of temperature $d\Theta$ necessary to be applied to maintain the shape of the body without change—that is, with zero strains. These are usually pressures—that is, negative tensions.

The last equation of (3.121) can be expressed in terms of the strains by using Eq. (3.128). The result is

$$d\sigma = \lambda_{ij}S_{ij} + \frac{[\rho C_p - \Theta\alpha_{ij}\lambda_{ij}]}{\Theta}\, d\Theta \qquad (3.130)$$

Since the last term is the relation between the entropy and the temperature when the strains S_{ij} are zero, the term in brackets is the specific heat at constant volume C_v times the density ρ—that is, the specific heat per unit volume.

Finally, by eliminating $d\Theta$ from (3.130) and (3.128) by setting $d\sigma = 0$, the difference between the adiabatic and isothermal elastic moduli becomes

$$c_{ijkl}^{\sigma} = c_{ijkl}^{\Theta} + \frac{\lambda_{ij}\lambda_{kl}\Theta}{\rho C_v} \tag{3.131}$$

An example is of some interest. From Table 4, the thermal-expansion coefficients for quartz are

$$\alpha_{11} = \alpha_1 = \alpha_{22} = \alpha_2 = 14.3 \times 10^{-6} ; \qquad \alpha_{33} = \alpha_3 = 7.8 \times 10^{-6}$$

while the elastic constants given in Table 2 are

$$c_{11} = c_{1111} = 8.68 \times 10^{10} ; \qquad c_{33} = 10.59 \times 10^{10} ; \tag{3.132}$$
$$c_{12} = 0.909 \times 10^{10} ; \qquad c_{13} = 1.19 \times 10^{10} \text{ newtons/meter}^2$$

Hence from (3.129) the values of λ_i are

$$\lambda_1 = \alpha_1(c_{11} + c_{12}) + \alpha_3 c_{13} = 1.43 \times 10^6 \text{ newtons/meter}^2 \text{ degree}; \qquad \lambda_4 = 0$$
$$\lambda_2 = \alpha_1(c_{12} + c_{11}) + \alpha_3 c_{13} = 1.43 \times 10^6 \text{ newtons/meter}^2 \text{ degree}; \qquad \lambda_5 = 0$$
$$\lambda_3 = 2\alpha_1 c_{13} + \alpha_3 c_{33} \quad = 1.16 \times 10^6 \text{ newtons/meter}^2 \text{ degree}; \qquad \lambda_6 = 0$$
$$\tag{3.133}$$

The specific heat at constant stress measured for quartz is 7.37×10^6 ergs per gram, or 1.95 Joules per cubic centimeter since the density of quartz is 2.65. In MKS units this is 1.95×10^6 joules per cubic meter. Hence the specific heat at constant volume is, at $300°$ K,

$$\rho C_v = 1.95 \times 10^6 - 300(14.8 \times 10^{-6} \times 1.38 \times 10^6 \times 2$$
$$+ 7.8 \times 10^{-6} \times 1.14 \times 10^6) \tag{3.134}$$

$$\rho C_v = [1.95 - 0.0015] \times 10^6 \tag{3.135}$$

Hence, the difference between the specific heat at constant stress and at constant volume is only about 0.1 per cent.

The differences between the adiabatic and isothermal moduli $c_{11} = c_{1111}$ and $c_{33} = c_{3333}$ are given by

$$c_{11}^{\sigma} - c_{11}^{\Theta} = \frac{\lambda_1^2 \Theta}{\rho C_v} = 2.94 \times 10^8 \text{ newton/meter}^2 \qquad \text{or} \qquad 0.34 \text{ per cent}^{[21]}$$
$$\tag{3.136}$$

$$c_{33}^{\sigma} - c_{33}^{\Theta} = \frac{\lambda_3^2 \Theta}{\rho C_v} = 2 \times 10^8 \text{ newton/meter}^2 \qquad \text{or} \qquad 0.187 \text{ per cent}$$

[21] See American Institute of Physics Handbook [25].

It is obvious from the fact that $\lambda_4 = \lambda_5 = \lambda_6 = 0$ that there is no change between adiabatic and isothermal values for shear moduli.

The effect is somewhat larger in a metal on account of the larger temperature-expansion coefficients. For an isotropic material the values of λ_i are all the same for $i = 1, 2,$ or 3, and are equal to

$$\lambda = \alpha(\lambda + 2\mu + 2\lambda) = 3\alpha B \qquad (3.137)$$

where B is the bulk modulus. For example, for zinc the constants are $\rho = 7.1 \times 10^3$ kilograms/meter^3 ; $C_v = 382$ joules/kilogram; $\lambda^\Theta = 4.2 \times 10^{10}$ newtons/meter^2 ; $B = 7 \times 10^{10}$ newtons/meter^2 ; and $\alpha = 29.7 \times 10^{-6}$. Hence

$$\lambda^\sigma - \lambda^\Theta = \frac{(3\alpha B)^2 \times 300}{2.71 \times 10^6} = 4.3 \times 10^9 \text{ newtons/meter}^2 \qquad (3.138)$$

If we compare this with $\lambda^\Theta + 2\mu$—which is the equivalent of c_{11}—the change is 3.4 per cent.

RESUME CHAPTER 3

I. Electrical Variables

Electrical variables are defined with respect to a parallel-plate condenser with major surfaces perpendicular to the x_3 axis. The electric field is defined in terms of the voltage gradient $\partial V/\partial x_3$, and the unit is volts/meter. The electric displacement represents the separation of charges on the two plates, and the unit is coulombs/square meter.

$$D_3 = \varepsilon_0 E_3 ; \qquad \varepsilon_0 = \frac{D_3}{E_3} = \frac{\text{coulombs}}{\text{meter}^2} \bigg/ \frac{\text{volts}}{\text{meter}} = \frac{\text{farads}}{\text{meter}}$$

For a vacuum dielectric the value of $\varepsilon_0 = 8.85 \times 10^{-12}$ farads/meter.

When a crystal is inserted between the plates of the condenser, a polarization is developed that in general is not parallel to E_3. This adds to $\varepsilon_0 E_3$. By Gauss' theorem the component normal to the surface is the only one measured. Hence $D_3 = \varepsilon_0 E_3 + P_3 = \varepsilon_{33} E_3$; $\varepsilon_{33}/\varepsilon_0 = K_{33}$, the relative dielectric constant. The rate of change of D_3—that is, $\partial D_3/\partial t = \dot{D}_3$—is known as the displacement current.

The measurement of energy $dW = v\varepsilon_{ij}E_i \, dE_j$ where v = volume. The total energy stored is

$$W = \tfrac{1}{2}v\varepsilon_{ij}E_iE_j ; \qquad \varepsilon_{ij} = \varepsilon_{ji}$$

2. Magnetic Variables

The magnetic field is defined in terms of the ampere-turns

$$H = Ni; \qquad N = \text{number of turns per meter, } i = \text{current in amperes}$$

H has the dimensions of ampere-turns per meter. The magnetic flux is determined from Faraday's law that the integral

$$\int_C E_i \, dS_i = -\frac{d\Phi}{dt}$$

where Φ is the total flux through the area determined by the path C. If the integral, the electromotive force, is expressed in volts, the flux is in webers. The flux per unit area B is expressed in webers/meter^2. In a vacuum, the relation between the flux and the field is

$$B_3 = \mu_0 H_3 ; \qquad \mu_0 = \frac{4\pi}{10^7} \text{ henries/meter}$$

With a magnetic material in the solenoid the flux is

$$B_3 = \mu_0 H_3 + I_3 = \mu_{33} H_3 ; \qquad \frac{\mu_{33}}{\mu_0} = \text{relative permeability}$$

Energy stored is

$$W = \tfrac{1}{2} v \mu_{ij} H_i H_j ; \qquad \mu_{ij} = \mu_{ji}$$

For a polycrystalline or ferrite sintered material the flux follows the applied field but is nonlinear. If the sample is polarized by a field H_S, the energy stored is

$$W = \tfrac{1}{2} v \mu_R (H_3 - H_S)^3 \qquad \text{where } \mu_R \text{ is reversible permeability.}$$

The applied field has to have subtracted from it the demagnetizing field H_D. The field in the energy equations is the $H = (H_A - H_D)$.

3. Maxwell's Equations

Maxwell's equations follow from Ampere's law and Faraday's law, together with the divergence equations connected with the electric displacement and the magnetic field. These equations are

Ampere's Law

$$\int_C H_i \, ds_i = \int_S \left(\frac{\partial D_3}{\partial t} + I_3 \right) dS = \int_S \varepsilon_{ijk} \frac{\partial H_k}{\partial x_j} \, dS;$$

Hence $\varepsilon_{ijk}(\partial H_k / \partial x_j) = \partial D_3 / \partial t + I_3$;

Faraday's Law

$$\int_C E_i \, ds_i = - \int_S \frac{\partial B}{\partial t} \, dS = \int_S \varepsilon_{ijk} \frac{\partial E_k}{\partial x_j} \, dS$$

Hence $\varepsilon_{ijk}(\partial E_k / \partial x_j) = - \partial B / \partial t$; $\partial D_i / \partial x_i = \rho$; $\partial B_i / \partial x_i = 0$; The force equation is $F_i = qE_i + q\varepsilon_{ijk}v_j B_k$ where F_i is the force on a unit whose charge is q, and velocity is v_j.

4. Stress and Strain

Stress is defined with respect to the forces exerted on the sides of a unit cube. There are in general nine components, denoted by T_{ij}. However, except in the case of a ferromagnetic material in a magnetic field or a ferroelectric material in an electric field

$$T_{ij} = T_{ji}$$

Infinitesimal strain is defined in terms of the change in length of a line as a function of strain. Tensor strains are defined as

$$S_{ij} = \frac{1}{2} \left(\frac{\partial u_j}{\partial x_i} + \frac{\partial u_i}{\partial x_j} \right)$$

where u_i are the displacements of the line along the three axes.

A reduced notation is in use for which the stresses are given by

$$T_1 = T_{11} ; \qquad T_2 = T_{22} ; \qquad T_3 = T_{33} ;$$
$$T_{23} = T_{32} = T_4 ; \qquad T_{13} = T_{31} = T_5 ; \qquad T_{12} = T_{21} = T_6$$

Engineering shearing strains are twice the tensor terms. According to the convention used, the reduced-notation terms refer to the engineering strains. Hence

$$S_1 = S_{11} ; \qquad S_2 = S_{22} ; \qquad S_3 = S_{33} ;$$
$$\frac{S_4}{2} = S_{23} = S_{32} ; \qquad \frac{S_5}{2} = S_{13} = S_{31} ; \qquad \frac{S_6}{2} = S_{12} = S_{21}$$

Stresses are related to strains through the elastic constants

$$T_m = c_{mn}S_n ; \qquad \text{or} \qquad T_{ij} = c_{ijkl}S_{kl} \quad \text{(tensor)}$$

The strains are related to the stresses through the compliance constants

$$S_n = s_{mn}T_m ; \qquad S_{kl} = s_{ijkl}T_{ij} \quad \text{(tensor)}$$

$c_{mn} = c_{nm} = c_{ijkl}$; $s_{mn} = s_{ijkl}$ if m and n are 1, 2, or 3. If $n = 4$, 5, or 6, $m = 1$, 2, or 3, $s_{mn}/2 = s_{ijkl}$; If both m and n have values 4, 5, or 6, $s_{mn}/4 = s_{ijkl}$. The energy stored in a body is

$$W = \tfrac{1}{2}vS_{ij}T_{ij} = \tfrac{1}{2}vS_iT_i = \tfrac{1}{2}vc_{ijkl}S_{ij}S_{kl} = \tfrac{1}{2}vc_{mn}S_mS_n$$

5. Thermal Terms

Quantity of heat is measured in Joules/M³. One Joule = 4.186 calories. Specific heat $C = dQ/(\Theta_2 - \Theta_1)$; entropy σ is defined by

$$d\sigma = \frac{d\Theta}{\Theta} = \frac{C}{\Theta} d\Theta$$

6. Interaction Between Thermal and Mechanical Variables

Using the thermal and elastic relations derived in Chapter 4 it is shown that relations between adiabatic and isothermal elastic compliances and stiffnesses are given by

$$s^\sigma_{ijkl} - s^\Theta_{ijkl} = - \frac{\alpha_{ij}\alpha_{kl}\Theta}{\rho C_p} \qquad c^\sigma_{ijkl} - c^\Theta_{ijkl} = \frac{\lambda_{ij}\lambda_{kl}\Theta}{\rho C_v}$$

The relation between the specific heat at constant stress and at constant strain is

$$\rho[C_p - C_v] = \Theta\alpha_{ij}\lambda_{ij}$$

PROBLEMS CHAPTER 3

1. How many crystal orientations and what orientations are required to measure the six dielectric components of a triclinic crystal?

2. Using the relation between the elastic compliances and elastic constants given by (3.106), determine the elastic compliances of LiF.

3. What are the principal axes of the expansion coefficients of the EDT crystal whose measured temperature expansion coefficients are shown by Fig. 3.14?

REFERENCES CHAPTER 3

1. S. A. Schelkunoff, "Electromagnetic Waves," p. 77. Van Nostrand, Princeton, New Jersey, 1943.
2. J. F. Nye, "Physical Properties of Crystals," p. 56. Oxford Univ. Press (Clarendon), London and New York, 1957.
3. R. M. Bozorth, "Ferromagnetism," p. 846. Van Nostrand, Princeton, New Jersey, 1951.
4. J. J. Kyame, *J. Acoust. Soc. Am.* **21,** 159 (1949); **26,** 990 (1954).
5. A. R. Hutson and D. L. White, *J. Appl. Phys.* **33,** 40 (1962).
6. S. A. Schelkunoff, "Electromagnetic Fields." Random House (Blaisdell), New York, 1963.
7. W. P. Mason, *Phys. Rev.* **82,** 715 (1951).
8. G. A. Alers, J. R. Neighbors, and H. Sato, *Phys. Chem. Solids* **9,** 21 (1958).
9. B. A. Auld, R. E. Tokheim, and D. K. Winslow, *J. Appl. Phys.* **34,** 2281 (1963).
10. R. C. LeCraw and R. L. Comstock *in* "Physical Acoustics" (Warren P. Mason, ed.), Vol. IIIB, Chapter IV. Academic Press, New York, 1965.
11. J. Laval, *Compt. Rend. Acad. Sci. (Paris)* **232,** 1947 (May 21, 1951).
12. Congr. Solvay, Stoops, Brussels, pp. 273–313 (1952).
13. C. V. Raman and K. S. Viswanathan, *Proc. Indian Acad. Sci.* **A42,** 1, 51 (1955).
14. (a) R. D. Mindlin, "Report US Signal Corps Contract DA 36–039 SC–87414." (March 29, 1961).
14. (b) M. Lax, "The relation between microscopic and macroscopic theories of elasticity." *Proceedings International Conference on Lattice Dynamics*, Copenhagen, 1963. (R. F. Wallace, ed.), pp. 583–596. Macmillan (Pergamon), New York, 1965.
15. H. Jaffe and C. S. Smith, *Phys. Rev.* **121,** 1604 (1961).
16. R. Beckmann, A. D. Ballato, and T. J. Lukaszek, *Proc. IRE* **50,** No. 8, 1912 (1962).
17. T. B. Bateman, W. P. Mason, and H. J. McSkimin, *J. Appl. Phys.* **32,** 928 (1961).
18. R. N. Thurston *in* "Physical Acoustics" (Warren P. Mason, ed.), Vol. IA, Chapter I. Academic Press, New York, 1964.
19. A. E. H. Love, "The Mathematical Theory of Elasticity," 4th Edition, p. 49. Cambridge Univ. Press, London and New York, 1934.
20. O. L. Anderson *in* "Physical Acoustics" (Warren P. Mason, ed.), Vol. IIIB, Chapter II. Academic Press, New York, 1965.
21. George Joos, "Theoretical Physics," Chapter 34. G. E. Steckert Co., 1934.
22. R. D. McCammon and G. K. White, *Phys. Rev. Letters* **10,** 234 (1963).
23. G. K. White, *Cryogenics* **1,** 151 (1961).
24. Warren P. Mason, "Piezoelectric Crystals and Their Application to Ultrasonics," Fig. 9.12, p. 178. Van Nostrand, Princeton, New Jersey, 1950.
25. "American Institute of Physics Handbook," pp. 3–91. McGraw-Hill, New York, 1963.

CHAPTER 4 • Equilibrium Thermodynamics, Thermodynamic Functions, and Maxwell Relations

It has already been remarked in Chapter 1, in connection with Eq. (1.6), that a relation exists between the piezoelectric constants that express the ratio between the electric displacement and stress—the direct piezoelectric effect—and those expressing the ratio between the strains and the applied electric fields—the converse piezoelectric effect. Another example in Chapter 3, Eq. (3.121), is the equality between the ratio of strain to temperature increment—the temperature-expansion coefficients—and the relations between the entropy and the applied stresses—the piezocaloric effect. Such relations can be demonstrated most easily from equilibrium thermodynamics. They are the result of applying the Maxwell relations to particular forms of thermodynamic functions.

4.1 Thermodynamic Functions for Mechanical and Thermal Variables

It follows directly from the first law of thermodynamics that the total energy U of a body is the sum of all the different types of energy. Four forms of energy have been discussed in Chapter 3: magnetic energy, electric energy, mechanical energy, and thermal energy. For these types of energy the increment of total energy dU can be written

$$dU = E_i \, dD_i + H_j \, dB_j + T_{kl} \, dS_{kl} + \Theta \, d\sigma \tag{4.1}$$

It rarely happens that all forms of energy occur in any interaction effect, so that some of these energies can be dropped from consideration. For example, for piezoelectric or electrostrictive effects the magnetic energy can be neglected, whereas for piezomagnetic materials the electrical energy can be neglected. This is fortunate, for otherwise the resulting relations would be exceedingly complicated.

In the present chapter, consideration is given to a combination of the electric energy, the mechanical energy, and the thermal energy, since all three forms are involved in piezoelectric, pyroelectric, and electrostrictive interactions. Similar results, however, can be obtained by substituting magnetic energy for electric energy when dealing with piezomagnetic and magnetostrictive materials.

If we consider first only the mechanical and thermal variables, a direct differentiation of Eq. (4.1) gives

$$\left.\frac{\partial U}{\partial S_{kl}}\right)_{\sigma} = T_{kl}\,; \qquad \left.\frac{\partial U}{\partial \sigma}\right)_{S_{kl}} = \Theta \tag{4.2}$$

Since both the stresses and the temperature are functions of the strains and the entropy, we can perform the further differentiation

$$dT_{kl} = \left.\frac{\partial T_{kl}}{\partial S_{ij}}\right)_{\sigma} dS_{ij} + \left.\frac{\partial T_{kl}}{\partial \sigma}\right)_{S_{ij}} d\sigma$$

$$d\Theta = \left.\frac{\partial \Theta}{\partial S_{ij}}\right)_{\sigma} dS_{ij} + \left.\frac{\partial \Theta}{\partial \sigma}\right)_{S_{ij}} d\sigma \tag{4.3}$$

Since the partial differentials are constants of the material that are assumed not to vary for any values of the strains considered here, these equations can be integrated into the forms

$$T_{kl} = \left.\frac{\partial T_{kl}}{\partial S_{ij}}\right)_{\sigma} S_{ij} + \left.\frac{\partial T_{kl}}{\partial \sigma}\right)_{S_{ij}} \delta\sigma\,; \qquad \delta\Theta = \left.\frac{\partial \Theta}{\partial S_{ij}}\right)_{\sigma} S_{ij} + \left.\frac{\partial \Theta}{\partial \sigma}\right)_{S_{ij}} \delta\sigma \tag{4.4}$$

The subscripts indicate which variables are held constant. The partial differentials are readily recognized from the interaction diagram of Fig. 1.2 as being

$$\left.\frac{\partial T_{kl}}{\partial S_{ij}}\right)_{\sigma} = \frac{\partial^2 U}{\partial S_{ij}\,\partial S_{kl}} = c^{\sigma}_{ijkl} = \text{adiabatic elastic constants}$$

$$\left.\frac{\partial T_{kl}}{\partial \sigma}\right)_{S_{ij}} = \frac{\partial^2 U}{\partial S_{kl}\,\partial \sigma} = \frac{\partial^2 U}{\partial \sigma\,\partial S_{kl}} = \left.\frac{\partial \Theta}{\partial S_{kl}}\right)_{\sigma} = \gamma_{kl}\,; \qquad \left.\frac{\partial \Theta}{\partial \sigma}\right)_{S_{ij}} = \left(\frac{\Theta}{\rho C_v}\right) \tag{4.5}$$

In these equations the γ_{kl} constants are either the increase in stresses due to an increment of entropy at constant strain or the increase in temperature due to strains at constant entropy. These two effects have similar constants since U is a state variable with a perfect differential. Hence, as first shown by Maxwell, it is immaterial which derivative is taken first and hence the relations for γ_{kl} follow by inverting the order of differentiation. Such relations are known as Maxwell relations. It was previously shown by Eqs. (3.102) and (3.103) that c^{σ}_{ijkl} is a symmetrical fourth-rank tensor by virtue of the Maxwell relation, making it immaterial which set of strains is used first in the differentiation of U.

Equation (4.5) can be used to discuss all the stress-strain thermal relations, but the constants γ_{kl} are not the usually measured relationships between the mechanical and thermal variables. To obtain such relations as the thermal stress coefficients, we have to use another thermodynamic variable known as the free energy, designated by A. As shown in books dealing with thermodynamics,[1] the free energy represents the amount of the internal energy U that can be converted into mechanical or electrical energy at constant temperature. It is defined as

$$A = U - \Theta\sigma \tag{4.6}$$

The differential form of the free energy is

$$dA = dU - \Theta\,d\sigma - \sigma\,d\Theta = T_{kl}\,dS_{kl} - \sigma\,d\Theta, \tag{4.7}$$

after incorporating the value of dU from (4.1), the electrical and magnetic energy being neglected.

From (4.7) we see that

$$\left.\frac{\partial A}{\partial S_{kl}}\right)_\Theta = T_{kl}\,; \qquad \left.\frac{\partial A}{\partial \Theta}\right)_{S_{kl}} = -\sigma \tag{4.8}$$

where S_{kl} and Θ are now the independent variables and T_{kl} and σ the dependent variables. But the dependent variables are functions of the independent variables, so that

$$dT_{kl} = \left.\frac{\partial T_{kl}}{\partial S_{ij}}\right)_\Theta dS_{ij} + \left.\frac{\partial T_{kl}}{\partial \Theta}\right)_{S_{ij}} d\Theta\,; \qquad d\sigma = \left.\frac{\partial \sigma}{\partial S_{ij}}\right)_\sigma dS_{ij} + \left.\frac{\partial \sigma}{\partial \Theta}\right)_{S_{ij}} d\Theta \tag{4.9}$$

Since the partial derivatives are assumed to be constant for all values of the strains and temperatures considered, the mechanical differentials can be replaced by the variables T_{kl} and S_{ij} and the thermal differentials can be replaced by $\delta\sigma$ and $\delta\Theta$.

The partial derivatives have the meanings

$$\left.\frac{\partial T_{kl}}{\partial S_{ij}}\right)_\Theta = c_{ijkl}^\Theta\,; \qquad \frac{\partial T_{kl}}{\partial \Theta} = \frac{\partial^2 A}{\partial \Theta\,\partial S_{kl}} = \frac{\partial^2 A}{\partial S_{kl}\,\partial \Theta} = -\frac{\partial \sigma}{\partial S_{kl}} = -\lambda_{kl}\,;$$

$$\frac{\partial \sigma}{\partial \Theta} = \left(\frac{\rho C_v}{\Theta}\right) \tag{4.10}$$

where the elastic constants are the isothermal constants, λ_{kl} are the temperature-stress constants, and ρC_v is the specific heat at constant

---

[1] See Joos [1] and Slater [2].

strain per unit volume. By virtue of the Maxwell relation given by the second equation of (4.10), we can write the relations

$$T_{kl} = c^{\Theta}_{ijkl} S_{ij} - \lambda_{kl}\,\delta\Theta; \qquad \delta\sigma = +\lambda_{ij}S_{ij} + \frac{\rho C_v}{\Theta}\,\delta\Theta \qquad (4.11)$$

By setting $\delta\sigma = 0$, the relation between the adiabatic and isothermal elastic moduli given by Eqs. (3.131) is determined directly.

Two other thermodynamic functions are of interest. These are the enthalpy H and the Gibbs function G. These are defined as shown by Table 5, which lists also the differential forms.

Table 5

THERMODYNAMIC FUNCTIONS FOR MECHANICAL AND THERMAL VARIABLES

Thermodynamic Functions	Definition	Independent Variables	Differential Relations
Total energy U	U	S_{ij}, σ	$dU = T_{kl}\,dS_{kl} + \Theta\,d\sigma$
Free energy A	$U - \Theta\sigma$	S_{ij}, Θ	$dA = T_{kl}\,dS_{kl} - \sigma\,d\Theta$
Enthalpy H	$U - T_{kl}S_{kl}$	T_{kl}, σ	$dH = -S_{kl}\,dT_{kl} + \Theta\,d\sigma$
Gibbs function G	$U - T_{ij}S_{ij} - \Theta\sigma$	T_{ij}, Θ	$dG = -S_{ij}\,dT_{ij} - \sigma\,d\Theta$

For example, the Gibbs function G can be used to derive directly Eqs. (3.121) of the preceding chapter. For this case

$$S_{ij} = -\left.\frac{\partial G}{\partial T_{ij}}\right)_{\Theta}; \qquad \sigma = -\left.\frac{\partial G}{\partial\Theta}\right)_{T_{ij}} \qquad (4.12)$$

Then

$$S_{ij} = \left.\frac{\partial S_{ij}}{\partial T_{kl}}\right)_{\Theta} T_{kl} + \left.\frac{\partial S_{ij}}{\partial\Theta}\right)_{T_{kl}} \delta\Theta; \qquad \delta\sigma = \left.\frac{\partial\sigma}{\partial T_{kl}}\right)_{\Theta} T_{kl} + \left.\frac{\partial\sigma}{\partial\Theta}\right)_{T_{kl}} \delta\Theta \qquad (4.13)$$

But

$$\left.\frac{\partial S_{ij}}{\partial T_{kl}}\right)_{\Theta} = -\frac{\partial^2 G}{\partial T_{kl}\,\partial T_{ij}} = s^{\Theta}_{ijkl};$$

$$\left.\frac{\partial S_{ij}}{\partial\Theta}\right)_{T_{kl}} = -\frac{\partial^2 G}{\partial\Theta\,\partial T_{ij}} = -\frac{\partial^2 G}{\partial T_{ij}\,\partial\Theta} = \left.\frac{\partial\sigma}{\partial T_{ij}}\right)_{\Theta} = \alpha_{ij}; \qquad (4.14)$$

$$\left.\frac{\partial\sigma}{\partial\Theta}\right)_{T_{kl}} = \frac{\rho C_p}{\Theta}$$

which reduces to the Eq. (3.121).

4.2 Thermodynamic Functions for Mechanical, Electrical, and Thermal Variables for Piezoelectric Crystals

For piezoelectric, electrostrictive, or pyroelectric material, it is necessary to add electric energy to the mechanical and thermal energy considered in the last section. Eight possible thermodynamic functions are of interest, depending on the independent variables that are chosen. They have been discussed previously, and Table 6 lists the names, definitions, independent variables, and differential forms.

It is obvious that eight types of equations—all related to each other—can be obtained by using the eight thermodynamic functions. The forms having the greatest use for piezoelectric crystals are the Gibbs function G and the electric Gibbs function G_2. For these the independent variables are respectively the stresses, fields, and temperature, and the strains, fields, and temperature. For ferroelectric crystals and ceramics, however, the electric displacement turns out to be a better independent variable than does the electric field. The two potentials of interest are then the free energy A and the elastic Gibbs function G_1. In the present section, these four forms will be discussed in detail and their interrelationships determined.

The most common relations are the ones for which the strains, electric displacements, and entropy are the dependent variables while the stresses, fields, and temperature are the independent values. From the differential form of the Gibbs function G we have

$$S_{kl} = -\frac{\partial G}{\partial T_{kl}}; \qquad D_m = -\frac{\partial G}{\partial E_m}; \qquad \sigma = -\frac{\partial G}{\partial \Theta} \qquad (4.15)$$

Since the dependent variables are functions of the independent variables, they can be developed in the form of the partial differential equations

$$dS_{kl} = \frac{\partial S_{kl}}{\partial T_{kl}}\bigg)_{E,\Theta} dT_{ij} + \frac{\partial S_{kl}}{\partial E_m}\bigg)_{T_{ij},\Theta} dE_m + \frac{\partial S_{kl}}{\partial \Theta}\bigg)_{T_{ij},E} d\Theta$$

$$dD_n = \frac{\partial D_m}{\partial T_{ij}}\bigg)_{E,\Theta} dT_{ij} + \frac{\partial D_n}{\partial E_m}\bigg)_{T_{ij},\Theta} dE_m + \frac{\partial D_n}{\partial \Theta}\bigg)_{T_{ij},E} d\Theta \qquad (4.16)$$

$$d\sigma = \frac{\partial \sigma}{\partial T_{ij}}\bigg)_{E,\Theta} dT_{ij} + \frac{\partial \sigma}{\partial E_m}\bigg)_{T_{ij},\Theta} dE_m + \frac{\partial \sigma}{\partial \Theta}\bigg)_{T_{ij},E} d\Theta$$

The terms on the parentheses indicate the variables that are held constant during the differentiation. In order to differentiate between the stress and the temperature, the use of the subscript Θ denotes constant temperature, E denotes constant field, and T_{ij}, or sometimes T without subscripts, denotes constant stress. The subscript or superscript σ denotes constant entropy, and S_{kl}, or sometimes S, denotes constant strain.

Since the partial derivatives are regarded as being constants over the range of variables used, these equations can be integrated, with the result that dS_{ij}

Table 6

THERMODYNAMIC FUNCTIONS FOR MECHANICAL, ELECTRICAL, AND THERMAL VARIABLES

Thermodynamic Function	Definition	Independent Variables	Differential Relations
Internal energy U	U	S_{ij}, D_m, σ	$dU = T_{ij}\,dS_{ij} + E_m\,dD_m + \Theta\,d\sigma$
Free energy A	$U - \Theta\sigma$	S_{ij}, D_m, Θ	$dA = T_{ij}\,dS_{ij} + E_m\,dD_m - \sigma\,d\Theta$
Enthalpy H	$U - T_{ij}S_{ij} - E_m D_m$	T_{ij}, E_m, σ	$dH = -S_{ij}\,dT_{ij} - D_m\,dE_m + \Theta\,d\sigma$
Elastic enthalpy H_1	$U - T_{ij}S_{ij}$	T_{ij}, D_m, σ	$dH_1 = -S_{ij}\,dT_{ij} + E_m\,dD_m + \Theta\,d\sigma$
Electric enthalpy H_2	$U - E_m D_m$	S_{ij}, E_m, σ	$dH_2 = T_{ij}\,dS_{ij} - D_m\,dE_m + \Theta\,d\sigma$
Gibbs function G	$U - S_{ij}T_{ij} - E_m D_m - \Theta\sigma$	T_{ij}, E_m, Θ	$dG = -S_{ij}\,dT_{ij} - D_m\,dE_m - \sigma\,d\Theta$
Elastic Gibbs function G_1	$U - S_{ij}T_{ij} - \Theta\sigma$	T_{ij}, D_m, Θ	$dG_1 = -S_{ij}\,dT_{ij} + E_m\,dD_m - \sigma\,d\Theta$
Electric Gibbs function G_2	$U - E_m D_m - \Theta\sigma$	S_{ij}, E_m, Θ	$dG_2 = T_{ij}\,dS_{ij} - D_m\,dE_m - \sigma\,d\Theta$

is replaced by S_{ij}, and so on. The partial derivates have special names as follows:

$$\frac{\partial S_{kl}}{\partial T_{ij}}\bigg)_{E,\Theta} = s_{ijkl}^{E,\Theta} = \text{elastic compliances at constant fields and temperatures}$$

$$\frac{\partial S_{kl}}{\partial E_m}\bigg) = -\frac{\partial}{\partial E_m}\frac{\partial G}{\partial T_{kl}} = -\frac{\partial}{\partial T_{kl}}\frac{\partial G}{\partial E_m} = \frac{\partial D_m}{\partial T_{kl}} = d_{mkl} = \text{piezoelectric constants}$$

$$\frac{\partial S_{kl}}{\partial \Theta} = -\frac{\partial}{\partial \Theta}\bigg(\frac{\partial G}{\partial T_{kl}}\bigg) = -\frac{\partial}{\partial T_{kl}}\bigg(\frac{\partial G}{\partial \Theta}\bigg) = \frac{\partial \sigma}{\partial T_{kl}} = \alpha_{kl} = \text{temperature-expansion coefficients}$$

$$\frac{\partial D_n}{\partial E_m}\bigg)_{T,\theta} = \varepsilon_{mn}^{T,\theta} = \text{dielectric constants at constant stress and temperature}$$

$$\frac{\partial D_n}{\partial \Theta} = -\frac{\partial}{\partial \Theta}\bigg(\frac{\partial G}{\partial E_n}\bigg) = -\frac{\partial}{\partial E_n}\bigg(\frac{\partial G}{\partial \Theta}\bigg) = \frac{\partial \sigma}{\partial E_n} = p_n = \text{pyroelectric constants}$$

$$\frac{\partial \sigma}{\partial \Theta}\bigg)_{T,E} = \frac{\rho C^{T,E}}{\Theta} = \text{specific heat per unit volume divided by the absolute temperature}$$

Introducing these values in Eqs. (4.16), these can be written

$$S_{kl} = s_{ijkl}^{E,\Theta} T_{ij} + d_{mkl}^{\Theta} E_m + \alpha_{kl}^E \delta\Theta$$

$$D_n = d_{nij}^{\Theta} T_{ij} + \varepsilon_{mn}^{T,\Theta} E_m + p_n{}^T \delta\Theta \qquad (4.17)$$

$$\delta\sigma = \alpha_{ij}^E T_{ij} + p_m{}^T E_m + \frac{\rho C^{T,E}}{\Theta} \delta\Theta$$

The interaction terms d_{mkl}^{Θ}, α_{kl}, $p_n{}^T$ need only one superscript because, for example, the ratio between the strain and the applied electric field—that is, the converse piezoelectric effect—is always measured with zero or constant stress, and so on. The relation between the electric displacement and the applied stress is always measured with a zero applied electric field. This is the direct piezoelectric effect. Since for no applied electric field there is no difference between the electric displacement and the polarization, we could replace D_n by P_n. Since D_n is the external variable, however, it is more convenient to retain the equation in this form.

The form of the equations of (4.17) is useful for static measurements. For example, the compliance moduli can be measured by means of extension or bending experiments, carried out at constant field or temperature. The piezoelectric constants d_{mij} have been measured by observing the strains produced by a static electric field or the charge produced by an applied stress. Similarly, the temperature-expansion coefficients are usually measured by the change in displacement or the increased strain caused by a temperature change $\delta\Theta$. The so-called direct pyroelectric effect is determined by the charge developed on the crystal surface for an increment of temperature in the absence of an applied stress. If we multiply the last equation through by the absolute temperature Θ, the left-hand side $\Theta\, d\sigma$ represents an increment of heat δQ. The first term represents the increment of heat caused by an applied stress, and the constants $\Theta\alpha_{ij}$ are known as the piezocaloric constants. The next term, $\Theta p_m{}^T$, represents the heat caused by an applied electric field, and the constants are called the electrocaloric constants. The last term represents the specific heat per unit volume measured at constant stress and field.

The greatest use for piezoelectric crystals is in exciting mechanical vibrations in the material or in an attached medium.[2] For this case the alterations occur so fast that there is no time to interchange heat between the various parts, and the entropy $\delta\sigma$ can be set equal to zero. If we solve the last equation of (4.17) for $\delta\Theta$ and substitute the result in the first two equations, the relations reduce to two equations of the form

$$S_{kl} = s_{ijkl}^{E,\sigma} T_{ij} + d_{mkl}^{\sigma} E_m$$
$$D_n = d_{nij}^{\sigma} T_{ij} + \varepsilon_{mn}^{T,\sigma} E_m \tag{4.18}$$

where the superscript σ indicates that the constants are adiabatic constants. Because only adiabatic conditions prevail, it is usually possible to leave off the superscript σ. By direct substitution we find

$$s_{ijkl}^{E,\sigma} = s_{ijkl}^{E,\Theta} - \frac{\alpha_{ij}^{E}\alpha_{kl}^{E}\Theta}{\rho C^{T,E}}$$

$$d_{nij}^{\sigma} = d_{nij}^{\Theta} - \frac{p_n{}^T\alpha_{ij}^{E}\Theta}{\rho C^{T,E}} \tag{4.19}$$

$$\varepsilon_{mn}^{T,\sigma} = \varepsilon_{mn}^{T,\Theta} - \frac{p_m{}^T p_n{}^T \Theta}{\rho C^{T,E}}$$

Hence unless the crystal is pyroelectric—which occurs in only 10 of the 32 crystal classes—there is no difference between the adiabatic and isothermal values of the piezoelectric and dielectric constants.

---

[2] See Mason [3].

Three other forms of the piezoelectric equations are in common use. By using the thermodynamic function G_2 it is readily found that

$$T_{ij} = c_{ijkl}^{E,\Theta} S_{kl} - e_{mij}^{\Theta} E_m - \lambda_{ij}^E \delta\Theta$$

$$D_n = +e_{nkl}^{\Theta} S_{kl} + \varepsilon^{S,\Theta} E_m + p_n{}^S \delta\Theta \qquad (4.20)$$

$$\delta\sigma = +\lambda_{kl} S_{kl} + p_m{}^S E_m + \frac{\rho C^{S,E}}{\Theta} \delta\Theta$$

The term $c_{ijkl}^{E,\Theta}$ represents the elastic constants measured at constant field and temperature, e_{mij}^{Θ} is a piezoelectric constant relating a stress to an applied field or the negative of an electric displacement to a strain, λ_{ij} is the thermal stress constant relating an increase in temperature to a stress at constant strain or field, $p_n{}^S$ is a pyroelectric constant relating an increase in polarization to an increase in temperature when the strain is constant. By multiplying the last equation through by the absolute temperature Θ it is seen that $\Theta\lambda_{kl}$ represents the heat of deformation and $\Theta p_m{}^S$ is the electro-caloric effect at constant strain. For adiabatic conditions, $\delta\Theta$ can be eliminated from these equations by setting $\delta\sigma = 0$, and one finds

$$T_{ij} = c_{ijkl}^{E,\sigma} S_{kl} - e_{mij}^{\sigma} E_m \; ; \qquad D_n = +e_{nkl}^{\sigma} S_{kl} + \varepsilon_{mn}^{S,\sigma} E_m \qquad (4.21)$$

where

$$c_{ijkl}^{E,\sigma} = c_{ijkl}^{E,\Theta} + \frac{\lambda_{ij}^E \lambda_{kl}^E \Theta}{\rho C^{T,E}} \; ; \qquad e_{mij}^{\sigma} = e_{mij}^{\Theta} - \frac{\lambda_{ij}^E p_m{}^S \Theta}{\rho C^{T,E}} \; ;$$

$$\varepsilon_{mn}^{S,\sigma} = \varepsilon_{mn}^{S,\Theta} - \frac{p_m{}^S p_n{}^S \Theta}{\rho C^{T,E}}$$

By comparing (4.18) with (4.21) it is obvious that several relations exist between the various constants. If we specify that all the interactions are adiabatic, we can leave off the superscript σ. If we multiply the first of Eqs. (4.21) by s_{ijkl}^E, the resulting equation is

$$s_{ijkl}^E T_{ij} = (s_{ijkl}^E c_{ijkl}^E) S_{kl} - (e_{mij} s_{ijkl}^E) E_m \qquad (4.22)$$

Since s_{ijkl} is the reciprocal tensor of c_{ijkl}—as shown by Eqs. (3.107)—it follows that the product is a unitary matrix (having 1's in the leading diagonal and zeros elsewhere). It follows that the product with the tensor S_{kl} equals S_{kl} and hence

$$S_{kl} = s_{ijkl}^E T_{ij} + (e_{mij} s_{ijkl}^E) E_m \qquad (4.23)$$

Comparing this equation with the first of (4.18) we see that

$$d_{mkl} = (e_{mij}s^E_{ijkl}) \tag{4.24}$$

By multiplying the first equation of (4.18) by c^E_{ijkl}, it is readily seen that

$$e_{mij} = (d_{mkl}c^E_{ijkl}) \tag{4.25}$$

Another relation that can be obtained is between the dielectric constants measured at constant stress and constant strain. From the first of Eqs. (4.21) the stresses in the absence of any strain S_{kl} can be written

$$T_{ij} = -e_{mij}E_m \tag{4.26}$$

Substituting this equation in the last of Eqs. (4.18), the relation between the electric field and the electric displacement becomes

$$D_n = (\varepsilon^T_{mn} - d_{nij}e_{mij})E_m = \varepsilon^S_{mn} \tag{4.27}$$

Hence

$$\varepsilon^T_{mn} - \varepsilon^S_{mn} = d_{nij}e_{mij} \tag{4.28}$$

Another relation that has received some discussion is the relation between the pyroelectric effect at constant stress, denoted by the constant $p_n{}^T$, and the pyroelectric constant at constant strain, $p_n{}^S$. This latter constant has sometimes[3] been called the primary (or true) pyroelectric effect and that due to a change in dimension the secondary effect. It seems more appropriate to call them the pyroelectric effects at constant strain and constant stress. The relation between the pyroelectric constants at constant stress and at constant strain is obtained by setting T_{ij} and E_m equal to zero in the first of Eqs. (4.17) and substituting the resulting expression for the strain in the second equation of (4.20). The resulting relation is

$$D_n = (\alpha^E_{kl}e^\Theta_{nkl} + p_n{}^S)\delta\Theta = p_n{}^T\,\delta\Theta$$

or

$$p_n{}^T - p_n{}^S = \alpha^E_{kl}e^\Theta_{nkl} \tag{4.29}$$

The other two forms of the piezoelectric equations make use of the electric displacement D_n as the fundamental variable rather than the electric field E_m. The equivalent thermodynamic functions are the free energy A and the elastic Gibbs function G_1. Because the second of these functions is important in the discussion of ferroelectric crystals and ceramics, it is discussed in detail here. By developing the strains, fields, and entropy in terms of the

---

[3] See Cady [4].

stresses, electric displacements, and the temperature, the three equations can be written in the forms

$$
\begin{aligned}
S_{kl} &= \frac{\partial S_{kl}}{\partial T_{ij}}\bigg)_{D,\Theta} T_{ij} + \frac{\partial S_{kl}}{\partial D_n}\bigg)_{T,\Theta} D_n + \frac{\partial S_{kl}}{\partial \Theta}\bigg)_{T,D} \delta\Theta \\
&= s_{ijkl}^{D,\Theta} T_{ij} + g_{nkl}^{\Theta} D_n + \alpha_{kl}^{D}\,\delta\Theta \\[1ex]
E_m &= \frac{\partial E_m}{\partial T_{ij}}\bigg)_{D,\Theta} T_{ij} + \frac{\partial E_m}{\partial D_n}\bigg)_{T,\Theta} D_n + \frac{\partial E_m}{\partial \Theta}\bigg)_{T,D} \delta\Theta \\
&= -g_{mij}^{\Theta} T_{ij} + \beta_{mn}^{T,\Theta} D_n + q_m{}^T\,\delta\Theta \\[1ex]
\delta\sigma &= \frac{\partial \sigma}{\partial T_{ij}}\bigg)_{D,\Theta} T_{ij} + \frac{\partial \sigma}{\partial D_n}\bigg)_{T,\Theta} D_n + \frac{\partial \sigma}{\partial D_n}\bigg)_{T,D} \delta\Theta \\
&= \alpha_{ij}^{D} T_{ij} - q_n{}^T D_n + \frac{\rho C^{T,D}}{\Theta}\,\delta\Theta
\end{aligned}
\tag{4.30}
$$

g_{mkl} is a new piezoelectric constant relating the strain to the electric displacement, β_{mn} is the inverse tensor to the dielectric-constant tensor and is called the impermeability tensor, and q_m is a pyroelectric constant that measures the field generated in a crystal by a change in temperature when the stress and electric displacements are held constant.

As before, an adiabatic piezoelectric equation can be obtained by setting $\delta\sigma = 0$ in the last of Eq. (4.29) and eliminating $\delta\Theta$. The result is

$$
S_{kl} = s_{ijkl}^{D} T_{ij} + g_{nkl} D_n \; ; \qquad E_m = -g_{mij} T_{ij} + \beta_{mn}^{T} D_n
\tag{4.31}
$$

where

$$
s_{ijkl}^{D,\sigma} = s_{ijkl}^{D,\Theta} - \frac{\alpha_{ij}^{D}\alpha_{kl}^{D}\Theta}{\rho C^{S,D}} \; ; \qquad
g_{nkl}^{\sigma} = g_{nkl}^{\Theta} + \frac{\alpha_{kl}^{D}q_n{}^T \Theta}{\rho C^{S,D}} \; ;
$$

$$
\beta_{mn}^{T,\sigma} = \beta_{m,n}^{T,\Theta} + \frac{q_m{}^T q_n{}^T \Theta}{\rho C^{S,D}}
$$

The final set of equations, obtained by using the free energy A, can be written in the form of (4.32) for the adiabatic case

$$
T_{ij} = c_{ijkl}^{D} S_{kl} - h_{nij} D_n \; ; \qquad E_m = -h_{mkl} S_{kl} + \beta_{mn}^{S} D_n
\tag{4.32}
$$

There are obviously many relations between the various constants employed in the four forms of the piezoelectric equations. Table 7 gives a number of these relations, which the reader can verify in a manner similar to that employed for the other relations.

Table 7

RELATIONS BETWEEN PIEZOELECTRIC CONSTANTS

$c_{ijpq}^{E}s_{pqkl}^{E} = I_{ijkl}$;	$d_{nkl} = \varepsilon_{mn}^{T}g_{mkl} = e_{nij}s_{ijkl}^{E}$
$c_{ijpq}^{D}s_{pqkl}^{D} = I_{ijkl}$;	$e_{nkl} = \varepsilon_{mn}^{S}h_{mkl} = d_{nij}c_{ijkl}^{E}$
$\beta_{mp}^{T}\varepsilon_{pn}^{T} = I_{mn}$;	$g_{nkl} = \beta_{mn}^{T}d_{mkl} = h_{nij}s_{ijkl}^{D}$
$\beta_{mp}^{S}\varepsilon_{pn}^{S} = I_{mn}$;	$h_{nkl} = \beta_{mn}^{S}e_{mkl} = g_{nij}c_{ijkl}^{D}$
$\varepsilon_{mn}^{T} - \varepsilon_{mn}^{S} = d_{nkl}e_{mkl}$;	$\alpha_{ij}^{E} - \alpha_{ij}^{D} = g_{lij}^{\Theta}p_{l}^{T}$
$\beta_{mn}^{S} - \beta_{mn}^{T} = h_{nkl}g_{mkl}$;	$\lambda_{ij}^{D} - \lambda_{ij}^{E} = e_{mij}^{\Theta}q_{m}^{T}$
$c_{ijkl}^{D} - c_{ijkl}^{E} = e_{mij}h_{mkl}$;	$\rho(C^{E} - C^{D}) = \Theta p_{i}^{T}p_{j}^{T}\beta_{ij}^{T,\Theta}$
$s_{ijkl}^{E} - s_{ijkl}^{D} = d_{mij}g_{mkl}$;	$\rho(C^{T,\Theta} - C^{S,\Theta}) = \Theta\alpha_{ij}^{E}\alpha_{kl}^{E}c_{ijkl}^{E,\Theta}$

The last two equations represent respectively the pyroelectric and thermo-elastic contributions to the specific heat capacity per unit volume.

4.3 Numerical Values for Interaction Effects

In many books and papers the piezoelectric equations and the interaction factors are expressed in terms of cgs units rather than the MKS units used in this book. Hence it is desirable to give conversion factors between the two systems. Table 8 gives the conversion factors of interest for piezo-electric crystals. These values may be inserted as factors in an equation without changing its validity. Insertion of the appropriate conversion factors into an equation in which cgs electrostatic units are explicit converts the equation to a form that is explicit in MKS rationalized units.

The magnitudes of the interaction effects of Table 7 are of interest. As an example let us consider X-cut quartz, for which all the values are well known. The following matrices give the relative values in MKS units.

	T_{11}	E_1	Θ		T_{11}	E_1	Θ
S_{11}	s_{1111}	d_{111}	α_{11}	S_{11}	1.28×10^{-11}	2.25×10^{-12}	1.43×10^{-5}
D_1	d_{111}	ε_{11}	0	D_1	2.25×10^{-12}	4×10^{-11}	0
σ	α_{11}	0	$\rho C^{E,S}/\Theta$	σ	1.43×10^{-5}	0	6.5×10^{3}

$$(4.33)$$

Quartz has no pyroelectric effect, so that $p^{S} = p^{T} = 0$. In order to evaluate the various coupling effects it is necessary to know the various piezoelectric constants. There are two independent constants, which have for the d

Table 8

CONVERSION FACTOR BETWEEN cgs ELECTROSTATIC AND MKS RATIONALIZED UNITS

Quantity	Symbol	Conversion Factor
Mechanical force	F	10^{-5} newton per dyne
Elastic strain	S	1, numeric = relative deformation
Elastic stress	T	10^{-1} newton/meter^2 per dyne/cm^2
Elastic displacement	u	10^{-2} meter per centimeter
Elastic compliance	s	10 meters^2/newton per cm^2/dyne
Elastic stiffness	c	10^{-1} newton/meter^2 per dyne/cm^2
Electric potential	V	300 volts per statvolt
Electric field	E	3×10^4 volts/meter per statvolt/cm
Electric charge	Q	$\frac{1}{3} \times 10^{-9}$ coulomb per statcoulomb
Electric displacement	D	$\dfrac{1}{12\pi \times 10^5}$ coulomb/meter^2 per statcoulomb/cm^2
Dielectric permittivity	ε	$\dfrac{1}{36\pi \times 10^9}$ farad/meter per statfarad/cm
Dielectric impermeability	β	$36\pi \times 10^9$ meters/farad per cm/statfarad
Relative dielectric constant	K	1, numeric = $\varepsilon/\varepsilon_0$
Dielectric polarization	P	$\frac{1}{3} \times 10^{-5}$ coulomb/meter^2 per statcoulomb/cm^2
Piezoelectric constant	d	$\frac{1}{3} \times 10^{-4}$ coulomb/newton per statcoulomb/dyne
Piezoelectric constant	e	$\frac{1}{3} \times 10^{-5}$ coulomb/meter^2 per statcoulomb/cm^2
Piezoelectric constant	g	3×10^5 meter^2/coulomb per cm^2/statcoulomb
Piezoelectric constant	h	3×10^4 newtons/coulomb per dyne/statcoulomb
Pyroelectric constant	p	$\dfrac{1}{12\pi \times 10^5}$ coulomb/meter$^2 \times$ °K per statcoulomb/cm$^2 \times$ °K
Pyroelectric constant	q	3×10^4 volts/meter $\times$ °K per statvolt/cm $\times$ °K
Temperature-expansion coefficient	α	1, relative deformation/°K
Thermal-stress constant	λ	10^{-1} newton/meter$^2 \times$ °K per dyne/cm$^2 \times$ °K
Energy per unit volume	U	10 joules/meter^3 per erg/cm^3
Specific heat per unit volume	ρC	10 joules/meter$^3 \times$ °K per erg/cm$^3 \times$ °K
Entropy	σ	10 joules/meter$^3 \times \Theta$ per erg/cm$^3 \times \Theta$

constants the values

$$d_{111} = -d_{122} = 2.25 \times 10^{-12} \, ;$$

$$d_{123} = \frac{d_{14}}{2} = -4.25 \times 10^{-13} \, \text{coulombs/newton}$$

(4.34)

To calculate the e_{111} and e_{123} values use is made of the equation $e_{nkl} = d_{nij}c^E_{ijkl}$. For the e_{111} and e_{123} values, by taking all the combinations of the ij terms, we find

$$e_{111} = d_{111}(c_{1111} - c_{1122}) + 2d_{123}c_{1123} \, ;$$

(4.35)

$$e_{123} = d_{111}c_{1123} + d_{122}c_{2223} + 2(d_{123}c_{2323})$$

But

$$c_{1111} = c_{11} = 8.68 \times 10^{10} \, ; \qquad c_{1122} = c_{12} = 0.709 \times 10^{10} \, ;$$

$$c_{1123} = c_{14} = 1.80 \times 10^{10}$$

(4.36)

$$c_{2223} = c_{24} = -1.80 \times 10^{10} \, ; \qquad c_{2323} = c_{44} = 5.82 \times 10^{10} \, \text{newtons/meter}^2$$

Hence the values of the e piezoelectric constants are

$$e_{111} = 0.164; \qquad e_{123} = 0.031 \, \text{coulombs/meter}^2$$

(4.37)

From the relation $g_{nkl} = \beta^T_{mn}d_{nkl}$ one can obtain directly

$$g_{111} = \frac{1}{4 \times 10^{-11}} \, d_{111} = 0.056;$$

$$g_{123} = \frac{1}{4 \times 10^{-11}} \, d_{123} = -0.0107 \, \text{meters}^2/\text{coulomb}$$

(4.38)

and from the last relation of the table

$$h_{111} = \frac{1}{4 \times 10^{-11}} \, e_{111} = 4.1 \times 10^9 \, ;$$

$$h_{123} = \frac{1}{4 \times 10^{-11}} \, e_{123} = 7.75 \times 10^8 \, \text{newtons/coulomb}$$

(4.39)

Here we have assumed that $\beta^T_{11} \doteq \beta^S_{11}$, which is true to about 1 per cent, as will be evident from Table 9. A check of these values can be obtained by using the last form for h_{nkl} from the table. With these values, the differences shown by Table 7 become the values in Table 9.

The differences between constant-displacement and constant-field moduli can be obtained for all the other elastic moduli from the same formulas, but they are not considered here. It should be noted that these formulas take account of all of the strains that can be generated by fields applied along the x axis. If the motion is limited to the thickness mode by resonance, the differences for both the dielectric and elastic moduli are about 1 per cent.

Another interaction effect of some interest is the difference between true $(p_n{}^S)$ and false pyroelectricity, or between pyroelectricity measured at constant strain and that measured at constant stress. The latter effect includes a piezoelectric polarization due to strains generated by temperature.

Table 9

EVALUATION OF INTERACTION EFFECTS FOR QUARTZ

$$\frac{\varepsilon_{11}^T - \varepsilon_{11}^S}{\varepsilon_{11}^T} = \frac{7.27 \times 10^{-13}}{4 \times 10^{-11}} = 0.0181; \qquad \frac{s_{1111}^E - s_{1111}^D}{s_{1111}^E} = \frac{2.52 \times 10^{-13}}{1.279 \times 10^{-11}} = 0.0197$$

$$\frac{\beta_{11}^S - \beta_{11}^T}{\beta_{11}^T} = \frac{4.53 \times 10^8}{2.5 \times 10^{10}} = 0.0181; \qquad \frac{\rho(C^T - C^S)}{\rho C^T} = \frac{1.5 \times 10^3}{2 \times 10^6} = 7.5 \times 10^{-4}$$

$$\frac{c_{1111}^D - c_{1111}^E}{c_{1111}^E E} = \frac{1.41 \times 10^9}{8.605 \times 10^{10}} = 0.0164$$

This calculation is given for tourmaline, probably the first crystal for which pyroelectricity was observed. From Eq. (4.40) this difference is

$$p_n{}^T - p_n{}^S = \alpha_{kl}^E e_{nkl}^\Theta \tag{4.40}$$

Pyroelectricity is observed along the trigonal axis and for a constant stress has the value[4]

$$p_n{}^T = 4.4 \times 10^{-6} \text{ coulombs/meter}^2 \text{ degree} \tag{4.41}$$

The piezoelectric e constants have been measured for tourmaline by Voigt and Röntgen, and the preferred values selected by Cady[5] are

$$e_{311} = 3_{322} = 0.103; \qquad e_{333} = 0.32 \text{ coulombs/meter}^2 \tag{4.42}$$

The expansion coefficients are[6]

$$\alpha_{11} = \alpha_{22} = 3.6 \times 10^{-6} ; \qquad \alpha_{33} = 9.05 \times 10^{-6} \tag{4.43}$$

If we expand Eq. (4.40) we find

$$p_3{}^T - p_3{}^S = (\alpha_{11}e_{311} + \alpha_{22}e_{322} + \alpha_{33}e_{333}) = 2\alpha_{11}e_{311} + \alpha_{33}e_{333} \tag{4.44}$$

Hence the so-called true pyroelectric effect is

$$p_3{}^S = 4.4 \times 10^{-6} - [(2 \times 3.6 \times 10^{-6} \times 0.103) + 9.05 \times 10^{-6} \times 0.32] \tag{4.45}$$

$$= (4.4 - 3.63) \times 10^{-6} = 0.77 \times 10^{-6} \text{ coulombs/meter}^2 \times \text{degree}$$

and hence most of the pyroelectric effect is due to the piezoelectric effect working through the temperature-expansion coefficients.

[4] See *American Institute of Physics Handbook* [5].
[5] See Cady [6].
[6] See Krishnan [7].

4.4 Electrostrictive Properties of Ferroelectric Materials

The ferroelectric crystals have some similarity to pyroelectric crystals in that they become spontaneously polarized as the temperature changes. However, this spontaneous polarization increases from zero to a rapidly increasing value at a specific temperature called the Curie temperature. Figure 4.1 shows the spontaneous polarization of a barium titanate ($BaTiO_3$) single crystal as a function of the temperature measured along a cube axis.

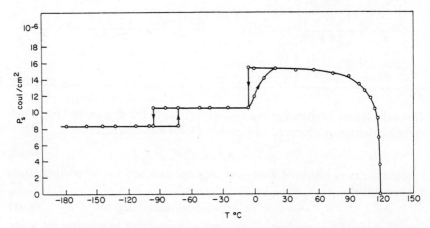

FIG. 4.1 Spontaneous polarization along a cube axis for barium titanate, plotted as a function of temperature (after W. J. Merz).

The temperature of 120° C is known as the Curie temperature. Two other transition temperatures occur at −5° C and −90° C, for which the spontaneous polarization changes from being along a ⟨001⟩ axis to being along a face diagonal (−5° C) and then along a cube diagonal (−90° C). These changes in polarization are accompanied by changes in the crystal structure. Above the Curie temperature, the crystal is cubic ($m3m$) with a center of symmetry, so that it is not piezoelectric. In the first ferroelectric phase it becomes tetragonal ($4mm$) with the axis of polarization along the c axis. As shown by Fig. 4.2 this axis becomes larger and the other two axes become smaller. Below −5° C the crystal structure is again changed to an orthorhombic form ($mm2$); the three axes are shown by the Fig. 4.2. Finally, at −90° C the crystal changes to a trigonal form ($3m$) and all three axes are equal although not at right angles. Figure 4.3 shows the successive shapes[7] taken by the unit cell.

[7] See Mason and Wick [8].

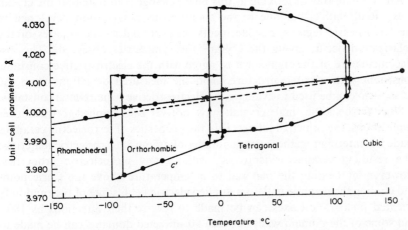

FIG. 4.2 Lengths of edges of cube of pseudocubic unit cell for BaTiO₃ (after Kay and Vousden).

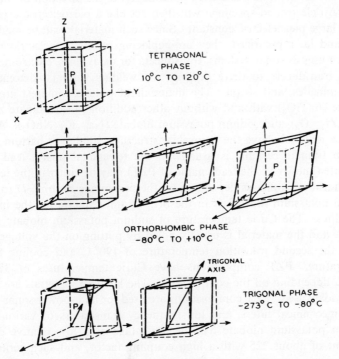

FIG. 4.3 Successive shapes taken by the unit cell of BaTiO₃ as a function of temperature.

An alternate way of looking at this phase change is to note that the crystal is essentially cubic, but due to the spontaneous polarization developed in the ferroelectric regions, displacements proportional to the squares of the polarization occur, giving the crystals the symmetries observed. The displacements are of the right form to agree with the electrostrictive constants consistent with a cubic structure,[8] and the measured deviations from the cubic form can be used to evaluate the electrostrictive (square-law) constants.

Since ferroelectric single crystals have not yet been applied to transducer applications, the principal interest is in the properties of ferroelectric ceramics made by sintering together a large number of grains of ferroelectric materials. The resulting ceramic is isotropic, without any piezoelectric properties. However, by heating the material to a temperature above the Curie point and cooling the ceramic under an applied electric field, some of the randomly oriented polarization axes can be made to reverse their directions by 180°, and some of the domain walls between 90°-directed domains can be made to move in such a direction as to increase the polarization in the direction of the field. When the temperature returns to normal and the electric field is taken off, the ceramic has a fixed polarization in the direction of the applied field. An electroded specimen will then act like a piezoelectric crystal with a very large piezoelectric constant. Since such materials can be made in any form and later polarized, they are replacing other piezoelectric materials for such uses as sonar systems, transducers for delay lines, electromechanical filters, transducers to drive drilling and welding systems, detonators for hand grenades, and so on. The materials of practical interest are barium titanate ($BaTiO_3$) with and without other additions, lead titanate zirconate ($PbTi_xZr_{1-x}O_3$), and sodium potassium niobate ($Na_{0.5}K_{0.5}NbO_3$). All three[9] of these materials have the perovskite structure, similar to barium titanate. Barium titanate ceramic is usually in the tetragonal phase, lead titanate zirconate (known by the trade name of PZT) may be either in the tetragonal or orthorhombic phases depending on the percentage ratio of Ti to Zr, and sodium potassium niobate is in the orthorhombic phase for the individual crystallites. The Curie temperature of sodium potassium niobate is above 400° C, and the material is usually poled by putting on the voltage slightly above the second transition temperature of 190° C and cooling to room temperature. PZT compositions have Curie temperatures of 328° C to 370° C, depending on the composition. The last two materials are now the most widely used. PZT compositions are used for high-power sonar systems, electromechanical filters, and low-frequency transducers of various forms. Sodium potassium niobate has the advantage of a low relative dielectric constant of about 225 with a high coupling factor, and such ceramics are

---

[8] For a thermodynamic derivation of these terms see Devonshire [9].

[9] The properties of these ceramics are discussed by Berlincourt et al. [10].

widely used as delay line transducers. The mechanical properties of the compositions are excellent, and fundamental longitudinal and shear modes have been obtained as high in frequency as 100 megacycles/sec.

It is the purpose of the present section to give a phenomenological derivation of the elastic, piezoelectric, and dielectric constants. All the constants can be derived by considering that the unpoled ceramic is isotropic with a center of symmetry. In the unpoled state the elastic and dielectric moduli are independent of the orientation and there are no piezoelectric moduli. When the material is poled by means of an electric voltage at high temperatures, a unique axis called the x_3 or z axis is established. All directions perpendicular to this direction are similar, and hence the symmetry is transverse isotropic. All the properties of any tensor of rank less than six are the same for this symmetry as those for the hexagonal symmetry $6mm$, and hence the elastic, piezoelectric, and dielectric constants will be

Elastic $\quad s_{1111} = s_{2222}, s_{1122}, s_{1133} = s_{2233}; \quad s_{3333}, s_{2323} = s_{1313};$

$$s_{1212} = (s_{1111} - s_{1122}).$$

(4.46)

Piezoelectric $\quad e_{113} = e_{223}; \quad e_{311} = e_{322}; \quad e_{333}.$

Dielectric $\quad \varepsilon_{11} = \varepsilon_{22}; \quad \varepsilon_{33}$

The same set of constants can be derived by considering that the material is isotropic[10] with the addition of a superposed dielectric displacement $D_{3_0} = P_0$. Since the isotropic material has a center of symmetry, all odd-rank tensors are zero and there is no direct piezoelectric effect. The material can be considered hard elastically but soft electrically, so that only the first derivatives of the stress are considered but second and fourth derivatives of the electric displacements are taken. Since we are interested only in the adiabatic moduli, the thermodynamic function H_1 is used and any derivatives with respect to the entropy σ are neglected. From Table VI

$$S_{ij} = -\frac{\partial H_i}{\partial T_{ij}}; \quad E_m = \frac{\partial H_i}{\partial D_m} \qquad (4.47)$$

Out to six derivatives, the function H_1 can be expressed in the form

$$2H_1 = -[s_{ijkl}^I + R_{ijklmn}D_m D_n]T_{ij}T_{kl} + [q_{ijmn}D_m D_n + N_{ijmnop}D_m D_n D_o D_p]T_{ij}$$
$$+ [\varepsilon_{mn}^T + K_{mnop}^T D_o D_p + K_{mnopqr}^T D_o D_p D_q D_R]D_m D_n \qquad (4.48)$$

From (4.47) the two equations of (4.49) result:

$$S_{ij} = [s_{ijkl}^I + R_{ijklmn}^D D_m D_n]T_{kl} - [q_{ijmn}D_m + N_{ijmnop}D_m D_o D_p]D_n$$
$$E_m = [q_{ijmn}D_n + N_{ijmnop}D_n D_o D_p]T_{ij} \qquad (4.49)$$
$$+ [\varepsilon_{mn}^I + K_{mnop}^T D_o D_p + K_{mnopqr}D_o D_p D_q D_r]D_n$$

[10] See Mason [11].

Comparing this with Eq. (4.31) we see that the elastic, piezoelectric, and dielectric constants for the polarized ceramic become

$$s_{ijkl}^{D} = [s_{ijkl}^{I} + R_{ijklmn}^{D} D_m D_n]$$

$$g_{mij} = -[q_{ijmn} D_n + N_{ijmnop} D_n D_o D_p] \qquad (4.50)$$

$$\varepsilon_{mn} = [\varepsilon_{mn}^{I} + K_{mnop}^{T} D_o D_p + K_{mnopqr}^{T} D_o D_p D_q D_r]$$

Here the superscript I indicates that the constants are those for the unpoled isotropic state.

By employing the components of a sixth-rank tensor appropriate to an isotropic material[11] it is readily shown that the same number of constants results for the elastic and dielectric cases as the ones given in (4.46). The deviation from the isothermal constants is proportional to the square of the polarization $D_{3_0} = P_0$, and the deviation can be used to measure the sixth-order constants R_{ijklmn}, of which there are five for an isotropic material. The same is true for the dielectric constants $\varepsilon_{11}^{T} = \varepsilon_{22}^{T}$; ε_{33}^{T}.

However, for the piezoelectric constants g_{nij} an additional relationship occurs. For the fourth-rank tensor q_{ijmn} there are only two independent components for the isotropic case, as shown by Table B.4 of Appendix B. These are, in terms of two-index symbols,

$$q_{11}, \quad q_{12}, \quad \text{and} \quad q_{44} = q_{11} - q_{12} \qquad (4.51)$$

Since the polarization is along the $z = x_3$ axis, the strains are, from the first of Eqs. (4.49),

$$S_{33} = q_{11} D_3^2 + N_{333} D_3^4 ; \qquad S_{11} = S_{22} = q_{12} D_3^2 + N_{133} D_3^4 \qquad (4.52)$$

D_3 consists of a fixed part P_0 plus a small variable part δD_3. The variable strain is then

$$\delta S_{33} = [2q_{11} P_0 + 4N_{333} P_0^3] \, \delta D_3 = g_{33} \, \delta D_3$$
$$\delta S_{11} = \delta S_{22} = [2q_{12} P_0 + 4N_{133} P_0^3] \, \delta D_3 = g_{31} \, \delta D_3 \qquad (4.53)$$

The thickness shear mode in piezoelectric ceramics is described by the piezoelectric constants $g_{15} = g_{24}$. It results when a signal δD_1 is applied perpendicular to the ceramic polarization $D_{3_0} = P_0$. For this case the one-index strain $S_5 = 2S_{13}$ is expressed in the form

$$\delta S_5 = 2\delta S_{13} = 2[q_{55} + 2N_{553} P_0^2] P_0 \, \delta D_1 = g_{15} \, \delta P_1 \qquad (4.54)$$

Since $q_{55} = q_{44} = (q_{11} - q_{12})$, if third-order terms are neglected,

$$g_{33} - g_{31} = g_{15} \qquad (4.55)$$

[11] See Mason [11].

Table 10

EQUIVALENT PIEZOELECTRIC, ELASTIC, AND DIELECTRIC PROPERTIES
FOR SEVERAL CERAMICS

		Material	
Quantity	PZT-4[a]	$BaTiO_3$ (0.8); $PbTiO_3$ (0.12); $CaTiO_3$ (0.08)	$Na_{0.5}K_{0.5}NbO_3$
k_p	−0.58	−0.19	−0.36
k_{31}	−0.334	−0.113	−0.22
k_{33}	0.70	0.34	0.51
k_{15}	0.71	0.30	0.60
$\varepsilon_{33}^T/\varepsilon_0$	1300	450	290
$\varepsilon_{33}^S/\varepsilon_0$	635	350	—
$\varepsilon_{11}^T/\varepsilon_0$	1475	—	—
$\varepsilon_{11}^S/\varepsilon_0$	730	—	—
$d_{33} \times 10^{-12}$ C/N	289	60	80
d_{31}	−123	−20	−32
d_{15}	496	—	—
$g_{33} \times 10^{-3}$ Vm/N	26.1	17.0	31.5
g_{31}	−11.1	−5.7	−12.6
g_{15}	39.4	—	—
$s_{11}^E \times 10^{-12}$ m²/N	12.3	7.8	9.6
s_{33}^E	15.5	8.1	—
s_{12}^E	−4.05	−2.3	—
s_{13}^E	−5.31	—	—
s_{44}^E	39.0	—	24.4
s_{66}	32.7	20.2	—
Q_C	500	1200	240
Q_E	250	170	160
ρ kgm/meter³	7.5	5.4	4.46
f_1, cycles/meter/sec	1650	2430	2540
f_3	2000	—	—
Curie point	328° C	140° C	>400° C
Volume resistivity, ohm-m	>10^{10}	>10^{10}	>10^{10}
Heat capacity, J/kg° C	420	500	—
Thermal conductivity, W/m° C	1.25	2.5	—
Static tensile strength, psi	13000	12000	—
Rated dynamic tensile strength, psi	3500	3000	—

[a] Trademark, Clevite Corporation.

Taking account of the third-order terms

$$\frac{(g_{33} - g_{31}) - g_{15}}{P_0} = [4(N_{333} - N_{133} - N_{553})P_0^{\,2}] \qquad (4.56)$$

The right-hand side of the equation may be positive or negative, depending on the nature of the higher-order terms. According to reference [10] the difference is negative for BaTiO$_3$ ceramics and increases in proportion to the square of P_0, indicating that terms higher than the third can be neglected. With some lead titanate zirconate ceramics the difference is positive.[12]

Once a ceramic is poled, it acts like a piezoelectric crystal having the symmetry 6*mm*. Hence it will have the set of elastic, piezoelectric, and dielectric moduli of Eq. (4.46). The development in terms of an isotropic solid is interesting in that it shows that the properties depend on how much polarization P_0 is introduced by the poling process. Conversely, if the polarization decays with time for any reason, all of the properties are going to change. Such effects have been observed in piezoelectric ceramics. They are usually ascribed to relief of mechanical strain introduced by the poling process. Such strain can cause the motion of 90° walls between adjacent domains, and the motion is always in a direction to cause a decrease of polarization. Table 10 shows[13] the equivalent piezoelectric, elastic, and dielectric parameters for several widely used ceramics.

The first four entries are the planar coupling coefficients and the coupling coefficients for the longitudinal length mode, the thickness longitudinal mode, and the thickness shear mode. The definitions of these terms are discussed in Chapter 6. The frequencies f_1 and f_3 are respectively the longitudinal-length-mode frequency for a ceramic one meter long and the longitudinal-thickness-mode frequency for a crystal one meter thick. All of these quantities are useful in determining the performance of ceramic transducers. The columns Q_C and Q_E are respectively the mechanical and electrical Q's—that is, the ratio of reactance to resistance—for the mechanical and electrical properties of the ceramics.

4.5 Magnetostrictive and Piezomagnetic Properties of Ferromagnetic Materials

If we include the magnetic energy in the expression (4.1) for the total energy and delete the first term containing the electrical energy, a set of piezomagnetic relations can be derived similar to the piezoelectric relations of Section 4.2. In fact this was done by Voigt,[14] who showed that a true

[12] See Berlincourt *et al.* [12].
[13] See Berlincourt *et al.* [10].
[14] See Voigt [13].

piezomagnetic effect is possible in 29 of the 32 crystal classes. In determining these constants it is necessary to note that B and H are axial rather than polar vectors. However, the existence of this effect has not been demonstrated with certainty for any single-domain crystal.

Biased magnetostriction is, however, phenomenologically equivalent to piezomagnetism. The magnetized ferromagnetic material has the symmetry ∞/m with an infinite fold symmetry axis and a symmetry plane normal to it. This combination results in a center of symmetry that is compatible with piezomagnetism, since a center of symmetry does not reverse the sign of an axial vector.

Since any tensor less than a sixth-rank tensor will have the same components for a hexagonal system as for one with infinite symmetry, one can use the piezomagnetic constant matrix that Voigt lists as

$$
\begin{array}{c|cccccc}
 & S_1 & S_2 & S_3 & S_4 & S_5 & S_6 \\
\hline
H_1 & 0 & 0 & 0 & d_{14} & d_{15} & 0 \\
H_2 & 0 & 0 & 0 & d_{15} & -d_{14} & 0 \\
H_3 & d_{31} & d_{31} & d_{33} & 0 & 0 & 0
\end{array}
\tag{4.57}
$$

where, in agreement with the terminology proposed by the IEEE committee on piezoelectric and ferroelectric crystals, the d constants are defined by the relations

$$
\frac{\partial B_m}{\partial T_{ij}} = \frac{\partial S_{ij}}{\partial H_m} = d_{mij}
\tag{4.58}
$$

This definition preserves the analogy between piezoelectric and piezomagnetic equations. The latter can then be written in the tensor form

$$
S_{ij} = s_{ijkl}^{H,\Theta} T_{kl} + d_{mij}^{\Theta} H_m + \alpha_{ij}^{H}\,\delta\Theta
$$

$$
B_n = d_{nkl}^{\Theta} T_{kl} + \mu_{mn}^{T,\Theta} H_m + i_n{}^{T}\,\delta\Theta
\tag{4.59}
$$

$$
\delta\sigma = \alpha_{kl}^{H} T_{kl} + i_m{}^{T} H_m + \frac{\rho C^{H\,T}}{\Theta}\,\delta\Theta
$$

which is comparable with Eq. (4.17) for piezoelectric effects. The four forms of the piezomagnetic constants and the new terms defined by Eq. (4.59) are shown by Table 11. Conversion factors for converting into cgs electromagnetic units are also shown.

Although symmetry lets the piezomagnetic constant $d_{14} = -d_{25}$, it is zero in a polarized ferromagnetic polycrystalline or sintered material. This is obvious if we consider the material to be isotropic with an induced magnetic polarization, as was done for ferroelectric materials in the preceding section. Since the squares and even powers of an axial vector are always positive, a phenomenological derivation of the function H_1 will have the same terms as (4.46), with B replacing D and H replacing E. Hence, a polarized ferromagnetic material will have the same constants as (4.46), with μ_{mn} replacing ε_{mn}. In the case of a ferroelectric ceramic most of the effective piezoelectric constant is due to the piezoelectric effects in the individual crystals and only

Table 11

MAGNETIC QUANTITIES AND PIEZOMAGNETIC CONSTANTS

Quantity	Symbol	Unit	Conversion factor to cgs electromagnetic units
Magnetic field	H	(ampere-turn)/meter	$(10^3/4\pi)$ ampere turns/m per oersted
Magnetic flux	Φ	weber	10^{-8} weber per maxwell
Magnetic flux density	B	weber/meter^2	10^{-4} weber/m^2 per gauss/cm^2
Permeability—material	μ	henry/meter	$4\pi \times 10^{-7}$ henry/m per gauss/oersted
Permeability—vacuum	μ_0	henry/meter	$4\pi \times 10^{-7}$ henry/m per gauss/oersted
Relative permeability	K	μ/μ_0	1
Inductance	L	henry	10^{-9} henry
Magnetic polarization	I	weber/meter^2	$(\frac{1}{4}\pi \times 10^4)$ weber/m^2 per gauss/cm^2
Magnetomotive force	$\mathfrak{F}$	ampere-turn	$(10/4\pi)$ ampere turn per gilbert
Magnetic reluctance	R	ampere-turn/weber	$(10^9/4\pi)$ ampere turn/weber per gilbert/maxwell
Piezomagnetic constant	d	weber/newton	10^{-3} weber/newton per gauss/dyne
Piezomagnetic constant	e	weber/meter^2	10^{-4} weber/m^2 per gauss/cm^2
Piezomagnetic constant	g	meter^2/weber	10^4 m^2/weber per cm^2/gauss
Piezomagnetic constant	h	newton/weber	10^3 newton/weber per dyne gauss
Pyromagnetic constant	i	weber/meter$^2 \times {}^\circ$K	10^{-4} weber/m$^2 \times {}^\circ$K per gauss/cm$^2 \times {}^\circ$K

a smaller amount to the motion of domain walls. Since piezomagnetism has not been demonstrated for a single-domain ferromagnetic crystal, all of the effect must be due to domain-wall motion. Some of the properties have been measured for a number of polycrystalline metals and sintered ferrite ceramics. Since the electrical conductivity is very low for the ferrites made from oxides of iron (Fe_3O_4) with the addition of zinc, nickel, cobalt, and other metals, eddy current losses are not of importance. Hence such materials can be used to quite high frequencies, whereas the metal ferromagnetic materials have to be finely laminated to work as high in frequency as 100 kilocycles. For very low frequencies, however, as now used in underwater

Table 12

PROPERTIES OF SEVERAL PIEZOMAGNETIC MATERIALS

Quantity	Ferroxcube		Kearfoot N-51	Nickel	Alfenol 13 Fe 0.87, Al 0.13
	7A1	7A2			
k_{33} (opt)	0.25 to 0.30	0.21 to 0.25	0.32 to 0.40	0.15 to 0.31	0.25 to 0.32
d_{33} (opt) (10^{-9} Wb/N)	-2.8 to -4.4	-1.6 to -2.9	~-3.9	-3.1	~7.1
μ_{33}^S/μ_0 (opt)	15 to 25	8 to 15	12	22	58
$1/s_{33}^P$ (10^{10} N/m²)	16.6	17.2	16	21	15
$1/s_{33}^H$ (10^{10} N/m²)	15.1	16.1	~14	20	~14
$-s_{13}^B/s_{33}^B = \sigma^B$	~0.4	~0.4	—	0.3	—
$1/s_{44}^B$ (10^{10} N/m²)	5.9	6.2	—	7.7	—
Q_M^B	≥5000	≥5000	—	—	—
Q_M^H	2500 to 5000	2500 to 5000	260	50 to 250	—
tan δ (electrical)	0.001 to 0.002	0.001 to 0.002	—	—	—
H (opt) (10^2 A turns/m)	15 to 24	11 to 19	10 to 15	7 to 10	7 to 10
Biasing flux B (opt) (Wb/m²)	0.22 to 0.24	0.22 to 0.24	0.17	0.4	~-0.6
B_S (Wb/m²)	0.11 to 0.16	0.15 to 0.17	—	—	—
μ_{33}^S/μ_0 (rem)	30 to 45	30 to 50	45	20	—
k_{33} (rem)	0.15 to 0.20	0.15 to 0.19	0.29 to 0.33	0.14	—
d_{33} (rem)	-2.3 to -3.8	-2.2 to -3.7	-6.0	-1.5	—
Curie point (°C)	530	530	590	358	~500
B_S (Wb/m²)	~0.33	~0.33	~-0.25	0.6	14
DC resistivity (ohm-m)	≥10	≥10	1 to 100	7×10^{-8}	9×10^{-7}
Heat capacity (J/kg° C)	670	670	—	460	510
$\alpha(10^{-6}/°$ C)	6.5 to 7.0	6.5 to 7.0	—	13	11
Thermal conductivity (W/m° C)	5.8	~5.8	—	59	80
H_C (10^2 A turns/m)	2.5 to 5.0	2.0 to 4.0	—	~-0.3	0.1
Saturation magnetostriction (10^{-6})	-26 to -28	-26 to -28	-30	-33	+40
Ultimate compressional stress (psi)	~150,000	~150,000	—	—	—
Ultimate static tensile stress (psi)	10-15,000	10-15,000	—	9000 yield / 15-30,000 tensile	20,000 yield / ~40-100,000 tensile
Critical dynamic tensile stress (psi)	~1500	~1500	—	≥5000	≥7000
ρ (density) (10^3 kg/m³)	5.35	5.35	~5.1	8.8	6.5
Sound velocity V^B (10^3 m/sec)	5.65	5.75	~5.6	5.0	4.8
Sound velocity V^H (10^3 m/sec)	5.45	5.6	~5.25	4.85	4.55

transducers, they are very useful materials. The data on a number of materials, taken mostly from reference [10], are shown by Table 12. The coupling factors k_{33} (opt) and k_{33} (rem) refer to the values obtained at optimum polarization (for which the superposed value of B is about 0.7 times the saturation value of B) and the value obtained at remanent polarization. The value of H_C is the coercive field as defined in Fig. 3.5.

RESUME CHAPTER 4

1. Thermodynamic Potentials

All the relations between the mechanical, electrical, and thermal variables can be obtained as second derivatives of a set of thermodynamic functions. For mechanical and thermal variables the potentials of interest are the total energy U, the Helmholtz free energy A, the enthalpy H and the Gibbs function G. The free energy A represents the amount of mechanical energy that can be abstracted from the body at constant temperature. The Gibbs function is a thermodynamic potential that has equal values for a transition such as the ferroelectric transition in barium titanate.

If we limit consideration to mechanical and thermal variables, the relations obtained from the Gibbs function are

$$S_{ij} = -\frac{\partial G}{\partial T_{ij}} ; \qquad \sigma \equiv -\frac{\partial G}{\partial \Theta} ; \qquad dG = -S_{ij}\,dT_{ij} - \sigma\,d\Theta$$

Since S_{ij} and σ are functions of the stresses and the temperature, second derivatives give

$$S_{ij} = \frac{\partial S_{ij}}{\partial T_{kl}}\bigg)_{\Theta} T_{kl} + \frac{\partial S_{ij}}{\partial \Theta}\bigg)_{T_{kl}} \delta\Theta ; \qquad \delta\sigma = \frac{\partial \sigma}{\partial T_{kl}}\bigg)_{\Theta} T_{kl} + \frac{\partial \sigma}{\partial \Theta}\bigg)_{T_{kl}} \delta\Theta$$

where

$$\frac{\partial S_{ij}}{\partial T_{kl}}\bigg)_{\Theta} = -\frac{\partial^2 G}{\partial T_{kl}\,\partial T_{ij}} = s^{\Theta}_{ijkl} ; \qquad \frac{\partial S_{ij}}{\partial \Theta}\bigg)_{T_{kl}} = -\frac{\partial^2 G}{\partial \Theta\,\partial T_{ij}} = -\frac{\partial^2 G}{\partial T_{ij}\,\partial \Theta} = \frac{\partial \sigma}{\partial T_{ij}}\bigg)_{\Theta} = \alpha_{ij} ;$$

$$\frac{\partial \sigma}{\partial \Theta}\bigg)_{T_{kl}} = \frac{\rho C_p}{\Theta}$$

The interchange of the order of differentiation results in relations known as Maxwell relations. Inserting the values of the second partial derivatives in the relations, we have

$$S_{ij} = s^{\Theta}_{ijkl} T_{kl} + \alpha_{ij}\,\delta\Theta ; \qquad \delta\sigma = \alpha_{kl} T_{kl} + \frac{\rho C_p}{\Theta}\,\delta\Theta$$

The product of $\Theta\alpha_{kl}T_{kl}$ is known as the heat of deformation.

2. Piezoelectric Relations

When electrical energy in the form of $E_m\,dD_m$ is added to the mechanical and thermal energy, there are eight thermodynamic functions, given by Table 6. The forms of greatest use for ordinary piezoelectric crystals such as quartz are the Gibbs function G and the electric Gibbs function G_2. For the Gibbs function we have

$$S_{kl} = -\frac{\partial G}{\partial T_{kl}}; \qquad D_m = -\frac{\partial G}{\partial E_m}; \qquad \sigma = -\frac{\partial G}{\partial \Theta}$$

Taking second derivatives of G, making use of the Maxwell relations between the independent variables, and giving symbols to the partial derivatives, the three dependent variables can be expressed in terms of the independent variables by the equations

$$S_{kl} = s_{ijkl}^{E,\Theta} T_{ij} + d_{mkl}^{\Theta} E_m + \alpha_{kl}^{E} \delta\Theta$$

$$D_n = d_{nij}^{\Theta} T_{ij} + \varepsilon_{mn}^{T,\Theta} E_m + p_n^{T} \delta\Theta$$

$$\delta\sigma = \alpha_{ij}^{E} T_{ij} + p_m^{T} E_m + \frac{\rho C^{T,E}}{\Theta} \delta\Theta$$

where the interaction terms, piezoelectric constants d_{mkl}^{Θ}, temperature expansion coefficients α_{kl}^{E}, and pyroelectric coefficients p_n^{T} need only one superscript since they are measured with the other independent variable held constant.

By setting $\delta\sigma = 0$, the relations between the isothermal and adiabatic constants can be derived as shown by Eq. (4.19). By employing another thermodynamic function such as G_2 the stress rather than the strain can be related to the applied field. By eliminating between the two forms of the equations, relations can be obtained between the two sets of constants.

For ferroelectric ceramics the electric displacement turns out to be a better independent variable and the thermodynamic functions A and G_1 are the preferred ones. Relations between all the different forms of the piezoelectric constants can be obtained and are shown by Table 7. These equations form the starting point for the equation of waves in piezo-electric crystals as discussed in Chapter 6.

3. Electrostrictive Ferroelectric Ceramics

By sintering together crystals of the ferroelectric materials $BaTiO_3$, $PbTiO_3$-$PbZrO_3$. or $NaNbO_3$-$KNbO_3$, ceramics are formed that have very useful transducer properties. Such ceramics have their polarization axes in random directions, and the ceramics appear isotropic. By poling them at temperatures above the Curie temperature by means of a large electric field and cooling under the field, some domains can be reversed in the direction of the field and 90° domain walls can move in a direction to increase the polarization parallel to the field. Such poled ceramics have very large effective piezoelectric constants, and they are being increasingly used in many forms of transducers.

The symmetry induced by the field is transverse isotropic. This symmetry has the same elastic, piezoelectric, and dielectric constants as a hexagonal crystal of symmetry $6mm$. There are three independent piezoelectric constants,

$$e_{33}; \qquad e_{31} = e_{32} \qquad \text{and} \qquad e_{14} = e_{25}$$

These constants produce vibrations along the direction of polarization (e_{33}), a radial vibration perpendicular to the direction of poling (e_{31}), and a thickness shear mode when the alternating field is applied perpendicular to the poling field.

By considering the ceramic as essentially isotropic, with a superposed polarization, it can be shown that there is one relation between the three piezoelectric constants of the form

$$\frac{(g_{33} - g_{31}) - g_{15}}{P_0} = [4(N_{333} - N_{133} + N_{553})P_0^2]$$

where the N terms are third-order terms in the electrostrictive constants.

4. Magnetostrictive and Piezomagnetic Properties

Magnetic energy can be substituted for electric energy in the thermodynamic potentials, and equations can be obtained similar to the piezoelectric equations of 2. Following an IEEE standard, piezomagnetic constants d_{mij} are defined by the equations

$$\frac{\partial B_m}{\partial T_{ij}} = \frac{\partial S_{ij}}{\partial H_m} = d_{mij}$$

The resulting constants have been worked out by Voigt but have not been found experimentally. However, a magnetostrictive metal or ceramic can be given properties similar to a hexagonal crystal with symmetry $6/m$ by poling it with a magnetic field. This material will have the same effective constants as a polarized ferroelectric ceramic and can be used to generate longitudinal, radial, torsional, and shear modes of motion. Such materials are useful for a variety of transducer applications.

PROBLEMS CHAPTER 4

1. What relations between the mechanical variables and the thermal variables can be obtained from the thermodynamic function, enthalpy H?
2. Derive the relations between the four sets of piezoelectric constants shown by Table 7.
3. Using the e piezoelectric constants for tourmaline given by Eq. (4.42) determine the d, g, and h piezoelectric constants. The elastic constants are given by Table 2, while the dielectric constants are given by Table 1.
4. How can one excite a torsional vibration in a magnetostrictive rod (Wiedeman effect)?

REFERENCES CHAPTER 4

1. G. Joos, "Theoretical Physics," Chapter 4 p. 496. G. E. Steckert & Co., New York, 1934.
2. J. C. Slater, "Introduction to Chemical Physics," Chapter II. McGraw-Hill, New York, 1939.
3. Warren P. Mason, "Piezoelectric Crystals and Their Applications to Ultrasonics," p. 34. Van Nostrand, Princeton, New Jersey, 1950.
4. W. G. Cady, "Piezoelectricity," Revised Edition, Chapter XXIX. Dover, New York, 1964.
5. "American Institute of Physics Handbook," Table p. 9–103. McGraw-Hill, New York, 1963.
6. W. G. Cady, "Piezoelectricity," p. 227. Dover, New York, 1964.
7. R. S. Krishnan, "Progress in Crystal Physics," p. 33. S. Viswanathan, Madras, 1958.

8. W. P. Mason and R. F. Wick, "Ferroelectrics and the Dielectric Amplifier." *Proc. IRE*, **42**, 1606 (1954).
9. A. F. Devonshire, "Theory of Ferroelectrics." *Phil. Mag. Suppl.* **3**, 85, (1958).
10. D. A. Berlincourt, D. R. Curran, and H. Jaffe *in* "Physical Acoustics" (Warren P. Mason, ed.), Vol. IA, Chapter III. Academic Press, New York, 1964.
11. W. P. Mason, *Phys. Rev.* **82**, 715 (1951).
12. D. A. Berlincourt, C. Cmolek, and H. Jaffe, *Proc. IRE* **48**, 220 (1960).
13. W. Voigt, "Lehrbuch der Kristallphysik," p. 938. B. G. Teubner, 1928.

CHAPTER 5 • Crystal Symmetries and Their Effects on Tensor Properties

The equations given in the preceding chapter are general enough to represent the most unsymmetrical crystal. As indicated for polarized ceramics and for several crystals, however, the symmetry inherent in the crystal structure eliminates some of the possible constants and causes others to have numerical relations. Hence, in order to understand the relations existing for crystals, it is necessary to consider the crystal systems, the crystal classes, and the symmetries that exist in crystals. It is the purpose of this chapter to consider such relationships and how they affect crystal constant terms.

5.1 Crystal Systems, Crystal Classes, and Miller Indices

Crystals are classified into seven crystal systems and 32 point groups or crystal classes. The ideal crystal is referred to identical unit cells, any one of which can be made to coincide with its neighbors by simple translations along the three crystal axes. The ensemble of all the unit cells forms the crystal lattice. The unit cell is usually but not always chosen as the smallest parallelopiped out of which the crystal can be constructed. The edges of the unit cell are parallel to the crystallographic axes a, b, and c, and their relative dimensions are the unit distances along these axes.

Bravais showed that the number of types of polyhedra that will completely fill space is seven. When body-centered and face-centered polyhedra were added, the total number increased to fourteen. Each polyhedron can be considered a unit cell. From these simple lattices are evolved the seven crystal systems. The edges of the polyhedra are the crystallographic axes, and the faces are the pinacoids of the crystal. The seven crystal systems evolved from the Bravais lattices are, in order of increasing symmetry, the triclinic, monoclinic, orthorhombic, tetragonal, trigonal, hexagonal, and cubic (or isometric) systems. In addition, isotropic and transverse isotropic systems are of interest for polycrystalline and sintered ceramics.

All of these systems can be specified in terms of the directions of the crystallographic axes a, b, and c with respect to each other and by the dimensions of the unit cells along the three axes. For example, the most unsymmetrical system, the triclinic system, has all three axes at angles different from 90° and with all three lengths different. Although tensor

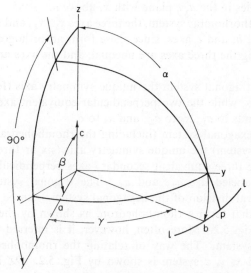

FIG. 5.1 Method for relating the crystallographic axes of a triclinic crystal to a set of rectangular axes.

notation can be applied to such a system by using covariant and contravariant notation, it is simpler to resort to the Cartesian tensors discussed in Chapter 2. In terms of the triclinic axes shown by Fig. 5.1, the piezoelectric crystals committee of the IEEE[1] has standardized the rectangular system as follows: The z or x_3 axis lies along the c crystallographic axis, the x_1 rectangular axis lies in the plane of the a and c crystallographic axes, and the y or x_2 axis is at right angles to x_1 and x_3 in a right-handed system of axes.

Certain conventions have been followed by the crystallographer in selecting the a, b, and c axes of the unit cell. Although they have not been universally observed, the following rules are in common use.

1. In the triclinic system the choice of axes is based on the lengths of the sides of the unit cell. The axes are chosen so that $c < a < b$.

[1] See *IEEE Standards on Piezoelectric Crystals* [1].

2. The monoclinic system is characterized by having one axis, usually taken as the b axis, perpendicular to a and c, which do not form a right angle. b is parallel to the symmetry axis or perpendicular to the symmetry plane. The positive directions of the a and c axes are outward from the obtuse angle between them, while the positive direction of the b axis is such as to make a right-handed system of axes with the a and c axis. The lengths of the unit cell are unequal on all three axes. By referring to the general case of Fig. 5.1 it is seen that x_3 lies along c and x_2 along b, while the x_1 axis lies in the a, c plane with x_1 above a.

3. In the orthorhombic system, the three axes x_1, x_2, and x_3 lie respectively along the a, b, and c axes, since these form an orthogonal system. The lengths along the three axes are unequal, and the axes are chosen so that $c < a < b$.

4. In the tetragonal system the unique symmetry axis (fourfold) is taken as the c axis, while the two perpendicular equivalent axes are a_1 and a_2. x_1 corresponds to a_1, x_2 to a_2, and x_3 to c.

5. In the hexagonal system (including the rhombohedral division or the "trigonal" system) the unique symmetry axis (six or threefold) is chosen as c and the three equivalent secondary axes perpendicular to c and $120°$ apart are labeled a_1, a_2, and a_3. The trigonal system—that is, the rhombohedral division of the hexagonal system—can be characterized by three axes that form a rhombohedron, as shown by the vectors a_1, a_2, and a_3 of Fig. 5.2. More often, however, it is referred to the axes of a hexagonal system. The way of relating the rhombohedral axes to the right-angled x, y, z system is shown by Fig. 5.2. OZ is a line making equal angles with all three equal crystallographic axes a_1, a_2, and a_3. If we extend these axes down to a plane perpendicular to OZ, the intersection points M_1, M_2, and M_3 form an equilateral triangle. If in this triangle we inscribe a hexagon $BCDEFG$, the x_1 axis is taken as OG (or OC or OE). The y axis is perpendicular to x and z in a right-handed system of axes. The hexagonal system is referred to the same axes—that is, OG, OC, or OE are the x_1 axes while the x_2 axis is perpendicular to one of these in a right-handed system.

6. In the cubic (or isometric) system the three equivalent mutually perpendicular axes are called a_1, a_2, and a_3.

The direction of a crystal axis is specified by the coordinates of the axis in terms of the unit-cell dimensions. For example, if a vector starts at the origin and passes through a point that is located one unit-cell displacement along the a axis, one unit cell along the b axis, and one along the c axis, the vector will have the notation $\langle 111 \rangle$. Square brackets are the crystallographers convention for indicating a direction.

A plane of atoms is specified by means of the intercepts along the a, b, and c axes. However, the Miller indices are reciprocals of the intercepts along the three axes. Thus, if the plane intersects the a axis at a distance of 4 unit cells, the b axis at a distance of 4 unit cells and the c axes at a distance of 2 unit cells the Miller indices will be $\frac{1}{4}$, $\frac{1}{4}$, and $\frac{1}{2}$. It is customary to multiply these indices by a number that makes all the indices whole numbers. Therefore, the Miller indices for this plane will be 112. If a, b, and c are the

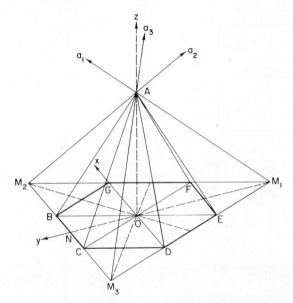

FIG. 5.2   Crystallographic axes of a hexagonal or trigonal crystal.

intercepts of the unit cell along the a, b, and c axes, respectively, the unit plane (the 111 plane) is a plane passing through these three points or is one parallel to it. Any plane drawn through three points having the coordinates a/h, b/k, c/l, where h, k, and l are integers (including zero), is parallel to a plane of the lattice and hence to a geometrically possible crystal face or one from which X rays can be reflected. The integers h, k, and l are known as the Miller indices. The integers h, k, and l are taken in the same order as the a, b, and c axes. They are usually small positive or negative integers, including zero. For example, the 001 face is one perpendicular to the c axis at its positive end and the $00\bar{1}$ face is one perpendicular to c at its negative end. For the trigonal and hexagonal systems, which are usually specified by the four axes (a_1, a_2, a_3, and c) of Fig. 5.2, it is common to use the Bravais-Miller symbols h, k, i, l, which are the reciprocals of the intercepts of the planes on all four axes. Since the sum of $h + k + i = 0$, it is common practice to write the symbol as $(h, k \cdot l)$, the dot signifying that $i = -(h + k)$.

5.2 Crystal Point-Group Symmetries

In general, the points that form the space lattices do not represent the positions of the atoms but merely serve to define the unit cell within which the atoms may be situated in a definite number of configurations. The space groups define the symmetry of the arrangement of the atoms throughout

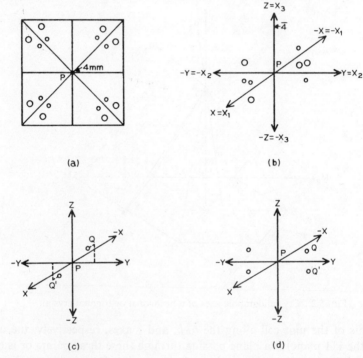

(a)   (b)

(c)   (d)

FIG. 5.3   Point symmetry operations in a crystal: (a) symmetry $4mm$; (b) symmetry $\bar{4}$; (c) symmetry $\bar{1}$; (d) symmetry $\bar{2} = m$.

the unit cell. Crystallographers have shown that there are a total of 230 possible space groups or ways the atoms can be arranged in the seven types of unit cells.

In order to determine the macroscopic properties of the crystal—such as the number of elastic constants—it is not necessary to know the space groups but rather the point groups, so called because all the elements of symmetry can be referred to a single point in the unit cell.

The elements of symmetry possible in the unit cell can be illustrated by the arrangement of atoms in Fig. 5.3. Figure 5.3a shows a pattern made up of two different kinds of atoms, as shown by the larger and smaller circles.

If the array is rotated through 90° about the point P, the elements reproduce themselves. Three other rotations of 90° produce the same patterns, and the axis perpendicular to the paper at the point P is known as a fourfold axis. The four full lines denote mirror planes or planes of symmetry perpendicular to the plane of the paper. The patterns on the two sides of these planes are related as the object and image in a plane mirror. In the Hermann-Mauguin system of designation the fourfold symmetry axis is designated by the number 4. In general, there are 1-fold, 2-fold, 3-fold, 4-fold, and 6-fold axes, which are designated respectively by the numbers 1, 2, 3, 4, and 6. Mirror planes are designated by the latter m.

Figure 5.3b designates another type of symmetry known as an inversion symmetry. The vertical axis through P is known as a fourfold inversion axis and is given the symbol $\bar{4}$. For this fourfold inversion axis, the operation is a rotation of 90°, followed by an inversion of all the elements through the central point P. This results in the positioning of all the atoms as shown. A rotation of 180° restores the elements on the positive end of the y axis to those at the negative end. A onefold inversion axis $\bar{1}$ is identical to a center of symmetry. In this symmetry any point Q having the coordinates x_0, y_0, z_0 has its counterpart at Q' having the coordinates $-x_0$, $-y_0$, $-z_0$, as shown by Fig. 5.3c. A twofold axis of inversion symmetry consists of a rotation of 180° plus an inversion through the center. This operation puts a point Q' of Fig. 5.3d under the point Q. It is equivalent to a mirror operation, or $\bar{2} \equiv m$. A threefold inversion axis, $\bar{3}$, is equivalent to a threefold rotation axis, 3, plus a center of symmetry. A fourfold inversion axis, $\bar{4}$, includes also the operation of a rotation diode axis—that is, rotation of 180° around z restores the symmetry. A sixfold inversion axis, $\bar{6}$, is equivalent to a threefold rotation axis plus a mirror plane normal to it, or $\bar{6} \equiv 3/m$. In general, a mirror perpendicular to a rotation axis is designated as above by the order of the rotation axis followed by a slant bar and the letter m. In general the table of symmetry symbols (Table 13) shows twofold, fourfold, and sixfold axes of this type. Combinations of these types of symmetries are the only ones possible for the point groups.

The 32 crystal classes result from possible combinations of these symmetry elements. The symbols are listed in Table 13. Some other elements of symmetry are sometimes inherent from the principal ones; these are listed in the table. The symbol $mm2$ designates two mirror planes intersecting in a twofold axis, The symbol mmm indicates three mirror planes at right angles to the a, b, and c axes. The first number always refers to an axis parallel to c, except in the monoclinic system, where the unique axis is b. The second number refers to an axis perpendicular to the first, except in the cubic system. The third number, if present, represents an axis perpendicular to the other two, except in the cubic system. The symbol $4mm$ denotes a fourfold

Table 13

THE 32 CRYSTALLOGRAPHIC POINT GROUPS

System	Point Group Symbol	Number of Symmetry Elements of Each Kind									
		m	2	3	4	6	$\bar{1}$	$\bar{2}$	$\bar{3}$	$\bar{4}$	$\bar{6}$
Triclinic	1	—	—	—	—	—	—	—	—	—	—
	$\bar{1}$	—	—	—	—	—	1	—	—	—	—
Monoclinic	2	—	1	—	—	—	—	—	—	—	—
	m	1	—	—	—	—	—	1	—	—	—
	2/m	1	1	—	—	—	1	1	—	—	—
Orthorhombic	222	—	3	—	—	—	—	—	—	—	—
	mm2	2	1	—	—	—	—	2	—	—	—
	mmm	3	3	—	—	—	1	3	—	—	—
Tetragonal (every point group has one 4 or $\bar{4}$ axis)	4	—	(1)	—	1	—	—	—	—	—	—
	$\bar{4}$	—	(1)	—	—	—	—	—	—	1	—
	4/m	1	(1)	—	1	—	1	1	—	1	—
	422	—	4 + (1)	—	1	—	—	—	—	—	—
	4mm	4	(2)	—	1	—	—	4	—	—	—
	$\bar{4}$2m	2	2 + (1)	—	—	—	—	2	—	1	—
	4/mmm	5	4 + (1)	—	1	—	1	5	—	1	—
Trigonal (every point group has one threefold axis)	3	—	—	1	—	—	—	—	—	—	—
	$\bar{3}$	—	—	(1)	—	—	(1)	—	1	—	—
	32	—	3	1	—	—	—	—	—	—	—
	3m	3	—	1	—	—	—	3	—	—	—
	$\bar{3}m$	3	3	(1)	—	—	(1)	3	1	—	—
Hexagonal (every point group has one 6 or $\bar{6}$ axis)	6	—	(1)	(1)	—	1	—	—	—	—	—
	$\bar{6}$	1	—	(1)	—	—	—	1	—	—	1
	6/m	1	(1)	(1)	—	1	1	1	1	—	1
	622	—	6 + (1)	(1)	—	1	—	—	—	—	—
	6mm	6	(1)	(1)	—	1	—	6	—	—	—
	$\bar{6}m$2	4	3	(1)	—	—	—	4	—	—	1
	6/mmm	7	6 + (1)	(1)	—	1	1	7	1	—	1
Cubic (every point group has four threefold axes)	23	—	3	4	—	—	—	—	—	—	—
	m3	3	3	4	—	—	1	3	4	—	—
	432	—	6 + (3)	4	3	—	—	—	—	—	—
	$\bar{4}$3m	6	(3)	4	—	—	—	6	—	3	—
	m3m	9	6 + (3)	4	3	—	1	9	4	3	—

rotation axis with two mirror planes parallel to it. The symbol 4/*mmm* denotes a fourfold rotation axis with a mirror perpendicular to it and two mirrors parallel to the rotation axis and perpendicular to each other. The cubic system poses a special problem, since it has symmetry axes that are neither parallel nor normal to the crystallographic axes. These are the three-fold axes along ⟨111⟩ directions present in every cubic point group. Therefore a 3 is always the second item in a cubic point-group symbol.

Axes that are inherent in the point-symmetry group are marked in parentheses in Table 13. To determine where all these axes are, it is usual to employ

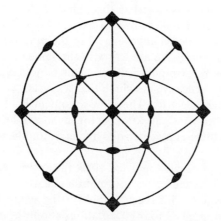

FIG. 5.4   Sterographic projection of a cubic crystal of symmetry 432.

a sterographic projection. To obtain such a projection[2] one first starts with a sphere and marks all the normals to the planes on the surface of the sphere—that is, ⟨100⟩, ⟨110⟩, ⟨111⟩ and so on. Next a great circle is put through the sphere in such a way that the ⟨100⟩, ⟨010⟩, ⟨$\bar{1}$00⟩, and ⟨0$\bar{1}$0⟩ points lie on the circumference. Starting at a point on the bottom of the sphere that marks the 00$\bar{1}$ axis, lines are drawn from this point to the top part of the sphere to the positions of the axes intersecting the sphere. Where these lines intersect the plane, a point marks the presence of an axis. Figure 5.4 shows the principal axes of a cubic crystal. The symmetries inherent in the 432 system are marked for each direction. Boat-shaped symbols indicate a binary axis, triangles threefold axes, and squares fourfold axes. Similar sterographic projections for all the other point groups are given in standard works,[3] but they will not be pursued farther here since they are not essential to the calculation of crystal properties.

² See Wood [2].
³ For example, see Wood [2].

5.3 Effects of Symmetries on the Macroscopic Properties of Crystals

Crystal symmetry eliminates some possible components of tensors of all ranks and establishes relations between others. Some of these relations are obvious without extended calculation, whereas others require a complete tensor calculation of the effect of symmetry on all the tensor components.

For example, consider the effect of a center of symmetry on the piezo-electric effect. In this effect, the application of a stress causes a polarization in a given direction. Now imagine that the whole system—that is, the crystal plus the stress—is inverted through a center of symmetry. This corresponds to using the reflected set of axes of Fig. 2.7b. The stress, being centrosymmetric, does not change sign; but the polarization, being a polar vector, does change sign. We are left with the same crystal under the same stress, but with the reverse polarization. This is possible only if the polarization is zero, and hence we conclude that a crystal with a center of symmetry cannot be piezoelectric. The same argument holds for any odd-rank tensor. On the other hand, an axial vector, as shown by Fig. 2.7c, does not change sign when there is a center of symmetry, and hence such a vector is not suppressed.

The normal way of calculating what effect the symmetry has on the tensor properties is to use the tensor transformation equation (2.17) and introduce the effect of symmetry. This is illustrated for the symmetry 4, for which a set of axes can be rotated by four 90° rotations about the $c = x_3$ axis without changing the properties. In particular, for a 90° rotation $x_1' = x_2$; $x_2' = -x_1$; $x_3' = x_3$. The direction cosines of Eq. (2.12) become

$$
\begin{array}{c|ccc}
 & x_1 & x_2 & x_3 \\
\hline
x_1' & \dfrac{\partial x_1'}{\partial x_1} = 0 & \dfrac{\partial x_1'}{\partial x_2} = 1 & \dfrac{\partial x_1'}{\partial x_3} = 0 \\[2ex]
x_2' & \dfrac{\partial x_2'}{\partial x_1} = -1 & \dfrac{\partial x_2'}{\partial x_2} = 0 & \dfrac{\partial x_2'}{\partial x_3} = 0 \\[2ex]
x_3' & \dfrac{\partial x_3'}{\partial x_1} = 0 & \dfrac{\partial x_3'}{\partial x_2} = 0 & \dfrac{\partial x_3'}{\partial x_3} = 1
\end{array}
\tag{5.1}
$$

Considering first a second-rank tensor; the transformation considered in (2.16) is

$$
W'_{jl} = \frac{\partial x_j'}{\partial x_i} \frac{\partial x_l'}{\partial x_k} W_{ik}
\tag{5.2}
$$

Considering the term W'_{11} this equation becomes

$$W'_{11} = \left(\frac{\partial x_1'}{\partial x_1}\right)^2 W_{11} + \frac{\partial x_1'}{\partial x_1}\frac{\partial x_1'}{\partial x_2} W_{12} + \frac{\partial x_1'}{\partial x_1}\frac{\partial x_1'}{\partial x_3} W_{13}$$

$$+ \frac{\partial x_1'}{\partial x_2}\frac{\partial x_1'}{\partial x_1} W_{21} + \left(\frac{\partial x_1'}{\partial x_2}\right)^2 W_{22} + \frac{\partial x_1'}{\partial x_2}\frac{\partial x_1'}{\partial x_3} W_{23} \qquad (5.3)$$

$$+ \frac{\partial x_1'}{\partial x_3}\frac{\partial x_1'}{\partial x_1} W_{31} + \frac{\partial x_1'}{\partial x_3}\frac{\partial x_1'}{\partial x_2} W_{32} + \left(\frac{\partial x_1'}{\partial x_3}\right)^2 W_{33}$$

Evaluating all these coefficients the only one different from zero is the one W_{22}, and we find

$$W'_{11} = W_{22} \qquad (5.4)$$

Since, however, the rotation leaves the properties unchanged, we have

$$W'_{11} = W_{11} = W_{22} \qquad (5.5)$$

Carrying out the same calculations for the other terms we find

$$W'_{12} = -W_{21} = W_{12} \,; \qquad W'_{13} = 0; \qquad W'_{21} = -W_{12} = W_{21} \,;$$
$$\qquad (5.6)$$
$$W'_{22} = W_{11} = W_{22} \,; \qquad W'_{23} = 0; \qquad W'_{32} = 0 \qquad W'_{33} = W_{33}$$

Hence the effect of this symmetry 4 on the second-rank tensor is to reduce the number of constants from nine to four, and these are, as shown by Table B.2 of Appendix B,

$$W_{ik} = \begin{vmatrix} W_{11} & W_{12} & 0 \\ -W_{12} & W_{11} & 0 \\ 0 & 0 & W_{33} \end{vmatrix} \qquad (5.7)$$

If W_{ij} is symmetrical, $W_{12} = W_{21}$, which can occur only if $W_{12} = 0$.

The same process can be applied to higher-rank tensors. However, it can be simplified by noting that the axis $1'$ equals 2, the axis $2' = -1$, and the axis $3' = 3$. This is the basis for a direct-inspection method,[4] which can be applied to crystal systems that have a unique transformation but not to ones for which the rotated axis is a combination of the linear axes.

[4] See Fumi [3].

If we apply this method to the third-rank piezoelectric tensor, the 27 terms have the following relations:

$$d'_{111} = d_{222} = d'_{111} \; ; \qquad d'_{112} = -d_{221} = d_{112} \; ; \qquad d'_{113} = d_{223} = d_{113}$$

$$d'_{121} = -d_{212} = d_{121} \; ; \qquad d'_{122} = d_{211} = d_{122} \; ; \qquad d'_{123} = -d_{213} = d_{123}$$

$$d'_{131} = d_{232} = d_{131} \; ; \qquad d'_{132} = -d_{231} = d_{132} \; ; \qquad d'_{133} = d_{233} = d_{133}$$

$$d'_{211} = -d_{122} = d_{211} \; ; \qquad d'_{212} = d_{121} = d_{212} \; ; \qquad d'_{213} = -d_{123} = d_{213}$$

$$d'_{221} = +d_{112} = d_{221} \; ; \qquad d'_{222} = -d_{111} = d_{222} \; ; \qquad d'_{223} = d_{113} = d_{223}$$

$$d'_{231} = -d_{132} = d_{231} \; ; \qquad d'_{232} = d_{131} = d_{232} \; ; \qquad d'_{233} = -d_{133} = d_{233}$$

$$d'_{311} = d_{322} = d_{311} \; ; \qquad d'_{312} = -d_{321} = d_{321} \; ; \qquad d'_{313} = d_{323} = d_{313}$$

$$d'_{321} = -d_{312} = d_{321} \; ; \qquad d'_{322} = d_{311} = d_{322} \; ; \qquad d'_{323} = -d_{313} = d_{323}$$

$$d'_{331} = d_{332} = d_{331} \; ; \qquad d'_{332} = -d_{331} = d_{332} \; ; \qquad d'_{333} = d_{333} \qquad (5.8)$$

If we compare these sets of relations we have such results as

$$d_{222} = d_{111} \; ; \qquad d_{222} = -d_{111} \qquad\qquad (5.9)$$

This can be satisfied only if $d_{111} = d_{222} = 0$. The number of constants and the relations between them are

$$d_{113} = d_{223} \; ; \qquad d_{123} = -d_{213} \; ; \qquad d_{131} = d_{232} \; ; \qquad d_{132} = -d_{231} \; ;$$
$$d_{331} = d_{322} \; ; \qquad d_{312} = -d_{321} \; ; \qquad d_{333} \qquad\qquad (5.10)$$

The piezoelectric constants represent a relation between an applied voltage and a resulting strain. The first number refers to the electrical variable, the field; the last two numbers refer to the strain, which is a symmetrical second-rank tensor. Hence, the last two numbers can be interchanged. This relation eliminates the next-to-last term. Furthermore, it is usual to represent the two symmetrical indices by a single number running from 1 to 6, as discussed in connection with Eq. (3.86). Since the shearing strains (engineering) are twice the tensor strains, the piezoelectric constants having the index 4, 5, or 6 are half the corresponding three-index symbols. The relations of Eq. (5.10) can be written in the form of the matrix

	S_1	S_2	S_3	S_4	S_5	S_6
E_1	0	0	0	$d_{123} = \dfrac{d_{14}}{2}$	$d_{113} = \dfrac{d_{15}}{2}$	0
E_2	0	0	0	$d_{223} = \dfrac{d_{15}}{2}$	$d_{231} = -\dfrac{d_{14}}{2}$	0
E_3	$d_{311} = d_{31}$	$d_{322} = d_{31}$	$d_{333} = d_{33}$	0	0	0

$$(5.11)$$

The same methods can be applied to higher-rank tensors. For a fourth-rank tensor with the same symmetry 4, if the number 2 occurs an odd number of times and the number 1 an even number, the term is always zero. For example,

$$M_{1123} = -M_{2213} \tag{5.12}$$

after a 90° rotation. But the term

$$M_{2213} = M_{1123} \tag{5.13}$$

and hence it is seen that this term equals zero.

If, however, the numbers 1 and 2 each occur an odd number of times, the result is that the transformed term is the negative of the original. For example,

$$M_{1112} = -M_{2221} \; ; \qquad M_{2221} = -M_{1112} \tag{5.14}$$

In terms of the two-index symbols

$$M_{16} = -M_{26} \tag{5.15}$$

Hence the number of terms in a fourth-rank tensor can be obtained by inspection. The result is shown by Table B.4. Appendix B.

As long as a simple transformation law for axes results from the symmetry, the remaining terms can be obtained by inspection methods. These methods can be applied to the symmetries $\bar{1}$, 2, m, $2/m$, 222, $mm2$, mmm, 4, $\bar{4}$, $4/m$, 422, $4mm$, $\bar{4}2m$, $4/mmm$, 23, $m3$, 432, $\bar{4}3m$, and $m3m$.

In general, however, it is a rather formidable task to calculate the components of an unknown tensor. Group theory has also been applied to this problem. However, since all tensors up to rank 4 have been calculated, and some up to rank 6, they will not be discussed further. The resulting components are shown in Appendix B.

RESUME CHAPTER 5

I. Crystal Systems

The seven Bravais lattices provide the basis for the seven crystal systems. These systems in order of increasing symmetry are the triclinic, monoclinic, orthorhombic, trigonal, tetragonal, hexagonal, and cubic systems. The triclinic system is based on a set of axes that do not intersect at right angles and that have different lengths along the three axes. These are usually specified by the inequality $c < a < b$. The method of relating right-angled coordinates to these is given by Fig. 5.1.

The monoclinic system has one axis, the b axis, at right angles to the other two. The orthorhombic system has three mutually perpendicular axes, with different cell lengths, along each axis with the convention $c < a < b$. For both the trigonal and hexagonal systems, the unique symmetry axis (three or sixfold) is taken along the c axis, and three equivalent a_1, a_2, and a_3 axes, 120° apart, lie in a plane perpendicular to c. The tetragonal system has three mutually perpendicular axes with two axes, a_1 and a_2, of equal length and a third, the c axis, having a different length. The cubic system has again three mutually perpendicular axes with three equal axes designated as a_1, a_2, and a_3.

2. Miller Indices

Crystal-axis directions are specified by vectors starting at the origin and passing through a point h unit cells in the a direction, k unit cells in the b direction, and l unit cells in the c direction. The Miller indices of a plane are determined by the reciprocals of the intercepts of the plane on the a, b, and c axes. These reciprocals are multiplied by a number that makes all the indices small whole numbers, and the indices are associated with the crystal axes in the relation $h \rightarrow a$, $k \rightarrow b$, $l \rightarrow c$. For the trigonal and hexagonal systems the three axes a_1, a_2, and a_3 in the plane perpendicular to c have the Bravais-Miller indices h, k, i, l where $h + k + i = 0$.

3. Crystal Point-Group Symmetries

All the macroscopic properties of a crystal—such as the number of elastic constants— are determined by the symmetries inherent in the arrangement of atoms in the unit cell. These symmetries, which are specified with respect to a point at the center of the unit cell, involve rotation axes 1, 2, 3, 4, and 6 for which a rotation of 1, $\frac{1}{2}$, $\frac{1}{3}$, $\frac{1}{4}$, or $\frac{1}{6}$ of a complete revolution will cause all the elements to coincide with their original positions inversion axes $\bar{1}$, $\bar{2}$, $\bar{3}$, $\bar{4}$, and $\bar{6}$ for which a rotation of 1, $\frac{1}{2}$, $\frac{1}{3}$, $\frac{1}{4}$, or $\frac{1}{6}$ of a complete revolution will cause all the elements to invert their positions with respect to the original arrangement, and mirror planes m for which all the elements on one side of the mirror are a mirror image of those on the other side. The thirty-two crystal classes are formed by various combinations of the different symmetry operations. The first number refers to an axis parallel to c, except for monoclinic crystals, where the unique axis is b. The second number refers to the symmetries inherent for an axis perpendicular to c, except for cubic crystals. The third number, if present, represents an axis perpendicular to the other two, except in cubic systems. Special rules apply to the cubic system.

4. Effect of Symmetries on Crystal Properties

A center of symmetry has the effect of causing all odd-rank polar tensors to vanish. This includes the phenomenon of piezoelectricity. Odd-rank axial tensors do not vanish in general for a center of symmetry, but even-rank axial tensors do. To obtain the components for a general symmetry, use can be made of the tensor transformation equation (2.17). By introducing the element of symmetry, such as a fourfold axis of rotation, and specifying that the crystal properties shall remain unchanged after such a rotation, it is found that some of the constants must be zero and that relations exist between other constants. For a number of crystal symmetries an inspection method is possible, which greatly reduces the amount of calculation. The effect of crystal symmetry on all polar tensors up to rank four and one to rank six are given in the tables in Appendix B. The first three ranks of axial tensors are shown by Tables B.5, B.6, and B.7.

PROBLEMS CHAPTER 5

1. Prove that the [112] crystal axis is perpendicular to the (112) crystal plane.

2. Prove the relation between the Bravais-Miller h, k, i indices for trigonal and hexagonal systems of axes, namely that $h + k + i = 0$.

3. Noting that the symmetry 3 for the second symbol 432 of the cubic crystal indicates a threefold axis around a [111] direction, this causes x_1 to go to x_2, x_2 to x_3, and x_3 to x_1 on a rotation of 120° about this axis. What effect does this have on the constants already calculated in Eq. (5.10) for the fourfold axis of symmetry 4?

4. A second-rank axial tensor transforms according to the equation

$$g_{ij} = \pm a_{ik} a_{jl} g_{kl}$$

where the plus sign refers to transformations between axial systems having the same handedness while the negative sign refers to a transformation to axes of different handedness. For the class m perpendicular to x_2, show that the terms of the tensor are

$$\begin{vmatrix} 0 & g_{12} & 0 \\ g_{12} & 0 & g_{23} \\ 0 & g_{23} & 0 \end{vmatrix}$$

as listed in Table B.7.

REFERENCES CHAPTER 5

1. "IEEE Standards on Piezoelectric Crystals, 1949." *Proc. IRE* **37**, No. 12, 1378–1395 (1949).
2. Elizabeth A. Wood, "Crystals and Light" (Momentum Book #5), p. 44. Van Nostrand, Princeton, New Jersey, 1964.
3. F. G. Fumi, *Acta Cryst. 5*, **44**, 691 (1952).

Part II

Applications of Equilibrium
Interaction Processes

CHAPTER 6 • Applications of Piezoelectric and Piezomagnetic Materials to Transducers

6.1 Introduction

Piezoelectric crystals and ceramics and piezomagnetic polycrystalline metals and ceramics are among the principal detectors and generators of acoustic and mechanical power. The earliest application was in the generation of underwater sound energy by quartz—carried out by P. Langevin during World War I—and the use of piezoelectric crystals and ceramics for this purpose is still one of the largest applications.

Shortly after World War I, Professor W. G. Cady[1] showed that oscillators can have their frequencies closely controlled by quartz crystal plates in mechanical vibration. In this use the piezoelectric effect drives the crystal at its mechanical resonance frequency. Crystals have the advantage of a very small internal dissipation, which causes the variations of the vacuum tube, transistor, or external circuit to contribute very little change in the frequency of oscillation of the crystal-controlled oscillator. Coupled with this advantage is the fact that certain crystal cuts—namely, the GT and AT cuts—have very little change in frequency of resonance over a wide temperature range. Crystals are widely used in the control of frequency and time standards.

Another and even larger use[2] for vibrating crystals and piezoelectric ceramics is in obtaining very selective wave filters for separating a given telephone conversation from a number of others transmitted over the same pair of wires, cable, or microwave radio system. For this use the low internal friction and high temperature stability are the prime requisites. For this use quartz crystals are almost universally employed. Each year about 500,000 new filters employing 2,000,000 crystals are required to meet the

[1] See Cady [1].
[2] See Mason [2].

125

needs of the long-distance carrier systems, the submarine cable systems, and the microwave telephone systems.

It is the purpose of this chapter to describe the properties of the principal transducer materials used for these purposes.

6.2   Properties of Principal Piezoelectric Crystals

As mentioned previously, a necessary condition for the occurrence of piezoelectricity is the absence of a center of symmetry. Hence, piezoelectric media are intrinsically anisotropic. This anisotropy may be connected with the structure of the crystal, or it may be supplied synthetically by the application of a poling voltage. Since piezoelectricity is a coupling between elastic and dielectric phenomena, piezoelectric properties are discussed with reference to all three sets of properties.

The properties of a large number of piezoelectric crystals have been measured and may be found in several reference books. The most complete table containing crystals measured up to 1956 is the one prepared by Dr. R. Bechmann of the Signal Corps for the Landolt-Börnstein Tables.[3] Shorter convenient tables complete to 1963 are found in the *American Institute of Physics Handbook*.[4] Not many of these crystals are of technical interest at the present time. The present chapter is limited to a discussion of the most used crystals.

6.21   Properties of Quartz and Tourmaline Crystals

The most widely used piezoelectric crystal is probably quartz, which belongs to the 32 trigonal class. The elastic constants in units of newtons/meter^2 × 10^{10} are given by Table 2, Chapter 3. These are the constant-field adiabatic constants discussed in Chapter 4. The relative dielectric constants are given by Table 1, Chapter 3, the thermal-expansion constants by Table 4; the two independent piezoelectric constants for the four forms are given by Eqs. (4.34) to (4.39) inclusive. Since most of the uses of quartz depend on a low temperature coefficient of frequency, or on a mode free from the interfering effects of other modes of motion, considerable work has been done on investigating crystal plates cut with various orientations to the crystal axes. Figure 6.1 shows the various orientations that have been used for transducers and for zero-coefficient crystals. The series +5° X cut, AT, BT, CT, DT, and GT have the property of a very small change in frequency over a relatively wide temperature range. Figure 6.2 shows the frequency change in parts per million obtained with these crystal cuts. Of these the AT

---

³ See Landolt and Börnstein [3].
⁴ See reference [4].

and GT have the smallest changes over the widest temperature range, and these cuts are used principally to control the frequency of oscillators. The +5° X cut is the crystal cut used in the construction of very selective filters and is the most widely used crystal cut. It has the frequency constant shown

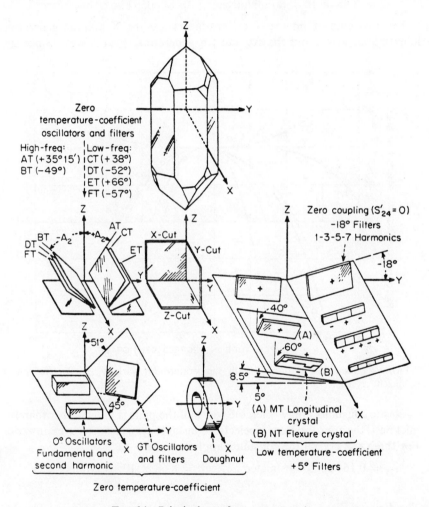

FIG. 6.1 Principal cuts for quartz crystals.

by Fig. 6.3. In this figure the lower part of the line shows the resonance frequency of the crystal while the width of the line is proportional to the difference between the resonance and antiresonance frequencies, as discussed in Section 6.3. In general this crystal is usable in the ratio of width to length of from 0 to 0.2 and from 0.28 to 0.6. The values pertaining to this cut are

$$d'_{12} = -2.26 \times 10^{-12} \text{ coulombs/newton};$$

$$s^{E'}_{22} = 1.22 \times 10^{-11} \text{ meters}^2/\text{newton};$$

$$\varepsilon^T_{11} = 3.98 \times 10^{-11} \text{ farads/meter}; \qquad \rho = 2650 \text{ kilograms/meter}^3$$

(6.1)

The two cuts of interest for transducers are the X cut for generating longitudinal waves and the AC cut for generating shear waves. Since the

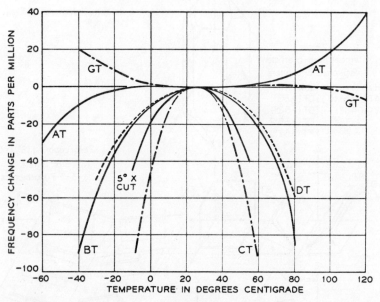

FIG. 6.2 Frequency variations with temperature for various "zero" temperature coefficient crystals.

c_{ij} elastic constants apply when the width of the plate is much larger than the thickness, the appropriate piezoelectric constants are e piezoelectric constants. For the X-cut crystal the constants applicable are

$$e_{11} = 0.164 \text{ coulombs/meter}^2; \qquad c^E_{11} = 8.68 \times 10^{10} \text{ newtons/meter}^2$$

(6.2)

The dielectric constant and the density are the same as in Eq. (6.1). The AC cut is a rotated cut with one edge along the x axis and the other making an angle of $+31°$ from the z axis. For this mode there is no coupling to the face shear, and energy propagation is normal to the crystal face. From the equations for the transformations of second-, third-, and fourth-rank tensors—Eqs. (2.16) and (2.17)—it is readily shown that the dielectric constants, the piezoelectric constants, and the elastic constants have the values

$$\varepsilon'_{22} = \varepsilon_{33} \sin^2 \theta + \varepsilon_{22} \cos^2 \theta = 4.0 \times 10^{-11} \text{ farads/meter}$$

$$e'_{212} = e'_{26} = -[e_{111} \cos^2 \theta - e_{123} \sin \theta \cos \theta]$$

$$= 0.1064 \text{ coulombs/meter}^2 \tag{6.3}$$

$$c'_{66} = c_{66} \cos^2 \theta + c_{44} \sin^2 \theta - 2c_{14} \sin \theta \cos \theta$$

$$= 2.88 \times 10^{10} \text{ newtons/meter}^2$$

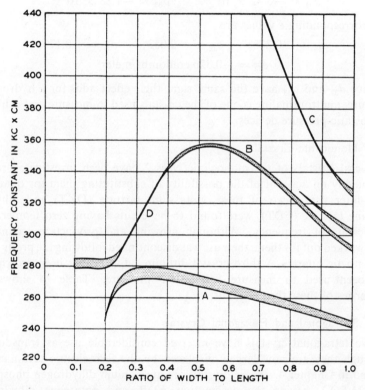

FIG. 6.3 Frequency spectrum and relative separation of resonant and antiresonant frequencies for +5° X cut crystals.

This cut is widely used in measuring the shear properties of crystals and other solids. The particle motion, which determines the plane of polarization, is along the x axis.

Another crystal that has been used to some extent to measure the longitudinal wave properties of low-expansion-coefficient solids is tourmaline. The expansion coefficients along the x and y axes are equal to 3.7 parts per million per degree C, as shown by Table 4 of Chapter 3. This is a fair match for such materials as silicon and germanium. The principal cut for this material is the Z cut—that is, a cut normal to the z axis—which has a longitudinal driving constant e_{333} that is appreciable. The complete set of elastic constants, relative dielectric constants, and temperature-expansion coefficients are given by Tables 1, 2, and 4 of Chapter 3. The piezoelectric constants are in units of coulombs/newton.

$$d_{15} = -3.63 \times 10^{-12} ; \qquad d_{22} = -3.33 \times 10^{-13} ;$$
$$d_{31} = -3.42 \times 10^{-13} ; \qquad d_{33} = -1.83 \times 10^{-12} \qquad (6.4)$$

The corresponding e values are

$$e_{15} = -0.238; \qquad e_{22} = -0.052; \qquad e_{31} = -0.170;$$
$$e_{33} = -0.326 \text{ coulombs/meter}^2 \qquad (6.5)$$

Since d_{31} and d_{33} have the same sign, their effect adds for a hydrostatic pressure, and tourmaline is one of the crystals used to measure and calibrate hydrostatic-pressure devices.

6.22 Monoclinic Crystals

Several crystals of the monoclinic class 2 have been investigated rather thoroughly on account of the possibility of substituting them for quartz in wave-filter use. Two of these, potassium tartrate (DKT) and ethylene diamine tartrate (EDT), were found to have cuts having zero temperature coefficients of frequency. Although growing and production techniques were worked out for them, their use was discontinued following the production of synthetic quartz. A third crystal, lithium sulphate, has been grown and has been used to measure hydrostatic pressure. Table 14 shows the properties of these three crystals.

6.23 Tetragonal and Hexagonal Crystals

Two tetragonal crystals have received considerable use as transducers, although they are now largely displaced by the ferroelectric ceramics described in Chapter 4. These crystals are ammonium dihydrogen phosphate ($NH_4H_2PO_4$), having the designation ADP, and potassium dihydrogen phosphate (KH_2PO_4), which has the designation KDP. Although no longer used in transducers, they are still of interest on account of their electro-optic constants, which are as large as those for any other nonferroelectric crystal. The relative dielectric constants and the elastic constants have already been given in Tables 1 and 2 of Chapter 3, and the temperature-expansion coefficients are given in Table 4. The compliance elastic constants are also of interest, and Table 15 gives these values as well as the densities and piezoelectric constants.

Table 14

PROPERTIES OF THREE MONOCLINIC (CLASS 2) CRYSTALS*

Crystal	Density kg/m³	Elastic Compliances s_{ij}^E (10^{-12} m²/N)												
		s_{11}	s_{22}	s_{33}	s_{12}	s_{13}	s_{23}	s_{44}	s_{55}	s_{66}	s_{15}	s_{25}	s_{35}	s_{46}
Potassium Tartrate (DKT)	1987	23.6	35.3	47.7	−6.1	−4.2	−17.5	122.7	96.1	113.3	−13.4	7.7	−7.4	−6.7
Ethylene Diamine Tartrate (EDT)	1538	33.4	36.5	100.2	−3.0	−32.8	−18.0	191.8	122	191.4	−17.0	15.0	−26.5	3.8
Lithium Sulphate	2052	26.2	22.5	23.9	−7.0	−9.8	−3.0	38.4	50.6	69.0	4.0	−7.3	−0.5	−8.8

Crystal	Piezoelectric Constants (10^{-12} C/N)							
	d_{14}	d_{25}	d_{36}	d_{16}	d_{34}	d_{21}	d_{22}	d_{23}
Potassium Tartrate (DKT)	7.9	−6.5	−23.2	3.5	−12.3	−0.8	4.5	−5.3
Ethylene Diamine Tartrate (EDT)	−10.0	−17.9	−18.4	−12.2	−17.0	10.1	2.2	−11.3
Lithium Sulphate	0.8	−5.0	−4.2	−2.0	−2.1	−3.6	18.3	1.7

	Dielectric Constants Relative			
	K_{11}	K_{22}	K_{33}	K_{13}
Potassium Tartrate (DKT)	6.44	5.8	6.49	0.005
Ethylene Diamine Tartrate (EDT)	4.8	7.74	5.25	0.25
Lithium Sulphate	5.6	10.3	6.5	0.07

* Data from Landolt-Börnstein Tables

Two hexagonal crystals, cadmium sulphide and zinc oxide, have properties of considerable interest for amplifiers of acoustic waves. These crystals are semiconductors of the II-VI type and usually have such a low electrical resistance that piezoelectric responses cannot be excited in them. By introducing a material such as copper, which forms strong bonds with the electrons, these can be captured and the resistivity becomes very high. These materials have been used as diffused-layer transducers and in the

Table 15

ELASTO-ELECTRIC CONSTANTS OF ADP AND KDP[a]

Crystal	Density kg/m^3	s_{11}	s_{33}	s_{12}	s_{13}	s_{44}^E	s_{66}^E	d_{14}	d_{36}
		Elastic Compliances (10^{-12} m^2/N)						Piezoelectric Constants (10^{-12} C/N)	
NH$_4$H$_2$PO$_4$ (ADP)	1803	18.1	43.5	1.9	−11.8	116	166	−1.5	+48
KH$_2$PO$_4$ (KDP	2338	17.5	20.0	−4.0	−7.5	77.7	161	+1.3	+21

[a] Data from Landolt-Bornstein Tables

conducting state as amplifiers for acoustic waves. Table 16 gives the elasto-electric properties of four crystals of this type.

6.24   Ferroelastic Crystals—Rochelle Salt and Barium Titanate

Two other crystals are of some interest for transducer applications. These are the ferroelectric crystals Rochelle salt and single-crystal barium titanate, which, as discussed in Section 4.4, becomes ferroelectric—that is, develops a spontaneous polarization—for all temperatures below about 120° C. Ferroelectric crystals are characterized by very large changes in the mechanical, dielectric, and piezoelectric properties, particularly near the Curie temperature or a transition temperature of any kind.

Rochelle salt was once widely used as a transducer material because it has a high electromechanical coupling factor, which—as discussed in the next section—allows a large percentage of the applied electrical energy to be transformed into mechanical energy and vice versa. Although the original use in underwater sound transducers has now largely been taken over by the ferroelectric ceramics, Rochelle salt still has uses as phonograph pickups and other voltage-generating devices for which the temperature variation is not too serious.

Table 16

ELASTO-ELECTRIC CONSTANTS OF SEVERAL HEXAGONAL CRYSTALS

Crystal	Density kg/m³	Elastic Compliances s_{ij}^E (10^{-12} m²/N)						Piezoelectric Constants (10^{-12} C/N)			Relative Dielectric Constants	
		s_{11}	s_{33}	s_{12}	s_{13}	s_{44}	s_{66}	d_{15}	d_{33}	d_{31}	K_{11}	K_{33}
BeO	3009	2.30	—	—	—	—	—	—	+0.24	−0.12	—	7.66
ZnO	5675	7.9	7.5	−2.35	−2.35	23.4	20.5	−12.0	+12	−4.7	8.1	8.5
CdS	4819	20.69	16.97	−9.99	−5.81	66.49	61.36	−14.0	+10.3	−5.2	9.35	10.33
CdSe	5684	23.38	17.35	−11.22	−5.72	75.95	69.2	−10.5	+7.8	−3.9	9.7	10.65

[a] $s_{66} = 2(s_{11} - s_{12})$

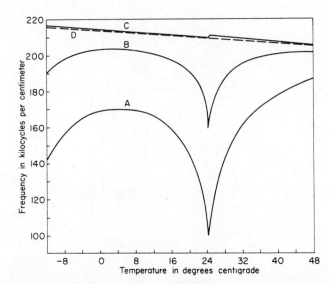

Fig. 6.4   Measured frequencies for a longitudinally vibrating Rochelle salt crystal measured as a function of the temperature. Curves A and B are, respectively, the measured resonance and antiresonance frequencies of a fully plated crystal. Curve C is the measured resonance for a crystal lightly plated at the center. Curve D is the calculated natural mechanical resonance (determined by c_{11}^{D}) of a fully plated crystal.

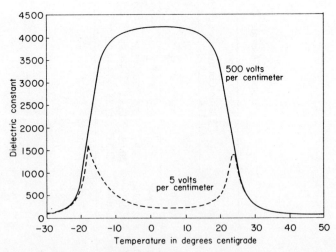

Fig. 6.5   Free dielectric constant for Rochelle salt for fields of 5 and 500 volts per centimeter.

The effect of the large, variable electromechanical coupling inherent in the ferroelectric effect of Rochelle salt is shown by the curves of Figs. 6.4 and 6.5. Figure 6.4 shows the resonant frequency of a crystal cut with its length 45° between the $y = b$ and $z = c$ crystallographic axes. This cut, known as the 45° X cut, is the one that has received the widest use. The resonant frequency, labeled A, shows a large drop at the upper Curie temperature of $+24°$ C and an equally large drop at the lower Curie temperature of $-18°$ C. The curve labeled B is the antiresonant frequency—that is, the frequency for which the crystal reaches its maximum impedance, as discussed in Section 6.3. If the crystal is plated over only a small area or is put in an air-gap holder with a large gap, the curve labeled C results. For both curves A and B the large electromechanical coupling factor reacts on the elastic modulus and lowers the two curves, especially near the Curie temperature. A similar effect occurs for the "free" dielectric constant ε_{11}^T shown by Fig. 6.5. For a low field strength the relative dielectric constant is about 200 between the Curie temperatures but rises to about 1,500 at these temperatures. By going to much higher field strengths the relative dielectric constant becomes much larger. The piezoelectric constants d_{14} and e_{14} show a similar variation, but the constants g_{14} and h_{14}, which relate the strain and stress to the polarization, have a very small variation. At 25° C these constants have the values

$$g_{14} = 0.189 \text{ meters}^2/\text{coulomb}; \quad h_{14} = 2.35 \times 10^9 \text{ newtons/coulomb} \quad (6.6)$$

The elastic compliance determined by the frequency constant of Fig. 6.4 is a combination of the compliances

$$s_{22}' = \frac{(s_{22} + s_{33} + 2s_{23} + s_{44}^E)}{4} \quad (6.7)$$

Table 17

ELASTO-ELECTRIC CONSTANTS OF ROCHELLE SALT AT 34°C

Density kg³m/	Elastic Compliances s_{ij}^E (10^{-12} m²/N)								
	s_{11}	s_{22}	s_{33}	s_{12}	s_{13}	s_{23}	s_{44}	s_{55}	s_{66}
1767	52.0	36.8	35.9	-16.3	-11.6	-12.2	150.2	350.3	104.2

Piezoelectric Constants g_{ij} in m²/C; h_{ij} in (N/C) $\times 10^9$						Relative Dielectric Constants		
g_{14}	g_{25}	g_{36}	h_{14}	h_{25}	h_{36}	K_{11}^T	K_{22}^T	K_{33}^T
0.189	0.57	0.144	2.35	-1.74	1.44	205	9.6	9.5

Table 18

ELASTO-ELECTRIC CONSTANTS OF SINGLE-CRYSTAL BARIUM TITANATE AT 25° C

Density kg/m²	Elastic Compliances s_{ij}^E (10^{-12} m²/N)						Piezoelectric Constants (10^{-12} C/N)			Relative Dielectric Constants	
	s_{11}	s_{33}	s_{12}	s_{13}	s_{44}	s_{66}	d_{15}	d_{33}	d_{31}	K_{11}^T	K_{33}^T
6020	8.05	15.7	−2.35	−5.24	18.4	8.84	392	85.6	−34.5	2920	168

Of these the only constant that varies appreciably with temperature is the elastic compliance s_{44}^E. The other compliance constants are shown by Table 17. There are two other piezoelectric constants, g_{25} and g_{36}, and these are independent of the temperature because the crystal is not ferroelectric along the y and z axes.

Single-crystal barium titanate shows similar effects in the ferroelectric region. However, such crystals cannot yet be obtained in large sizes. The elastic, piezoelectric, and relative dielectric properties are shown by Table 18.

6.3 Wave Transmission in Piezoelectric Crystals— Equivalent Circuits

6.31 Definition of Coupling Coefficient

The most important property of piezoelectric crystals and ceramics is the piezoelectric coupling factor k. If this factor is large—that is, if it approaches unity—electrical energy can be converted into mechanical energy over wide frequency ranges with good efficiency. The piezoelectric coupling factor may be defined as the ratio of the mutual elastic and dielectric energy density to the geometric mean of the elastic and dielectric self-energy densities. This definition parallels the definition of a coupling factor in a magnetic transformer. Neglecting magnetic and thermal terms, the internal energy of a linear system is given by (6.8), as seen by integrating Eq. (4.1):

$$U = \tfrac{1}{2}T_i S_i + \tfrac{1}{2}E_m D_m \tag{6.8}$$

Using the set of Eqs. (4.18) with the two-index symbols in place of the three- and four-index symbols,

$$
\begin{aligned}
U &= \tfrac{1}{2}T_i s_{ij}^E T_j + \tfrac{1}{2}T_i d_{mi}E_m + \tfrac{1}{2}E_m d_{mj}T_j + \tfrac{1}{2}E_m \varepsilon_{mn}^T E_n \\
&= U_e + 2U_m + U_d
\end{aligned} \tag{6.9}
$$

where the subscripts e, m, and d stand for elastic, mutual, and dielectric energies. From this the coupling factor can be written in the form

$$k = \frac{U_m}{\sqrt{U_e U_d}} \tag{6.10}$$

Equation (6.9) contains a large number of terms in general, but when the number of independent constants is reduced by symmetry and when most of the stresses are zero, considerable simplification results. For example, consider the case of an X-cut quartz crystal with its length $+5°$ with respect to the $y = x_2$ axis ($+5°$ X cut). From the data of Eq. (6.1) the static coupling factor for this cut is

$$k = \frac{|d'_{12}|}{\sqrt{s_{22}^{E'} \varepsilon_{11}^T}} = \frac{2.26 \times 10^{-12}}{\sqrt{1.22 \times 10^{-11} \times 3.98 \times 10^{-11}}} = 0.1025 \tag{6.11}$$

It should be noted that the definition of Eq (6.9) does not work for the e and g piezoelectric constants, since they appear with positive and negative signs for the elastic and electric equations. For these cases and for the h form, absolute values have to be taken.

6.32 Derivation of the Impedance of a Piezoelectric Crystal

In dynamic systems the coupling factors are dependent on the stress distribution and in general are less than the static ones. In certain cases, however, they are the same. These cases include a bar heavily mass-loaded, the fundamental mode of a spherical shell poled radially and a piezoelectric ceramic poled radially or axially, since this mode has no overtones. The value of the dynamic coupling can be obtained by inserting the piezoelectric equations of the form of (4.18) for a bar, or (4.21) for a plate, in Newton's equation, which takes the form for a wave transmitted along the y axis:

$$\rho \frac{\partial^2 u_2}{\partial t^2} = F_2 = \frac{\partial T_{2i}}{\partial x_i} \tag{6.12}$$

This relation follows directly from (3.51). Let us consider the case of a fully plated crystal as used, for example, in piezoelectric crystal filters. Figure 6.6 shows the construction. For quartz the relations between the strains and stresses and the field E_1—the only one applied—can be written in the form

$$S_1 = s_{11}^E T_1 + s_{12}^E T_2 + s_{13}^E T_3 + s_{14}^E T_4 + d_{11} E_1$$

$$S_2 = s_{12}^E T_1 + s_{22}^E T_2 + s_{13}^E T_3 - s_{14}^E T_4 - d_{11} E_1$$

$$S_3 = s_{13}^E T_1 + s_{13}^E T_2 + s_{33}^E T_3 \tag{6.13}$$

$$S_4 = s_{14}^E T_1 - s_{14}^E T_2 + s_{44}^E T_4 - d_{14} E_1$$

For the rotated axes for the $+5°$ X cut the same equations hold, with the elastic constants slightly changed by the rotation. For the long, thin bar of Fig. 6.6 the stresses T_1, T_3, and T_4 are zero on the long surfaces, and since the width is small it is a good approximation to set them zero throughout the crystal. We can neglect the strains introduced by the fields in the first

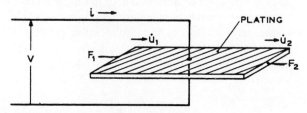

FIG. 6.6   Piezoelectric crystal driven longitudinally.

and last equations of (6.13), and hence the only equation of interest can be written in the form

$$T_2 = \frac{S_2}{s_{22}^E} + \frac{d_{11}}{s_{22}^E} E_1 \tag{6.14}$$

Introducing this equation in (6.12), $S_2 = \partial u_2/\partial x_2$ and noting that $\partial E_1/\partial x_2 = 0$, since the plating is an equipotential surface, Eq. (6.12) becomes, for simple harmonic motion,

$$-\omega^2 \rho s_{22}^E u_2 = \frac{\partial^2 u_2}{\partial x_2^2} \tag{6.15}$$

A solution of this equation with two arbitrary constants is

$$u_2 = A \cos \frac{\omega x}{v} + B \sin \frac{\omega x}{v} \tag{6.16}$$

where

$$v = \frac{1}{\sqrt{\rho s_{22}^E}}$$

To determine the constants A and B, use is made of (6.14). Differentiating (6.16)

$$\frac{du_2}{dx_2} = S_2 = \frac{\omega}{v}\left[-A \sin \frac{\omega x}{v} + B \cos \frac{\omega x}{v}\right] = s_{22}^E T_2 - d_{11}E_1 \tag{6.17}$$

When $x_2 = 0$ and $x_2 = l$, the crystal length, the stress $T_2 = 0$ for a free crystal. Under these conditions

$$\frac{\omega}{v} B = -d_{11}E_1 ; \qquad \frac{\omega}{v}\left[-A \sin \frac{\omega l}{v} + B \cos \frac{\omega l}{v}\right] = -d_{11}E_1 \tag{6.18}$$

Solving for the constants A and B and substituting in Eq. (6.17)

$$\frac{du_2}{dx_2} = S_2 = -d_{11}E_1\left[\frac{\sin\omega(l-x)/v + \sin\omega x/v}{\sin\omega l/v}\right] \qquad (6.19)$$

When a crystal is used as an electrical element in a filter or oscillator, it is desirable to know the electrical impedance of the element. This requires knowing the ratio of the applied voltage to the resultant current at all frequencies. This relation can be obtained from the last of (4.18), which for quartz can be written in the form

$$D_1 = d_{11}T_1 - d_{11}T_2 + d_{14}T_4 + \varepsilon_{11}^T E_1 \qquad (6.20)$$

T_1 and T_4 have been assumed zero, and T_2 can be obtained from (6.14) and (6.18). The current into the crystal is the rate of change of the surface charge, which is equal to the rate of change of D_1. For a simple harmonic voltage, i is given by

$$i = j\omega w\int_0^l D_1\,dx_2 = j\omega w\int_0^l\left[\left(\varepsilon_{11}^T - \frac{d_{11}^2}{s_{22}^E}\right)E_1 - \frac{d_{11}}{s_{22}^E}S_2\right]dx_2 \qquad (6.21)$$

The term

$$\left(\varepsilon_{11}^T - \frac{d_{11}^2}{s_{22}^E}\right) = \varepsilon_{11}^{LC} \qquad (6.22)$$

can be called the longitudinally clamped dielectric constant, since the completely clamped constant ε_{11}^S would require all of the motions to vanish. Introducing the value of S_2 from (6.19) and integrating from 0 to l, we have

$$i = j\omega wl\left[\varepsilon_{11}^{LC} + \frac{d_{11}^2}{s_{22}^E}\left(\frac{\tan\omega l/2v}{\omega l/2v}\right)\right]E_1 \qquad (6.23)$$

The admittance of a free crystal is then

$$\frac{1}{Z} = \frac{i}{V} = \frac{i}{E_1 l_t} = \frac{j\omega wl}{l_t}\varepsilon_{11}^{LC}\left[1 + \frac{d_{11}^2}{\varepsilon_{11}^{LC}s_{22}^E}\left(\frac{\tan\omega l/2v}{\omega l/2v}\right)\right] \qquad (6.24)$$

where w is the width, l the length, l_t the thickness, V the applied voltage, and i the current in amperes.

At very low frequencies this admittance reduces to the capacitative reactance

$$\frac{j\omega wl}{l_t}\left[\varepsilon_{11}^{LC} + \frac{d_{11}^2}{s_{22}^E}\right] = \frac{j\omega wl}{l_t}\varepsilon_{11}^T = j\omega C \qquad (6.25)$$

so that the low-frequency measurement of the capacitance C determines the free dielectric constant ε_{11}^T. When the tangent

$$\tan\frac{\omega l}{2v} = \infty \quad \text{or} \quad \frac{\omega l}{2v} = \frac{\pi}{2}; \quad f_R = \frac{v}{2l} = \frac{1}{2l\sqrt{\rho s_{22}^E}} \qquad (6.26)$$

This relation determines the resonant frequency f_R, which is the frequency for which the admittance of the crystal becomes infinite or the impedance equal to zero. The antiresonance occurs when the impedance becomes infinite—for a dissipationless crystal—or the admittance zero. Setting the term in brackets of (6.24) equal to zero at ω_A we find

$$\frac{\omega_A l}{2v} \cot \frac{\omega_A l}{2v} = -\frac{d_{11}^2}{\varepsilon_{11}^{LC} s_{11}^E} = -\frac{k^2}{(1-k^2)} \tag{6.27}$$

The last equality follows from the definition of the coupling factor—Eq. (6.11)—and Eq. (6.22) for the longitudinally clamped dielectric constant.

As shown by Table 7, Chapter 4, the difference between the elastic compliance at constant electric field and constant electric displacement is, for the single mode of motion considered,

$$s_{22}^E - s_{22}^D = \frac{d_{11}^2}{\varepsilon_{11}^T} \quad \text{or} \quad s_{22}^D = s_{22}^E\left[1 - \frac{d_{11}^2}{s_{22}^E \varepsilon_{11}^T}\right] = s_{22}^E(1-k^2) \tag{6.28}$$

Hence, as shown by Fig. 6.4, the ratio between the unplated mechanical resonance f_M—determined by the compliance s_{22}^D—and the resonant frequency for a fully plated crystal—determined by the compliance s_{22}^E—is given by

$$\frac{f_R}{f_M} = \sqrt{1-k^2} \tag{6.29}$$

The ratio of the antiresonance frequency to the mechanical resonance f_M is determined from Eq. (6.27) and is shown plotted by Fig. 6.7. These curves are in agreement with the values shown by the measurements of Fig. 6.4, which indicate that the coupling factor for this crystal of 0.88 at the Curie temperatures. Midway between the Curie temperatures, the coupling drops to 0.6.

For low-coupled crystals the separation Δf between the antiresonance frequency f_A and the resonance frequency f_R uniquely determines the coupling. Introducing the value

$$f_A = f_R + \Delta f; \quad \omega_A = \omega_R + 2\pi \Delta f \tag{6.30}$$

into Eq. (6.27) and expanding the cotangent function using the multiple-angle formula

$$\cot(A+B) = \frac{\cot A \cot B - 1}{\cot A + \cot B} \tag{6.31}$$

it is readily shown that

$$\frac{k^2}{1-k^2} = \frac{\pi^2}{4}\frac{\Delta f}{f_R} \tag{6.32}$$

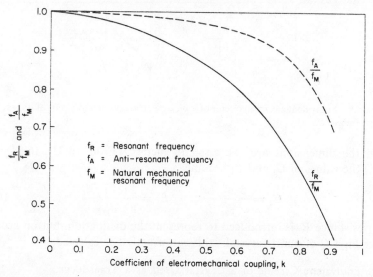

FIG. 6.7 Theoretical curves showing, respectively, the ratio between the electrical resonant and the natural mechanical resonant frequencies and the ratio between the electrical antiresonant frequency and the natural mechanical resonant frequency, each plotted as a function of the electromechanical coupling.

6.33 Equivalent Circuit of a Freely Vibrating Crystal

If the impedance Z of Eq. (6.24) is plotted as a function of the frequency it has the form shown by Fig. 6.8b. It starts out as a capacitative reactance, reaches zero at the resonant frequency f_R, is inductive and increases to infinity at f_A, and then becomes a capacitative reactance. The lumped circuit network of Fig. 6.8a can be given the same form if the elements are adjusted correctly. This equivalent circuit of a piezoelectric crystal is very useful in the design of piezoelectric filters and for other purposes.

To obtain an agreement for the resonant frequency f_R and the antiresonant frequency f_A we must have

$$f_R = \frac{1}{2\pi\sqrt{L_1 C_1}} \; ; \quad \frac{f_A{}^2 - f_R{}^2}{f_R{}^2} = \frac{C_1}{C_0} = \frac{1}{r} \tag{6.33}$$

where r is the ratio of C_0/C_1. Since $(f_A - f_R) \ll f_R$, the last equation reduces to

$$\frac{f_A - f_R}{f_R} = \frac{1}{2r} = \frac{4}{\pi^2}\frac{k^2}{1 - k^2} \quad \text{or} \quad r = \left(\frac{\pi^2}{8}\frac{(1 - k^2)}{k^2}\right) \tag{6.34}$$

from (6.32). The capacitance of the crystal C_0 is

$$C_0 = \frac{\varepsilon_{11}^{LC} l w}{l_t} \text{ farads} \tag{6.35}$$

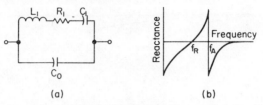

(a)   (b)

FIG. 6.8   Equivalent electrical circuit of a piezoelectric crystal and a plot of its reactance against frequency.

when the dimensions are expressed in meters. From (6.33), (6.34), and (6.11) the values of C_1 and L_1 become

$$C_1 = \frac{8}{\pi^2} \frac{d_{11}^2}{s_{22}^E} \frac{lw}{l_t} \; ; \qquad L_1 = \frac{1}{8} \frac{(s_{22}^E)^2 \rho l l_t}{d_{11}^2 w} \qquad (6.36)$$

A resistance R_1 is introduced to represent the dissipation, but on account of the high Q of quartz crystals, this can usually be neglected.

6.34 Equivalent Circuit of a Crystal Used as a Transducer

There are many applications for crystals in which they are used to convert electrical energy into mechanical energy and vice versa. Such devices are usually called transducers. One of the largest uses is for underwater sound or sonar transducers where large amounts of electrical energies are transferred into the mechanical energies associated with a sound wave in sea water.

For crystals or ceramics having the applied electrical field at right angles to the direction of particle motion, the equations for a transducer can be derived from (6.16) and (6.21) by introducing different boundary conditions. In general what is desired is the effect of terminating the two mechanical ends of the crystals in definite mechanical impedances—that is, definite ratios of the force F_1 or F_2 to the particle velocity $\dot{u}_1$ or $\dot{u}_2$—while working from an electrical impedance, usually a resistance R_E, which can represent the output impedance of a power amplifier. Two types of transducers are of special interest. One is the quarter-wave transducer, for which one end is terminated in a mechanical impedance much higher than the mechanical impedance of the transducer—ideally, an infinite impedance—whereas the other end actuates a mechanical load into which energy is to be delivered. This type of transducer is widely used in underwater sound applications. The other type of transducer is the half-wave type, which is terminated on one end by a zero impedance and delivers energy at the other end. This type is most widely used in high-frequency ultrasonic delay lines.

Since the force exerted by the transducer is the stress T_2 times the area, Eq. (6.14) can be written in the form

$$F = T w l_t = \frac{\partial u/\partial x}{s_{22}^E} w l_t + \frac{d_{11} w V}{s_{22}^E} \qquad (6.37)$$

where the subscript 2 has been dropped from T, u, and x. The last term follows since the voltage V is E_1/l_t. Introducing the strain $\partial u/\partial x$, this equation can be written

$$\left(F - \frac{d_{11}wV}{s_{22}^E}\right) = \frac{\omega w l_t}{v s_{22}^E}\left[-A\sin\frac{\omega x}{v} + B\cos\frac{\omega x}{v}\right] \tag{6.38}$$

When $x = 0$, $F = F_1$ and

$$\frac{w l_t \omega B}{s_{22}^E v} = \left(F_1 - \frac{d_{11}wV}{s_{22}^E}\right) \tag{6.39}$$

Similarly from (6.16)

$$A = u_1 \tag{6.40}$$

Introducing these terms in (6.36) we have one relation between the initial and final forces and the initial velocity $\dot{u}_1 = j\omega u_1$

$$\left(F_2 - \frac{d_{11}wV}{s_{22}^E}\right) = \left(F_1 - \frac{d_{11}wV}{s_{22}^E}\right)\cos\frac{\omega l}{v} + j\dot{u}_1 Z_0\sin\frac{\omega l}{v} \tag{6.41}$$

where $Z_0 = \sqrt{\rho/s_{22}}\,w l_t$ is the mechanical impedance of the crystal.

A second relation between the particle velocities and the forces is obtained by inserting the values of A and B in Eq. (6.16) expressed in terms of the particle velocities. The result is

$$\dot{u}_2 = \dot{u}_1\cos\frac{\omega x}{v} + j\frac{(F_1 - d_{11}wV/s_{22}^E)}{Z_0}\sin\frac{\omega l}{v} \tag{6.42}$$

The third relation involving the current i, the voltage V, and the particle velocities $\dot{u}_2$ and $\dot{u}_1$ can be obtained directly from Eq. (6.21)

$$i = \frac{j\omega\varepsilon^{LC}wl}{l_t}V - \frac{d_{11}w}{s_{22}^E}(\dot{u}_2 - \dot{u}_1) = j\omega C_0 V - \frac{d_{11}w}{s_{22}^E}(\dot{u}_2 - \dot{u}_1) \tag{6.43}$$

where C_0 is the total capacitance of the crystal.

One can use these three equations directly to solve any transducer problem. However, it is frequently more convenient to use the equivalent-circuit approach, where both the electrical and mechanical portions of the transducer are represented by electrical equivalents. This method has distinct advantages over the direct wave equations (6.41), (6.42), and (6.43) because it allows one to apply the powerful methods of network theory to the problem and also because the problem of arriving at the equivalent network has already been solved. Since the method of arriving at such a network has been described in detail,[5] only the final result is presented here. The equivalent circuit derived is shown on Fig. 6.9a, with the velocities, forces, and currents directed as shown. The transformer shown is an ideal transformer with a

[5] See Mason [5].

turns ratio $\varphi = d_{11}w/s_{22}^E$. For an ideal transformer the voltage on the output is the turns ratio φ times the input voltage, in this case V. Similarly, the current on the electrical side is the difference of the mechanical velocities $(\dot{u}_2 - \dot{u}_1)$ times the turns ratio φ. These considerations give directly

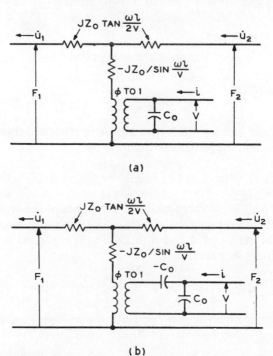

(a)

(b)

FIG. 6.9 (a) Complete equivalent circuit of a piezoelectric crystal or ceramic for which the particle motion is perpendicular to the applied field. (b) Complete equivalent circuit for a piezoelectric crystal or ceramic for which the direction of wave propagation is parallel to the applied field.

Eq. (6.43), for with the velocity and current directions shown the effective current in the capacitance is $i + \varphi(\dot{u}_2 - \dot{u}_1)$. This current times the reactance $(-j/\omega C_0)$ gives the voltage V. Similarly, by employing network theory the other two relations can be verified. Equation (6.42) results from the network equation

$$F_1 + \dot{u}_1\left(jZ_0 \tan \frac{\omega l}{2v}\right) + (\dot{u}_1 - \dot{u}_2)\left(-\frac{jZ_0}{\sin \omega l/v}\right) = \varphi V \qquad (6.44)$$

By taking the voltage-drop relation

$$F_1 + (\dot{u}_1 + \dot{u}_2)jZ_0 \tan \frac{\omega l}{2v} = F_2 \qquad (6.45)$$

and eliminating $\dot{u}_2$ by means of (6.42), the relations of (6.41) result. Hence the network gives directly the relations derived in (6.41), (6.42), and (6.43). The use of this network in designing transducers is considered in Section 6.4.

6.35 Thickness Vibrating Crystals as Transducers

Crystal and ceramic transducers for which the voltage is applied along the direction of wave propagation are also used. These include X-cut and AC-cut quartz crystals for generating respectively longitudinal and shear waves in solids, and ceramic transducers poled to generate longitudinal and shear waves. These types are widely used in ultrasonic delay lines.

For all these types the field can vary in the direction of the wave propagation—the thickness direction—and it is not legitimate to set $\partial E/\partial x = 0$. For these types of transducers, since there are no point charges present in the interior, the divergence of D is zero. Since only one displacement is usually excited, this results in the condition that $\partial D/\partial x = 0$.[6] Hence it is preferable to use the piezoelectric equations of the form given in (4.32). For a thickness longitudinal mode such as in X-cut quartz, the piezoelectric relations for the thickness mode can be written in the form

$$T_1 = c_{11}^D S_1 - h_{11} D_1 \; ; \qquad E_1 = -h_{11} S_1 + \beta_{11}^S D_1 \qquad (6.46)$$

Multiplying the first equation through by the cross-sectional area lw and integrating the last equation with respect to the thickness dimension, noting that $E_1 = \partial V/\partial x$, $S_1 = \partial u/\partial x$, and $\partial D_1/\partial x = 0$, we have

$$F = c_{11}^D S_1 lw - h_{11} D_1 lw; \qquad V = -h_{11}(u_2 - u_1) + \beta_{11}^S D_1 l_t \qquad (6.47)$$

Introducing the first of (6.46) in (6.12), noting that $\partial D/\partial x = 0$, one obtains a solution of the form

$$u = A \cos \frac{\omega x}{v} + B \sin \frac{\omega x}{v} \qquad (6.48)$$

where $v = \sqrt{c_{11}^D/\rho}$. Hence it is evident that the constant displacement elastic stiffness occurs for a thickness-vibrating crystal. Evaluating the constants as before, one obtains the three equations

$$(F_2 + h_{11} D_1 lw) = (F_1 + h_{11} D_1 lw) \cos \frac{\omega l}{v} - j \dot{u}_1 Z_0 \sin \frac{\omega l}{v}$$

$$\dot{u}_2 = \dot{u}_1 \cos \frac{\omega l}{v} - j \frac{(F_1 + h_{11} D_1 lw)}{Z_0} \sin \frac{\omega l}{v} \qquad (6.49)$$

$$i = j\omega C_0 V + \left(\frac{h_{11} lw}{\beta_{11}^S l_t}\right)(\dot{u}_2 - \dot{u}_1) = j\omega C_0 V + h_{11} C_0 (\dot{u}_2 - \dot{u}_1)$$

---

[6] However, if more than one is excited, it is preferable to start with the e piezoelectric constants as shown by Tierston (*J. Acoust. Soc. Am.* **35**, 53 [1963]). The elastic modulus is then not equal to c^D exactly, although close to it. The present treatment is valid for thickness-vibrating ferroelectric ceramics and for the X-cut quartz crystal. It is not valid for the AC cut.

The equivalent circuit for this type of crystal, as shown by Fig. 6.9b, differs from that of 6.9a principally in the electrical part. It involves a negative capacitance $-C_0$ as well as the positive capacitance C_0. The turns ratio of the electromechanical ideal transformer for this case is

$$-\frac{h_{11}lw}{\beta_{11}^{S}l_t} = -h_{11}C_0 \tag{6.50}$$

where C_0 is the capacitance of the crystal measured at constant strain. Since the mechanical velocity is $(\dot{u}_2 - \dot{u}_1)$ through the mechanical side of the transformer, this corresponds on the electrical side to the current directed as shown with the value

$$h_{11}C_0(\dot{u}_2 - \dot{u}_1) = i_E \tag{6.51}$$

which is in agreement with the last of Eqs. (6.49) since the current in the positive capacitance is $i - i_E$. The mechanical force on the mechanical side of the transformer is the voltage on the electrical side times the turns ratio. Hence

$$F_M = -h_{11}C_0\left[-h_{11}(\dot{u}_2 - \dot{u}_1)C_0\left(\frac{+j}{\omega C_0}\right) + V\right] \tag{6.52}$$

Since

$$-j\frac{h_{11}C_0(\dot{u}_2 - \dot{u}_1)}{\omega C_0} = +h_{11}(u_2 - u_1)$$

then from the second of Eqs. (6.47) the force F_M is

$$F_M = -\left(\frac{h_{11}lw}{\beta_{11}^{S}l_t}\right)(\beta_{11}^{S}D_1l_t) = -h_{11}D_1lw \tag{6.53}$$

Direct evaluation of the forces from the particle velocities and the impedances of the network in Fig. 6.9b reproduce the equations of (6.49).

For a ceramic shear thickness mode similar equations result, with the shear stiffness modulus c_{66}^{D} replacing c_{11}^{D}, and h_{26} replacing h_{11}.

6.4 Design of Transducers from Equivalent Circuits

The two types of transducers of particular interest are the quarter-wave transducer and the half-wave transducer. The quarter-wave transducer is largely used in underwater sound transducers and is usually of the type of Fig. 6.9a with the applied voltage at right angles to the particle motion. For this type of transducer the velocity $\dot{u}_2$ at the end attached to the high impedance—usually provided by a quarter-wave metal resonator—is zero, and the arm at the right can be neglected. The impedance of the two arms can be combined into one

$$jZ_0\tan\frac{\omega l}{2v} - \frac{jZ_0}{\sin \omega l/v} = -jZ_0\left[\frac{1 - \cos \omega l/v - 1}{\sin \omega l/v}\right] = -jZ_0\cot\frac{\omega l}{v} \tag{6.54}$$

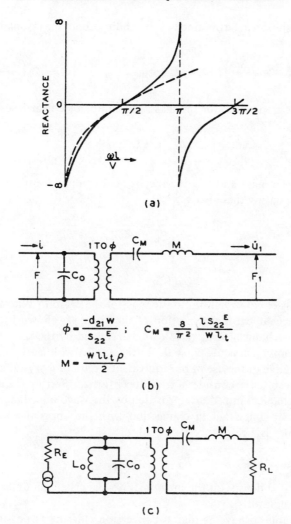

FIG. 6.10   Equivalent circuit for a quarter-wave crystal, used in the design of a transducer.

A plot of the negative cotangent function is shown by Fig. 6.10a. As a driving element the transducer is of use near the resonant frequency, which occurs when $\omega l/v = \pi/2$. A lumped mechanical circuit that has nearly the same impedance near the resonant frequency is a series coil and condenser, which has the reactance shown by the dashed line. To make the zero of both impedances coincide the coil must resonate the condenser at the frequency for which the cotangent function goes to zero. In addition we have to make the slopes of the two functions coincide. For the cotangent function the slope can be obtained by expanding about the resonant frequency f_R. Thus

$$-jZ_0 \cot\left(\frac{\omega_R + \Delta\omega}{v}\right)l = -jZ_0 \cot\left[\frac{\pi}{2}\left(1 + \frac{\Delta\omega}{\omega_R}\right)\right] = jZ_0\left(\frac{\pi}{2}\frac{\Delta\omega}{\omega_R}\right) \quad (6.55)$$

Similarly, the impedance for the mass and compliance near resonance is

$$-\frac{j}{\omega C_M}[1 - (\omega_R + \Delta\omega)^2 M C_M] = \frac{2j\,\Delta\omega}{\omega_R{}^2 C_M} \quad (6.56)$$

where M is the mass and C_M the mechanical compliance of the equivalent elements. Equating these two

$$C_M = \frac{4}{\pi\omega_R Z_0} = \frac{8l}{\pi^2 v Z_0} = \frac{8l s_{22}^E}{\pi^2 w l_t}$$

$$M = \frac{1}{\omega_R{}^2 C_M} = \frac{Z_0 l}{2v} = \frac{w l l_t \rho}{2} \quad (6.57)$$

or the mass is equal to half the static mass of the crystal. With these values the electromechanical representation of the crystal is shown by Fig. 6.10b. The electromechanical turns ratio φ is determined by the ratio of the force F_1 for the clamped crystal—$\dot{u}_1 = 0$—to the applied voltage.

As an example of the use of the equivalent circuit of a crystal in the design of a transducer, let us consider a transducer for transferring electrical energy into the mechanical impedance provided by the wave impedance of a liquid or solid. As pointed out in connection with the impedance of a crystal, this impedance is a resistance R_L equal to

$$R_L = \sqrt{\rho_M c_{11_M}}\,A \quad (6.58)$$

where ρ_M is the density of the medium, c_{11_M} the elastic modulus, and A the area of the radiator.

It is desirable to have as high a conversion efficient as possible over the widest frequency range. This usually requires designing the transducer as a filter circuit. The filter has two properties that make it of interest in electromechanical transducer systems. First, the filter is able to coordinate the action of several resonant elements to produce a device with a uniform transmission over a wide frequency range; and second, the dissipationless filter, with matched impedance terminations, is a device that delivers to its output all the energy impressed upon it over the widest possible frequency range consistent with the elements composing it.

The simplest wide-band filter structure is the one shown by Fig. 6.10c. For this system an electrical inductance L_0 tunes the capacitance C_0 at the midfrequency of the device. An all-electrical filter of this type has the design formulas[7]

$$L_0 = \frac{(f_B - f_A)Z_0}{2\pi f_A f_B}; \qquad C_0 = \frac{1}{2\pi(f_B - f_A)Z_0};$$

$$L_1 = \frac{Z_0}{2\pi(f_B - f_A)}; \qquad C_1 = \frac{f_B - f_A}{2\pi f_A f_B Z_0} \tag{6.59}$$

where f_A is the lower cutoff frequency, f_B the upper cutoff frequency, and Z_0 the characteristic impedance of the filter at midband. Such a transducer will deliver most of the input electrical energy into the output between the frequencies f_A and f_B if it is terminated on the electrical and mechanical ends by the equivalent resistances equal approximately to Z_0.

In order to use these formulas it is necessary to bring the mechanical elements through the electromechanical transformer, which results in the element values

$$L_1 = \frac{M}{\varphi^2} = \frac{l l_t}{2wv^2 k^2 \varepsilon_{11}^T}; \quad C_1 = C_M \varphi^2 = \frac{8}{\pi^2} \frac{lw}{l_t} k^2 \varepsilon_{11}^T; \quad C_0 = \frac{lw}{l_t} \varepsilon_{11}^T (1 - k^2) \tag{6.60}$$

$$L_0 = \frac{1}{(4\pi^2 f_R^2)C_0}; \quad R_T = \frac{\sqrt{\rho_M c_{11_M}}}{\varphi^2}(wl_t) = \frac{\sqrt{\rho_M c_{11_m}}}{wk^2 \varepsilon_{11}^T} s_{22}^E l_t$$

By taking the value of C_1/C_0 from (6.60) and comparing it with the value from (6.59) we find an expression for the fractional bandwidth defined as the actual bandwidth $f_B - f_A$ divided by the mean frequency $\sqrt{f_A f_B}$

$$\frac{f_B - f_A}{f_M} = \frac{1}{\sqrt{r}} = \frac{\sqrt{8}}{\pi} \frac{k}{\sqrt{1 - k^2}} \tag{6.61}$$

The products of $L_0 C_0$ and $L_1 C_1$ are equal to

$$L_1 C_1 = \frac{4l^2}{\pi^2 v^2} = \frac{1}{4\pi^2 f_R^2} = L_0 C_0 = \frac{1}{4\pi^2 f_A f_B} = \frac{1}{4\pi^2 f_M^2} \tag{6.62}$$

Hence the resonant frequency of the crystal system, which for the quarter-wave resonator is given by $f_R = v/4f_1$, and the resonant frequency of the coil and condenser occur at the mean frequency, f_M.

The other question is how well the mechanical and electrical impedances can be made to match their loads. For both ends the impedance of the device as a filter—that is, Z_0—is, from (6.59),

$$Z_0 = \left(\frac{f_B - f_A}{f_M}\right)\sqrt{\frac{L_0}{C_0}} = \left(\frac{f_B - f_A}{f_M}\right)\sqrt{\frac{L_1}{C_1}} \tag{6.63}$$

[7] See Mason [6].

Since the electrical amplifier has an output transformer, this device can be made to match Z_0 as closely as desired. For the mechanical output the only adjustable feature is the coupling factor k. Introducing the value of $\sqrt{L_1/C_1}$ and comparing Z_0 with R_T of Eq. (6.60) we find

$$Z_0 = \left(\frac{f_B - f_A}{f_M}\right)\sqrt{\frac{L_1}{C_1}} = \frac{f_B - f_A}{f_M}\left(\frac{l_t}{wvk^2\varepsilon_{11}^T}\right) \doteq R_T = \frac{\sqrt{\rho_M c_{11_M}}\, s_{22}^E l_t}{wk^2\varepsilon_{11}^T} \quad (6.64)$$

Cancelling out common terms,

$$\left(\frac{f_B - f_A}{f_M}\right)\sqrt{\frac{\rho}{s_{22}^E}} \doteq \sqrt{\rho_M c_{11_M}} \quad (6.65)$$

which states that the fractional bandwidth times the characteristic impedance of the crystal should equal the radiation impedance of the medium. For

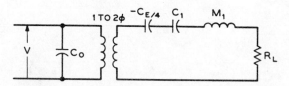

FIG. 6.11   Equivalent circuit of a half-wave transducer.

example if PZT-4 from Table 10, Chapter 4, is used in a length-expander mode, the coupling of 0.334 gives a fractional bandwidth of 0.32. The ratio of $\sqrt{\rho/s_{22}^E}$ times this ratio gives an impedance of

$$0.32 \times \sqrt{\frac{7.5 \times 10^3}{12.3 \times 10^{-12}}} = 7.9 \times 10^6 \text{ kgm meters/sec} \quad (6.66)$$

The impedance of sea water, for example, is

$$\sqrt{\rho C_M} = \rho V = 10^3 \times 1500 = 1.5 \times 10^6 \text{ kgm meters/sec} \quad (6.67)$$

This is somewhat lower, but since a wider band with only a small rise in the center results from terminating the filter in a low impedance, this arrangement should give a satisfactory result.

   The other type of transducer in wide use is the half-wave transducer, which is widely used on ultrasonic delay lines. Since the equivalent circuit requires considerable network manipulation to obtain the final value, the reader is referred to other sources[8] for the details. The final equivalent circuit is shown by Fig. 6.11. The negative capacitance is brought through

[8] See Mason [5].

the perfect transformer as a negative compliance, and when it is joined to the stiffness shown, it results in the term

$$\frac{2l_t}{\pi^2 c_{11}^D l w}\bigg/\left(1 - \frac{8}{\pi^2}\frac{h_{11}^2}{\beta_{11}^T c_{11}^D}\right) \doteq \frac{2l_t}{\pi^2 c_{11}^D l w}\bigg/(1 - k^2) \tag{6.68}$$

Hence the effective stiffness constant is equal to the elastic modulus c_{11}^E.

The method for designing such a transducer is the same as discussed above. If we use a $Na_{0.5}K_{0.5}NbO_3$ ceramic shear-vibrating crystal, as used in transducers to frequencies as high as 100 megacycles, the constants are, from Table 10,

$$\rho = 4.46 \times 10^3 \text{ kilograms/meter}^3; \qquad k_{15} = 0.60;$$
$$s_{44}^E = 24.4 \times 10^{-12} \text{ meters}^2/\text{newton} \tag{6.69}$$

On taking the mechanical elements through the transformer, which has a turns ratio of $2\varphi = (2h_{15}C_0)$, we have the element values

$$L_1 = \frac{\rho l_t^3 (\beta_{11}^S)^2}{8h_{15}^2 l w}; \qquad C_1 = \frac{8}{\pi^2}\frac{h_{15}^2 l w}{c_{44}^E (\beta_{11}^S)^2 l_t}; \qquad R_L = \frac{\sqrt{\rho_M c_{44M}}\, l_t^2 (\beta_{11}^S)^2}{4h_{15}^2 l w} \tag{6.70}$$

Comparing these equations with those given in (6.59) we find

$$\frac{f_B - f_A}{f_M} = \frac{1}{\sqrt{r}} = \frac{\sqrt{8}}{\pi}\frac{k}{\sqrt{1-k^2}} \qquad \text{where} \qquad k^2 = \frac{h_{15}^2}{c_{11}^E \beta_{11}^T}$$

$$\left(\frac{f_B - f_A}{f_M}\right)(\sqrt{\rho c_{55}^E}) = \sqrt{\rho_M c_{44M}} \tag{6.71}$$

Hence the bandwidth and matching conditions are the same as for the quarter-wave transducer. However, the crystal dimension in the direction of vibration is twice as large. For sodium potassium niobate in shear, the coupling is 0.60 so that the fractional bandwidth is

$$\frac{f_B - f_A}{f_M} = 0.56 \tag{6.72}$$

For a mean frequency of 100 megacycles, the bandwidth $f_B - f_A = 56$ megacycles. Fused silica in shear vibration has the values

$$\rho = 2.2 \times 10^3 \text{ kgm/meter}^3;$$
$$\sqrt{\rho c_{44}} = \rho\sqrt{\frac{c_{44}}{\rho}} = \rho v = 8.3 \times 10^6 \text{ kgm meters/sec} \tag{6.73}$$

Sodium potassium niobate has the value for shear waves

$$\rho = 4.46 \times 10^3 \text{ kgm/meter}^3; \qquad v = \frac{1}{\sqrt{\rho s_{44}^E}} = 3.03 \times 10^3 \text{ meters/sec}$$

$$\tag{6.74}$$

Hence the match obtained is

$$7.55 \times 10^6 \doteq 8.3 \times 10^6 \text{ kgm meters/sec} \tag{6.75}$$

and an excellent match is obtained.

6.5 Alteration of the Equivalent Circuit Caused by the Axial Nature of Magnetism

From the discussion of the equations for a piezomagnetic body given by Eqs. (4.59) it is evident that a solution for wave propagation in a biased piezomagnetic material will be the same as for a piezoelectric ceramic. Hence the equivalent circuit of Fig. 6.9 will hold, provided we replace V and i by

$$\int_0^l H_i \, dl = U; \qquad \dot{B}A = \dot{\Phi} \tag{6.76}$$

where U is the magnetomotive force and $\dot{\Phi}$ the rate of change of flux through the circuit. These hold exactly for a closed magnetic circuit or for a long, thin rod. If these conditions are not met, demagnetizing factors and additional reluctance factors have to be taken account of and $\dot{\Phi}$ is the average value determined by all of these factors.

In a transducer, however, it is not U and $\dot{\Phi}$ that we deal with, but rather the input voltage and current. These quantities are related by equations of the type

$$V = N \frac{d\Phi}{dt}; \qquad U = Ni \tag{6.77}$$

where N is the number of turns, and the voltage, current, flux, and magneto-motive force are directed as shown by Fig. 6.12a. Since a rate of change of flux is the analogue of a current and the magnetomotive force the analogue of a voltage it is seen that this element is different from an ideal transformer, which has the equations

$$i_1 = Ni_2; \qquad V_2 = NV_1 \tag{6.78}$$

This device, which converts extensive variables to intensive variables, is called a gyrator and represents a fifth independent network element, the other four being resistance, inductance, capacitance, and the ideal transformer. It was introduced by B. D. H. Tellegen.[9] In the equivalent circuit of Fig. 6.12a the gyrator is represented by the symbol on the left.

[9] See Tellegen [7].

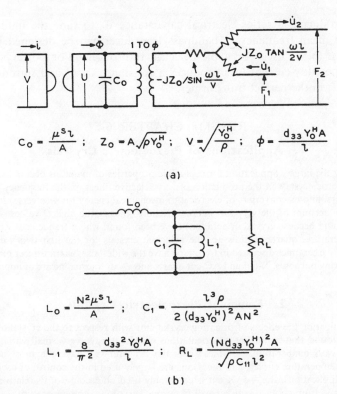

$$C_0 = \frac{\mu^S \ell}{A} \; ; \quad Z_0 = A\sqrt{\rho Y_0^H} \; ; \quad V = \sqrt{\frac{Y_0^H}{\rho}} \; ; \quad \phi = \frac{d_{33} Y_0^H A}{\ell}$$

(a)

$$L_0 = \frac{N^2 \mu^S \ell}{A} \; ; \quad C_1 = \frac{\ell^3 \rho}{2 (d_{33} Y_0^H)^2 A N^2}$$

$$L_1 = \frac{8}{\pi^2} \frac{d_{33}^2 Y_0^H A}{\ell} \; ; \quad R_L = \frac{(N d_{33} Y_0^H)^2 A}{\sqrt{\rho C_{11}} \ell^2}$$

(b)

FIG. 6.12   Equivalent circuit of a piezomagnetic device and its use in the design of a quarter-wave transducer.

If we call Z_M the magnetic impedance defined by

$$Z_M = \frac{U}{\dot{\Phi}} \tag{6.79}$$

it is evident that the gyrator inverts the impedance to the right of it into

$$Z_E = \frac{E}{i} = \frac{N^2}{Z_M} \tag{6,80}$$

For example, for the quarter-wave transducer with a cross-sectional area A and length l, the equivalent elements are shown by Fig. 6.12b. In these equations d_{33} is the piezomagnetic constant as defined in Eq. (4.59), Y_0^H the Young's modulus of the rod, measured at constant H, N the number of turns, A the cross-sectional area of the rod, l its length, and μ^S the average value of the reversible permeability as defined by Fig. 3.5, measured at constant strain.

By introducing a series electrical capacitance C_0 to tune the inductance L_0 at the mean frequency, a bandpass transducer can be designed by the same Eqs. (6.59) used for a piezoelectric transducer. For polycrystalline metals, the eddy-current loss produces a resistance in L_0 and introduces a phase shift in the perfect transformer.[10]

RESUME CHAPTER 6

1. Applications of Piezoelectric Crystals

Probably the largest application for crystalline properties of materials occurs in the use of piezoelectric crystals in the production of very selective filters, in the frequency control of oscillators, and as converters of electrical to mechanical energy (or vice versa) in transducers. On account of the very low internal friction—that is, the high Q values—present for quartz and because crystal orientations have been found whose frequencies vary little with temperature, quartz crystals are the principal crystals for the first two uses. The electrostrictive ceramics discussed in Chapter 4 have the widest uses as transducer materials. For measuring purposes, X-cut and AC-cut quartz and Z-cut tourmaline are of importance.

2. Principal Piezoelectric Crystals

By investigating the effects of orienting crystal cuts with respect to the crystallographic axes, it is found that a number of orientations exist that have very small variations of frequency with temperature. Of these cuts the AT and GT have the most constant frequency-temperature curves, and these cuts are largely used in the control of oscillators. For filter applications, the $+5°$ X cut is commonly used on account of its relatively pure frequency spectrum coupled with a small frequency variation with temperature.

The properties of a number of monoclinic, hexagonal, and tetragonal crystals are given. These have been or are used for a number of applications. The properties of the ferroelectric crystals rochelle salt and barium titanate are discussed, and it is shown that the properties have large variations near the Curie temperatures.

3. Wave Transmission in Piezoelectric Crystals

One of the most important properties of a piezoelectric crystal is the electromechanical coupling constant k. This is defined in terms of the energy stored in the mutual terms divided by the product of the energy stored in the electric and the mechanical terms. For quartz the coupling factor is about 0.1, but for ferroelectric ceramics it may be as high as 0.7.

Starting with Newton's laws of motion and inserting the converse piezoelectric relations, the equations of motion of a long, thin piezoelectric crystal are derived. By employing the direct piezoelectric relations and integrating the time rate of change of the electric displacement over the surface, the ratio of the voltage to the current can be determined, and hence, the electrical impedance of the piezoelectric element. There is a resonant frequency—current approaches infinity—and an antiresonant frequency—current approaches zero—for the fundamental mode and odd harmonic frequencies above this.

[10] See Hueter and Bolt [8].

By measuring the separation of the resonant and antiresonant frequencies, the electro-mechanical coupling factor can be determined. For a highly coupled crystal the set of curves of Fig. 6.7 have to be employed to determine the coupling factor. However, for low coupling we find

$$k = \frac{\pi}{2}\sqrt{\frac{\Delta f}{f_R}\Big/\left(1 + \frac{\pi^2}{4}\left(\frac{\Delta f}{f_R}\right)\right)}$$

where Δf is the separation of the resonant and antiresonant frequencies and f_R is the resonant frequency.

An equivalent circuit, consisting of a shunt condenser C_0 and a resonant circuit having the values L_1, R_1, and C_1, reproduces the impedance of the crystal near the resonant region.

4. Piezoelectric Crystals as Transducers

By including the force and particle velocity of each end of the crystal in the boundary conditions, we can obtain a set of three equations that can be solved for the conversion of electrical energy into mechanical energy or vice versa. If one end of the crystal is attached to the load, the two conditions for the other end are (1) large impedance compared to the crystal impedance (quarter-wave transducer) or (2) zero load on the second end (half-wavelength transducer). A considerable simplification occurs if a network is introduced that reproduces the three fundamental equations. In the neighborhood of the principal resonance such networks can be reduced to a set of equivalent masses, compliances, and resistances.

The equivalent circuit of a single-component thickness-vibrating transducer has a very similar equivalent network to the length-vibrating transducer, except for the addition of a negative capacitance. This is required to make the elastic modulus the C^D constant.

By considering the elements of the transducer as similar to a filter circuit, formulae are available for designing such transducers. High-coupling crystals can produce a transducer having a very good efficiency over a wide frequency range.

5. Piezomagnetic Transducer

If we replace the voltage and current of a piezoelectric transducer by the magnetomotive force U and the rate of change of magnetic flux through the transducer, the same equations and the same equivalent circuits result for a piezomagnetic transducer. However, it is the input current and voltage to the transducer that are the input variables. Since magnetic variables are axial vectors whereas electrical variables are polar, in order to couple them it is necessary to introduce a new circuit element, the "gyrator." This has the effect of inverting all the network elements—that is, causing the masses to appear as compliances and vice versa. Similar methods can be used to design transducers.

PROBLEMS CHAPTER 6

1. Calculate the piezoelectric constant d'_{12}, the elastic compliance s^E_{22}, and the dielectric constant ε^T_{11} for a $+5°$ X cut from the fundamental constants given in Tables 1, 2, and 4 and the tensor transformation Eq. (2.17).

2. Determine the g, e, and h piezoelectric constants of ADP ($NH_4H_2PO_4$) from the values of the d constants given in Table 15 and the dielectric and elastic constants given in Tables 1 and 2.

3. In the expansion of Eq. (6.27) in powers of $(f_A - f_R)/f_R$ determine the next correction term beyond the first term of Eq. (6.34).

4. If we replace the shunt coil L_0 of Fig. 6.10 by a coil L_0 in series with R_E, another type of band filter results, which has the element values

$$L_0 = \frac{Z_{0_1}}{2\pi(f_B - f_A)} \; ; \quad C_0 = \frac{1}{\pi(f_B + f_A)\sqrt{Z_{0_1}Z_{0_2}}}$$

$$L_1 = \frac{Z_{0_2}}{2\pi(f_B - f_A)} \; ; \quad C_1 = \frac{(f_B - f_A)(f_A{}^2 + f_B{}^2)}{4\pi Z_{0_2}f_A{}^2 f_B{}^2}$$

Here Z_{0_1} is the characteristic impedance on the input side and Z_{0_2} the characteristic impedance on the output side. Using the values of the inductance and capacitance L_1 and C_1 of Eq. (6.60)—that is, the mechanical elements taken through the electromechanical transformer—find out the bandwidth and the electrical input impedance Z_{0_1} for the same terminating impedance R_T of Eq. (6.60).

REFERENCES CHAPTER 6

1. W. G. Cady, "The Piezoelectric Resonator." *Phys. Rev.* **17**, 531 (1921).
2. Warren P. Mason *in* "Physical Acoustics" (Warren P. Mason, ed.), Vol. 1A, Chapter V. Academic Press, New York, 1964.
3. Landolt-Börnstein, "Numerical Values and Functions," 6th Edition, Vol. II/6, pp. 414–448. Springer, Berlin, 1959.
4. "American Institute of Physics Handbook," Section 9f–9, pp. 9–97 to 9–109. McGraw-Hill, New York, 1963.
5. Warren P. Mason, "Electromechanical Transducers and Wave Filters," 2nd Edition, Sections 6.3 and 6.32. Van Nostrand, Princeton, New Jersey, 1948.
6. Warren P. Mason, "Electromechanical Transducers and Wave Filters," 2nd Edition, Table I, No. 11, pp. 52, 53. Van Nostrand, Princeton, New Jersey, 1948.
7. B. D. H. Tellegen, *Phillips Res. Rep.* **3**, 31 (1948).
8. T. F. Hueter and R. H. Bolt, "Sonics," Chapter V. Wiley, New York, 1955.

CHAPTER 7 • Crystal Optics and the Electro-optic and Piezo-optic Effects

7.1 Introduction

The transmission of light in a crystalline medium has some interesting features not present for isotropic materials. In general, a light wave entering a crystal is broken up into two rays called the ordinary ray and the extraordinary ray. These waves have polarization vectors at right angles to each other, and in general they have different velocities and emerge at different angles from the crystal surface. These properties make possible the polarization of light in one direction, since the two rays can be separated in angle and one of the beams can be absorbed. The Nicol prism is one such device.

In all crystals there are directions called the optic-axis directions for which the two rays have the same velocities and the same paths. For cubic crystals every direction is an optic axis and the transmission of light is the same as in an isotropic substance, except that—as discussed in Chapter 8—the plane of polarization of a light ray can sometimes be rotated by an amount proportional to the thickness of the sample. For tetragonal, trigonal, and hexagonal crystals, there is one optic axis, which lies along one of the crystallographic axes, namely, the c axis. For triclinic, monoclinic, and orthorhombic crystals there are two such optic axes. These results follow directly from the form of the dielectric tensor given in Appendix B, Table B.2, which shows that cubic and isotropic materials have only one dielectric constant, tetragonal, trigonal, and hexagonal crystals have two dielectric constants, and in a plane perpendicular to the $c = z$ axis, the dielectric constant is independent of the orientation. As shown in Appendix A, this results in two waves with their polarization at right angles which have the same velocity. Finally for triclinic, monoclinic, and orthorhombic crystals there are three independent dielectric constants, and it is shown that there are two optic axes for which the dielectric constant in a plane perpendicular to these axes is independent of the orientation.

157

When light is transmitted along an optic axis, the two waves travel with the same velocity and emerge unchanged from the crystal. If, however, electric voltages or mechanical stresses are applied in definite directions in the crystal, the result is to shift the optic axis and to cause one of the rays transmitted in the direction of the original axis to have a velocity different from that of the other component. Hence birefringence is introduced by the voltage or stress. The first effect, the electro-optic effect, can be used to modulate light waves[1] by means of a modulating voltage applied in the appropriate direction. By employing a wave guide system for which the electrical wave velocity is the same as the velocity of light waves in the crystal, modulation can occur along the entire path. This is one method[2] used for modulating a laser. The principal use of the piezo-optic effect— sometimes called the photoelastic effect—is to simulate the stress pattern in a complicated structure by means of a photoelastic model. One can then analyze the stress pattern present by means of the light pattern generated by the photoelastic effect. This is a rather widely used technique in industry.[3]

It is the purpose of the present chapter to discuss the methods for deriving these effects.

7.2 The Indicatrix Ellipsoid

The principal tool for investigating the properties of crystal optics is the indicatrix ellipsoid. As shown in Appendix A and illustrated in Fig. A.1, light transmission in a given direction is determined by the wave direction n to which equal phase surfaces are perpendicular. The magnetic field H and the dielectric displacement D are in directions perpendicular to n. The electric field direction E can make an angle of θ degrees with D in the D, n plane, the angle being determined by the dielectric-constant matrix of the material. It was also shown that the energy-propagation direction—usually called the ray direction—makes the same angle θ with respect to the wave-normal direction n. When the relations between the electric vector and the dielectric displacement in the form

$$E_j = \beta_{ji}D_i = \beta_0 B_{ji}D_j \qquad (7.1)$$

are introduced in Maxwell's equations, it is shown that in general not one but two waves of different velocity will be propagated through the crystal for the same wave normal. Moreover, the two waves are plane polarized with the directions of polarization at right angles to each other. The value of V/v for each wave is called the refractive index for that wave. The refractive indices for the two waves as a function of the direction of their common wave normal are obtained from the ellipsoid called the indicatrix.

[1] See Billings [1] and Carpenter [2].

[2] See Kaminow [3], Rigrod and Kaminow [4], Didomenico and Anderson [5], and Sterzer et al. [6].

[3] See Coker and Filon [7] and Frocht [8].

The directions of the axes of the ellipse are determined by the directions for which the impermeability constants are maxima and minima—that is, the principal axes—and these three directions are at right angles to each other. As shown by equation (A.20), with the velocities a, b, and c being associated with B_1, B_2, and B_3 by the equations

$$a^2 = B_1 V^2 ; \qquad b^2 = B_2 V^2 ; \qquad c^2 = B_3 V^2 \qquad a > b > c \qquad (7.2)$$

the equation for the velocities takes the form

$$0 = (b^2 - v^2)(c^2 - v^2)l_1{}^2 + (a^2 - v^2)(c^2 - v^2)l_2{}^2 + (a^2 - v^2)(b^2 - v^2)l_3{}^2 \qquad (7.3)$$

where v is the velocity in any direction having the direction cosines l_1, l_2, and l_3 between the x_1, x_2, and x_3 axes. For example, if the transmission of light is along the $x_3 = z$ axis, $l_1 = l_2 = 0$ and $l_3 = 1$. Hence, the two velocities have the values a and b corresponding to the impermeability constants B_1 and B_2 at right angles to the direction of light transmission. If $B_1 = B_2$ or $a = b$, the two velocities are equal and $z = c$ is an optic axis. As pointed out previously, $z = c$ is an optic axis for tetragonal, trigonal, and hexagonal crystals. By introducing the direction cosines into this equation in terms of the angles of Fig. A.2 and solving for v^2, the result is given by Eq. (A.23b).

Another construction that has been shown to give the same result is the indicatrix ellipsoid defined by the equation

$$\frac{x_1{}^2}{n_1{}^2} + \frac{x_2{}^2}{n_2{}^2} + \frac{x_3{}^2}{n_3{}^2} = 1 = B_1 x_1{}^2 + B_2 x_2{}^2 + B_3 x_3{}^2 \qquad (7.4)$$

If we draw a line from the origin in the direction OP, as shown by Fig. 7.1,

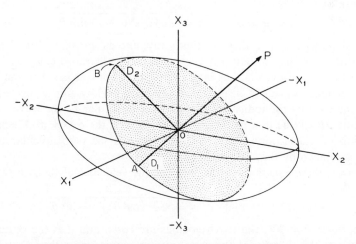

FIG. 7.1 Indicatrix ellipsoid and ellipse perpendicular to the light-propagation direction P showing the vibration directions D_1 and D_2.

and take a plane through the ellipsoid at right angles to OP, this section will be an ellipse. The semiaxes OA and OB of this ellipse will determine the direction of the electric displacement vectors D_1 and D_2, and the absolute values of OA and OB will give the two indices of refraction and hence will determine the two velocities of the waves.

7.21 Location of Optic Axes and Definition of Ordinary and Extraordinary Waves

When the radical of equation (A.23b) vanishes, the two velocities are equal and an optic axis exists. The expression inside the radical can be written

$$[(a^2 - b^2)(\cos^2 \theta \cos^2 \varphi + \sin^2 \varphi) - (b^2 - c^2) \sin^2 \theta]^2$$
$$+ 4(a^2 - b^2)(b^2 - c^2) \sin^2 \theta \sin^2 \varphi = 0 \quad (7.5)$$

Since the square is always positive and since $(a^2 - b^2) > 0$ and $(b^2 - c^2) > 0$, the equation can vanish only if $\varphi = 0$. But $\varphi = 0$ indicates that the two optic axes lie in a plane perpendicular to the direction of the intermediate velocity b. With $\varphi = 0$, the square vanishes when

$$\tan^2 \theta = \frac{a^2 - b^2}{b^2 - c^2} \quad \text{or} \quad \tan \theta = \pm \sqrt{\frac{a^2 - b^2}{b^2 - c^2}} \quad (7.6)$$

If $(a^2 - b^2) < (b^2 - c^2)$ the value of $\tan \theta$ is less than unity and the crystal is called a positive crystal. For this case the two axes approach more closely to the $z = c$ axis having the velocity c than they do to the x axis having the velocity a. If $(a^2 - b^2) > (b^2 - c^2)$ the crystal is called negative.

If $a = b$ or $b = c$ the crystal has a single optic axis and is respectively a positive or negative uniaxial crystal. For the positive crystal the two velocities are given by

$$v_1 = a = b; \quad v_2 = \sqrt{a^2 \cos^2 \theta + c^2 \sin^2 \theta} \quad (7.7)$$

The first velocity is that of the ordinary ray, and as shown by Fig. 7.2a, the wave surface of the ordinary ray is a sphere. The extraordinary ray has an ellipsoid of revolution that lies inside the sphere but touches it in the optic-axis direction. Since $c < a$, the maximum axis for any ellipse formed by a plane intersecting the indicatrix ellipsoid perpendicular to the wave normal direction OP will lie in a plane formed by OP and the c axis, and hence the direction of polarization of the ordinary ray will be perpendicular to this plane, as shown by Fig. 7.3.

If $b = c$, the a axis is the optic axis and the velocities of the two rays are again

$$v_1 = c \quad \text{and} \quad v_2 = \sqrt{a^2 \cos^2 \theta + c^2 \sin^2 \theta} \quad (7.8)$$

Hence, when $\theta = 90°$, the two velocities are equal and a is the optic axis. In this case the velocity of the extraordinary ray is larger than that of the

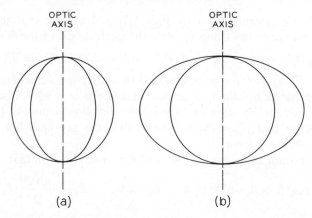

FIG. 7.2 (a) Wave surfaces of positive uniaxial crystals. (b) Wave surfaces of negative uniaxial crystals.

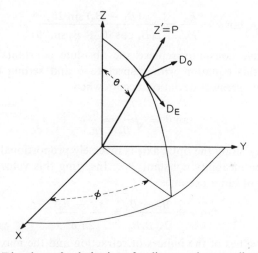

FIG. 7.3 Directions of polarization of ordinary and extraordinary rays.

ordinary ray, as shown by Fig. 7.2b. The polarization of the extraordinary ray lies again in the plane of the normal and the optic axis, while the polarization of the ordinary ray is perpendicular to this plane.

7.22 Direction of the Ray Path

As indicated by Fig. A.1 the path of energy propagation—the ray path— does not necessarily coincide with the normal to the wave surface. The angle between the two directions is determined by the angle between the electric-field vector E_i and the electric-displacement vector D_i. If we go to the coordinate system shown by Fig. A.2, and assume that the electric-displacement vector lies along x_1', then the electric-field vector E_i is given by

$$E_i = \beta_{i1}' D_1' \quad \text{or} \quad E_1' = \beta_{11}' D_1', \quad E_2' = \beta_{12}' D_1', \quad E_3' = \beta_{13}' D_1' \quad (7.9)$$

For the ordinary ray, the polarization is in a direction perpendicular to the plane determined by the θ, φ angles, and hence $\psi = 90°$. Inserting this direction cosine in equation (A.14) with the direction cosines of (A.26) it is readily shown that $\beta_{12}' = \beta_{13}' = 0$ and hence the ray path lies along the normal to the wave front.

For the extraordinary ray, $\psi = 0$, and it is readily shown that

$$E_1' = (\beta_1 \cos^2 \theta + \beta_3 \sin^2 \theta) D_1'; \quad E_2' = 0; \quad E_3' = \frac{\sin 2\theta}{2} (\beta_1 - \beta_3) D_1'$$

$$(7.10)$$

Hence, the angle α that the extraordinary ray deviates from the wave normal, or from the ordinary ray, is given by the formula

$$\tan \alpha = \frac{E_3'}{E_1'} = \frac{(B_1 - B_3) \sin 2\theta}{2(B_1 \cos^2 \theta + B_3 \sin^2 \theta)} \quad (7.11)$$

where the relative constants replace the absolute constants in the ratio. Differentiating this equation with respect to θ and setting the derivative equal to zero, the greatest deviation occurs when

$$\tan \theta = \sqrt{\frac{B_1}{B_3}} = \frac{n_e}{n_o} \quad (7.12)$$

since the velocity along the optic axis is inversely proportional to the square root of the impermeability constant B_1. Inserting this value of $\tan \theta$ into (7.11) the value of $\tan \alpha$ becomes

$$\tan \alpha = \frac{B_1 - B_3}{2\sqrt{B_1 B_3}} = \frac{n_e^2 - n_o^2}{2 n_o n_e} \quad (7.13)$$

Table 19 gives values of the indices of refraction and the maximum angle α for several uniaxial crystals. These indices of refraction are for the yellow sodium light.

Table 19

INDICES OF REFRACTION AND MAXIMUM ANGLE α FOR SEVERAL CRYSTALS

Material	System	n_e	n_o	Sign	Maximum α
ADP	tetragonal	1.5254	1.4798	positive	$+1° 46'$
Calcite	trigonal	1.486	1.658	negative	$-6° 16'$
KDP	tetragonal	1.510	1.4684	positive	$+1° 35'$
Quartz	trigonal	1.553	1.544	positive	$+20'$
Rutile	tetragonal	2.903	2.616	positive	$+5° 56'$

7.3 Uses for the Optical Properties of Crystals

If, as shown by Fig. 7.4, a polarized light ray is sent normal to the x, y plane along the z axis, and if the optic axis is in the y, z plane making an angle θ with the z axis, the ordinary ray will propagate without change in direction. However, the extraordinary ray is displaced by an angle α and will come out of the crystal parallel to the entry direction but at a distance d

$$d = t \tan \alpha \tag{7.14}$$

from the ordinary ray. The polarization of the extraordinary ray lies in the y, z plane, while that of the ordinary ray lies in the x, z plane. Hence, if a

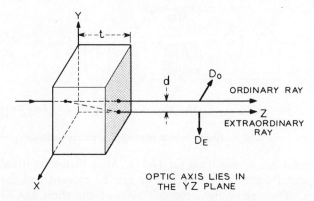

FIG. 7.4 Energy (ray) directions for the ordinary and extraordinary rays in a uniaxial crystal. Optic axis in y, z plane.

modulator of the electro-optic type is placed in front of the crystal and the plane of polarization is shifted from one direction to the other by 90°, the light spot can be given a displacement d. A system has been suggested[4] for storing a large amount of information on a photographic plate or screen by combining a number of modulators and uniaxial plates with their thicknesses and optic axis positions adjusted to give a series of spots displaced in both the x and y directions.

A crystal plate can also be used to polarize a nonpolarized beam of light, as in the Nichol prism. This device separates the extraordinary ray from the ordinary ray and absorbs the latter.[5] It is more usual to use polaroid sheets made from fibrous crystalline materials whose absorption coefficients are markedly different for the two polarization directions. An arrangement such as a Nichol prism or a polaroid sheet is called a polarizer. It automatically absorbs more than half the light energy. By using a polarizer as

[4] See Nelson [9].
[5] See Born and Wolf [10].

an analyzer, one can detect the degree of polarization by rotating the analyzer by 90° and observing the amount of light passed.

Other optical devices produced by birefringent crystals are quarter-wave plates and various types of compensators for turning elliptically polarized waves into plane polarized waves. These devices make use of the fact that light striking the crystal is broken up into two components whose polarization is at right angles and that these two components will be propagated with different velocities. In order that the ray direction shall coincide with the wave-normal direction, light is sent in a direction perpendicular to the optic axes, since the ray direction and wave-normal directions coincide. For a

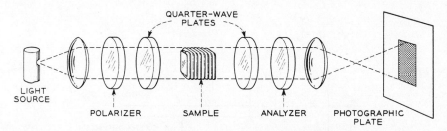

FIG. 7.5   Elements of a photoelastic polariscope.

uniaxial crystal this is seen from Eq. (7.11). Mica plates are usually used to produce quarter-wave plates, since they can be cleaved and form a very flat surface. For yellow light the thickness of the sheet has to be 0.032 inches; for green light 0.0333 inches is required. The angle of delay is given by

$$\varphi = \frac{\pi}{2} = \frac{2\pi}{\lambda} (n_1 - n_2)l \qquad (7.15)$$

Figure 7.5 shows the use of two quarter-wave plates in a polariscope for investigating the stresses in a photoelastic model. Without the quarter-wave plates, the polarizer and analyzer alone would produce a large cross in the center. The lines produced without the quarter-wave plates are called isoclinic lines; those with the quarter-wave plates are called isochromatic lines. Both sets are needed in the analysis of a stress present in a photoelastic sample. The isoclinic lines determine the directions of the principal stresses, whereas the isochromatic lines determine their difference.

As described by Wood in reference [11], another optical use for birefringent crystals is as an optical ring sight for determining the direction of a target. For this use a calcite crystal, with faces normal to the optic axis, is mounted between crossed polaroids and quarter-wave plates. Light coming in at an angle to the optic axis is broken up into two rays with slightly different velocities, which interfere at certain angles and produce a series of rings about an open center. If the target is centered in these rings, the direction is normal to the calcite crystal. Hence, if this device is mounted rigidly on a gun barrel, it serves as a directing sight.

7.4 Thermodynamic Derivation of Electro-optic and Piezo-optic Effects

When one applies an electrical variable or a mechanical variable to a crystal, the dielectric properties are slightly changed. Probably the most fundamental way of developing these properties is to express them in terms of the strains, the electric displacements, and the entropy. In terms of the electric displacements, the electro-optic constants do not vary much with temperature, whereas if they are expressed in terms of the fields, the constants of a ferroelectric type crystal such as KDP increase many fold near the Curie temperature. The entropy is considered as the fundamental thermal variable, since most measurements are carried out so rapidly that the entropy does not vary.

The thermodynamic potential that has the strains, electric displacements, and the entropy as the independent variables is the internal energy U, which has the differential form shown by line 1, Table 6, Chapter 4. From the differential form of dU we have the relations

$$T_{ij} = \frac{\partial U}{\partial S_{ij}}; \qquad E_m = \frac{\partial U}{\partial D_m}; \qquad \Theta = \frac{\partial U}{\partial \sigma} \qquad (7.16)$$

Since, for most conditions of interest, adiabatic conditions prevail, we can set $d\sigma = 0$ and can develop the dependent variables—the fields and the stresses—in terms of the independent variables—the strains and the electric displacements. We carry the development up to third derivatives in the electric displacements, in order to include the quadratic electro-optical effect (the Kerr effect), but to only first derivatives of the strains. This considers that the crystal is "soft" electrically and "hard" elastically.

Performing these differentiations, assuming that the partial differentials are constant over the range of the variables used, we have

$$E_m' = \frac{\partial E_m}{\partial S_{ij}} S_{ij} + \frac{\partial E_m}{\partial D_n} D_n + \frac{1}{2!} \left[2 \frac{\partial^2 E_m}{\partial S_{ij}\,\partial D_n} S_{ij} D_n + \frac{\partial^2 E_m}{\partial D_n\,\partial D_o} D_n D_o \right]$$
$$+ \frac{1}{3!} \frac{\partial^3 E_m D_n D_o D_p}{\partial D_n\,\partial D_o\,\partial D_p} + \cdots$$

$$T_{kl} = \frac{\partial T_{kl}}{\partial S_{ij}} S_{ij} + \frac{\partial T_{kl}}{\partial D_n} D_n + \frac{1}{2!} \left[2 \frac{\partial^2 T_{kl}}{\partial S_{ij}\,\partial D_n} S_{ij} D_n + \frac{\partial^2 T_{kl}}{\partial D_n\,\partial D_o} D_n D_o \right] \qquad (7.17)$$
$$+ \frac{1}{3!} \frac{\partial^3 T_{kl} D_n D_o D_p}{\partial D_n\,\partial D_o\,\partial D_p} + \cdots$$

For the linear and quadratic electro-optic effects and the piezo-optic effect, the tensors of interest are

$$\frac{\partial^2 E_m}{\partial D_n \, \partial D_o} = \frac{\partial^3 U}{\partial D_m \, \partial D_n \, \partial D_o} = \beta_0 r_{mno} = \text{linear electro-optic effect,}$$

$$\frac{\partial^3 E_m}{\partial D_n \, \partial D_o \, \partial D_p} = \beta_0 f_{mnop} = \begin{array}{l}\text{quadratic electro-optic} \\ \text{effect}\end{array}$$

$$\frac{\partial^2 T_{kl}}{\partial D_n \, \partial D_o} = \frac{\partial^3 U}{\partial S_{kl} \, \partial D_n \, \partial D_o} = \frac{\partial^2 E_n}{\partial S_{kl} \, \partial D_o} = \beta_0 m_{klno} = \text{piezo-optic effect}$$

(7.18)

where β_0 is introduced in order to give relations directly to the indices of refraction, n_i.

For the first partial derivatives we have the values

$$\frac{\partial T_{kl}}{\partial S_{ij}} = c^D_{ijkl}; \qquad \frac{\partial T_{kl}}{\partial D_n} = \frac{\partial^2 U}{\partial S_{kl} \, \partial D_n} = \frac{\partial E_n}{\partial S_{kl}} = -h_{nkl}; \qquad \frac{\partial E_m}{\partial D_n} = \beta^S_{mn} \quad (7.19)$$

The last term of the second equation of (7.17) is the second-order electrostrictive effect and is here neglected. However, we see from the last of Eqs. (7.18) that there is a relation between the electrostrictive stress from the first of the second derivatives of the last equation of (7.17) and the piezo-optic effect in the first equation of (7.17) through the tensor m_{klno}. However, the two effects would have to be measured at the same frequencies before equality can be obtained. The relation may be of some interest in accounting for the large strains experienced in lasers at high electric field values.

To obtain the changes in the optical properties caused by the strains and the electric displacements, we have to make use of the electric fields and displacements occurring at the high frequencies of optics. Even for microwave fields and piezoelectric vibrations occurring at as high frequencies as they can be driven by the piezoelectric effect, these frequencies are small compared to the optic frequencies and hence can be considered to be static displacements or strains. Hence, we can write

$$E_m = E_m{}^0 + E_m e^{j\omega t}; \qquad D_n = D_n{}^0 + D_n e^{j\omega t}; \qquad D_o = D_o{}^0 + D_o e^{j\omega t};$$

$$D_p = D_p{}^0 + D_p e^{j\omega t}; \qquad S_{ij} = S^0_{ij}$$

(7.20)

where ω pertains to the optical frequency. Inserting the tensors of (7.18) and (7.19) instead of the partial derivatives, Eq. (7.17) can be separated into a static or low-frequency form and an optical-frequency form

$$E_m{}^0 = -h_{mij}S_{ij} + D_n{}^0\left[\overline{\beta_{mn}^S} + \beta_0\left(\overline{m_{ijmn}}S_{ij} + \frac{\overline{r_{mno}^S D_o{}^0}}{2}\right)\right]$$

$$+ \tfrac{1}{6}\beta_0\overline{f_{mnop}}D_n{}^0D_o{}^0D_p{}^0 \qquad (7.21)$$

$$E_m e^{j\omega t} = D_n e^{j\omega t}\left[\beta_{mn}^S + \beta_0\left(m_{ijmn}S_{ij} + \frac{r_{mno}^S}{2}D_o{}^0\right)\right] + \beta_0\frac{r_{mno}}{2}D_n{}^0D_o e^{j\omega t}$$

$$+ \tfrac{1}{6}\beta_0[f_{mnop}D_o{}^0D_p{}^0D_n e^{j\omega t} + f_{mnop}D_n{}^0D_p{}^0D_o e^{j\omega t} + f_{mnop}D_n{}^0D_o{}^0D_p e^{j\omega t}]$$

In these equations a bar over the tensor term indicates that the low-frequency value is used, whereas the unbarred tensor refers to the optical-frequency term. By development in terms of the static terms and the optical terms, the optical equation can be written in the form

$$E_m e^{j\omega t} = D_n e^{j\omega t}\left[\beta_{mn}^S + \beta_0\left(m_{ijmn}S_{ij} + r_{mno}^S D_o{}^0 + \frac{f_{mnop}}{2}D_o{}^0D_p{}^0\right)\right] \qquad (7.22)$$

where the first number of r or f refers to the field, the second to the optical value of D, and the third and fourth to the static values of D.

From the definition of the tensors m_{ijno}, r_{mno}, and f_{mnop} given by Eqs. (7.18), it is obvious that there are relations between the various components of the tensors. For the first tensor, m_{ijmn}, since $S_{ij} = S_{ji}$ is a symmetrical tensor, the indices i and j can be interchanged. From the definition of the tensor in the form

$$\beta_0 m_{ijmn} = \frac{\partial}{\partial S_{ij}}\left(\frac{\partial U}{\partial D_m \partial D_n}\right) \qquad (7.23)$$

it is obvious that the order of m and n can be interchanged. Since i and j and m and n are reversible it is customary to abbreviate the tensor by writing one number in place of the two in the following form

$$11 = 1; \quad 22 = 2; \quad 33 = 3; \quad 12 = 21 = 6; \quad 13 = 31 = 5;$$

$$23 = 32 = 4 \qquad (7.24)$$

Since the convention is used that the reduced tensor is associated with the engineering strains, it is necessary to investigate the numerical relations between the four-index symbols and the two-index symbols. From Eq. (7.21) when $i \neq j$, the change in the impermeability constant β_{mn} is proportional to

$$m_{ijmn}S_{ij} + m_{jimn}S_{ji} = m_{rs}S_r \qquad (7.25)$$

Since $S_r = 2S_{ij} = 2S_{ji}$, we have the relation that

$$m_{ijmn} = m_{rs} \qquad (i, j, m, n = 1 \text{ to } 3; \quad r, s = 1 \text{ to } 6) \qquad (7.26)$$

In Eq. (7.18) we cannot interchange ij and mn since U does not contain product terms of strains and electric displacements, and hence in general

$$m_{rs} \neq m_{sr} \tag{7.27}$$

From the definitions of the tensor r^S_{mno} and the fact that both m and n refer to the values at the optical frequencies, we can interchange m and n. The same is true for the m and n values of f_{mnop}. r^S_{mno} is usually replaced by

$$r^S_{qo} = r^S_{mno} \qquad (m, n, o = 1 \text{ to } 3; \quad q = 1 \text{ to } 6) \tag{7.28}$$

The first letter refers to the optic variables; the last refers to the static or low-frequency component of the electric displacement.

Finally, for the term f_{mnop} as defined in (7.18), we can interchange m and n and o and p so that

$$f_{rs} = f_{mnop} \qquad (m, n, o, p = 1 \text{ to } 3; \quad r, s = 1 \text{ to } 6) \tag{7.29}$$

Since m and n refer to the optical frequencies and o and p to the low frequencies, in general we cannot interchange r and s. In the form of (7.20) the term

$$\left[\frac{f_{mnop}}{2} (D_o^{\ 0} D_p^{\ 0} + D_p^{\ 0} D_o^{\ 0}) \right] = f_{rs} D_p^{\ 0} D_o^{\ 0} \tag{7.30}$$

so that in terms of the two index symbols Eq. (7.22) can be written in the form

$$E_m e^{j\omega t} = D_n e^{j\omega t} [\beta^S_{mn} + \beta_0 (m_{rs} S_s + r^S_{qo} D_o^{\ 0} + f_{rs} D_o^{\ 0} D_p^{\ 0})] \tag{7.31}$$

The so-called "true" electro-optic constants are measured at constant strain. These can be measured for a piezoelectric crystal by applying an electric field of a high enough frequency so that the principal mechanical resonances and their harmonics cannot be excited by the applied field, and measuring the resulting birefringence along definite directions in the crystal. Figure 7.6[6] illustrates this process for ADP. On the other hand, if we apply a static field to the crystal, an additional effect occurs because the crystal is strained by the piezoelectric effect and this causes a piezo-optic effect in addition to the "true" electro-optic effect. A better designation for these effects is the electro-optic effect at constant strain ("true" effect) and at constant stress. This latter effect can be calculated from Eq. (7.17) by setting the stresses T_{kl} equal to zero and eliminating the S_{ij} strains. Neglecting second-order corrections,

$$r^T_{qo} = r^S_{qo} + \frac{m_{pq} h_{or}}{s^p_{pr}} = r^S_{qo} + m_{pq} g_{op} \tag{7.32}$$

[6] From Billings [1].

The last relation follows from the relation of Table 7 between the h and g piezoelectric constants. Measurements[7] given for ADP, reproduced in Fig. 7.6, show that the "true" effect is about two thirds of the total.

The linear electro-optic effect is usually measured in terms of the applied electric field. The electro-optic effect can be determined by replacing $D_n^{\ 0}$ by

$$D_n^{\ 0} = E_m^{\ 0}\varepsilon_{mn}^T \tag{7.33}$$

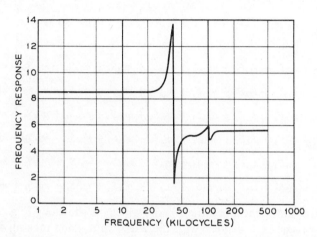

FIG. 7.6   High-frequency response of electro-optical effect in $NH_4H_2PO_4$ (ADP) (after R. O. B. Carpenter).

Introducing this in (7.29), the electro-optic term becomes

$$r_{qn}^S \varepsilon_{mn}^T = z_{qn}^S \tag{7.34}$$

Similarly, the difference between the electro-optic effect at constant stress and constant strain is

$$z_{qn}^T = z_{qn}^S + m_{pq}d_{np} \tag{7.35}$$

where d_{np} is a piezoelectric constant.

Finally, the photoelastic effect is sometimes expressed in terms of the stresses rather than the strains. This involves a new set of piezo-optic constants π_{pq} related to the strain piezo-optic constants by the equation

$$\pi_{pq} = m_{pr}s_{rq}^D \tag{7.36}$$

where s_{rq}^D are the elastic compliances measured at constant dielectric displacement.

[7] See Carpenter [2].

7.5 Birefringence along Any Direction in the Crystal and Determination of the Electro-optic and Piezo-optic Constants

If we take axes along the indicatrix ellipsoid when no stress or field is applied to the crystal, the result of the electro-optic and photoelastic effects is to change the impermeability constants to the values

$$\beta_{11} = \beta_1 + \beta_0\Delta_1 ; \qquad \beta_{22} = \beta_2 + \beta_0\Delta_2 ; \qquad \beta_{33} = \beta_3 + \beta_0\Delta_3$$
$$\beta_{23} = \beta_0\Delta_4 ; \qquad \beta_{13} = \beta_0\Delta_5 ; \qquad \beta_{12} = \beta_0\Delta_6 \tag{7.37}$$

where

$$\Delta_1 = z_{11}E_1 + z_{12}E_2 + z_{13}E_3 + m_{11}S_1 + m_{12}S_2 + m_{13}S_3 + m_{14}S_4 + m_{15}S_5 + m_{16}S_6$$

$$\Delta_2 = z_{21}E_1 + z_{22}E_2 + z_{23}E_3 + m_{21}S_1 + m_{22}S_2 + m_{23}S_3 + m_{24}S_4 + m_{25}S_5 + m_{26}S_6$$

$$\Delta_3 = z_{31}E_1 + z_{32}E_2 + z_{33}E_3 + m_{31}S_1 + m_{32}S_2 + m_{33}S_3 + m_{34}S_4 + m_{35}S_5 + m_{36}S_6$$

$$\Delta_4 = z_{41}E_1 + z_{42}E_2 + z_{43}E_3 + m_{41}S_1 + m_{42}S_2 + m_{43}S_3 + m_{44}S_4 + m_{45}S_5 + m_{46}S_6 \tag{7.38}$$

$$\Delta_5 = z_{51}E_1 + z_{52}E_2 + z_{53}E_3 + m_{51}S_1 + m_{52}S_2 + m_{53}S_3 + m_{54}S_4 + m_{55}S_5 + m_{56}S_6$$

$$\Delta_6 = z_{61}E_1 + z_{62}E_2 + z_{63}E_3 + m_{61}S_1 + m_{62}S_2 + m_{63}S_3 + m_{64}S_4 + m_{65}S_5 + m_{66}S_6$$

If we transmit light along the z' axis, which, as shown by Figs. A.2 or 7.3, makes an angle of θ degrees with the z axis in a plane making an angle φ with the x, z plane, the birefringence can be calculated as follows: Keeping z' fixed and rotating the other two axes about z' by varying the angle ψ, one light vector will occur when β'_{11} is a maximum and the other when β'_{11} is a minimum. Using the transformation equations for a second-rank tensor

$$\beta'_{11} = \alpha_{1i}\alpha_{1j}\beta_{ij} \tag{7.39}$$

and the direction cosines of Eqs. (A.26) we find that β'_{11} is given by the equation

$$\beta'_{11} = \beta_{11}\left[\cos^2\theta\cos^2\varphi\cos^2\psi - \frac{\sin 2\varphi\sin 2\psi\cos\theta}{2} + \sin^2\varphi\sin^2\psi\right]$$

$$+ \beta_{12}[\sin 2\varphi\cos 2\psi - \sin^2\theta\sin 2\varphi\cos^2\psi + \cos\theta\sin 2\psi\cos 2\varphi]$$

$$+ \beta_{13}[-\sin 2\theta\cos\varphi\cos^2\psi + \sin\varphi\sin\theta\sin 2\psi] \tag{7.40}$$

$$+ \beta_{22}\left[\cos^2\theta\sin^2\varphi\cos^2\psi + \frac{\cos\theta\sin 2\varphi\sin 2\psi}{2} + \cos^2\varphi\sin^2\psi\right]$$

$$+ \beta_{23}[-\sin 2\theta\sin\varphi\cos^2\psi - \sin\theta\cos\varphi\sin 2\psi] + \beta_{33}\sin^2\theta\cos^2\psi$$

Differentiating with respect to ψ and setting $\partial\beta'_{11}/\partial\psi = 0$, we find an expression for $\tan 2\psi$ in the form

$$\tan 2\psi = \frac{\begin{array}{c} -\beta_{11}\sin 2\varphi\cos\theta + 2\beta_{12}\cos\theta\cos 2\varphi + 2\beta_{13}\sin\varphi\sin\theta \\ + \beta_{22}\cos\theta\sin 2\varphi - 2\beta_{23}\sin\theta\cos\varphi \end{array}}{\begin{array}{c} \beta_{11}[\cos^2\theta\cos^2\varphi - \sin^2\varphi] + \beta_{12}[(1+\cos^2\theta)\sin 2\varphi] \\ - \beta_{13}\sin^2\theta\cos\varphi + \beta_{22}(\cos^2\theta\sin^2\varphi - \cos^2\varphi) \\ - \beta_{23}\sin 2\theta\sin\varphi + \beta_{33}\sin^2\theta \end{array}} \quad (7.41)$$

Inserting this value back in Eq. (7.40) we find that the two extreme values of β'_{11} are given by the equation

$$2\beta'_{11} = 2\beta_{22} + (\beta_{11} - \beta_{22})(\cos^2\theta\cos^2\varphi + \sin^2\varphi) + (\beta_{33} - \beta_{22})\sin^2\theta$$
$$- \beta_{12}\sin^2\theta\sin 2\varphi - \beta_{13}\sin 2\theta\cos\varphi - \beta_{23}\sin 2\theta\sin\varphi$$
$$\pm \{(\beta_{11} - \beta_{22})^2(\cos^2\theta\cos^2\varphi + \sin^2\varphi)^2 + 2(\beta_{11} - \beta_{22})(\beta_{33} - \beta_{22})\sin^2\theta$$
$$\times (\cos^2\theta\cos^2\varphi - \sin^2\varphi) + (\beta_{33} - \beta_{22})^2\sin^4\theta - 2(\beta_{11} - \beta_{22})$$
$$\times [\beta_{12}(\sin 2\varphi\sin^2\theta)(\cos^2\theta\cos^2\varphi + \sin^2\varphi) + \beta_{13}\sin 2\theta\cos\varphi$$
$$\times (\cos^2\theta\cos^2\varphi + \sin^2\varphi) - \beta_{23}\sin 2\theta\sin\varphi(1 + \cos^2\varphi\sin^2\theta)]$$
$$+ 2(\beta_{33} - \beta_{22})\sin^2\theta$$
$$\times [\beta_{12}\sin 2\varphi(1 + \cos^2\theta) - \beta_{13}\sin 2\theta\cos\varphi - \beta_{23}\sin 2\theta\sin\varphi]$$
$$+ (2\beta_{12})^2[\sin^4\theta\sin^2\varphi\cos^2\varphi + \cos^2\theta]$$
$$- 4\beta_{12}\beta_{13}\sin^2\theta\sin\varphi[\cos^2\theta\cos^2\varphi + \sin^2\varphi]$$
$$- 4(\beta_{12}\beta_{23})[\sin 2\theta\cos\varphi(\sin^2\varphi\cos^2\theta + \cos^2\varphi)]$$
$$+ (2\beta_{13})^2\sin^2\theta(\cos^2\theta\cos^2\varphi + \sin^2\varphi) - 4\beta_{13}\beta_{23}\sin 2\varphi\sin^4\theta$$
$$+ (2\beta_{23})^2\sin^2\theta(\cos^2\theta\sin^2\varphi + \cos^2\varphi)\}^{1/2} \quad (7.42)$$

The birefringence in any direction can be calculated from Eq. (7.42); since $B'_{11} = \beta'_{11}/\beta_0 = v_1^2/V^2$ it equals $1/n_1^2$, where n_1 is the index of refraction corresponding to a light wave with its electric displacement in the β'_{11} direction. Similarly, for the second solution at right angles to the first

$$B''_{11} = \frac{v_2^2}{V^2} = \frac{1}{n_2^2} \quad (7.43)$$

Hence if we designate the expression under the radical by K_2 and half the expression on the right of (7.42) by K_1, we have

$$\frac{1}{n_1^2} + \frac{1}{n_2^2} = \frac{K_1}{\beta_0}; \quad \frac{1}{n_1^2} - \frac{1}{n_2^2} = \sqrt{\frac{K_2}{\beta_0}} \quad (7.44)$$

Since n_1 and n_2 are very nearly equal, even in the most birefringent crystal, we have nearly

$$(n_2 - n_1)l = \Delta B = \frac{n^3 l}{2} \sqrt{\frac{K_2}{\beta_0'}} \qquad (7.45)$$

The crystals for which the electro-optic effect have been most widely used are ADP and KDP. With the symmetry $\bar{4}2m$ they have the electro-optic constants r_{63} and r_{41}. Another crystal for which the properties may be more advantageous is cuprous chloride (CuCl). This has a cubic structure ($\bar{4}3m$), which allows a birefringence at right angles to the applied field. All electro-optic crystals so far have used transmission of light along the optic axis. For ADP and KDP this is the $z = c$ crystalline axis. For this case $z' = z$, $\theta = 0$ and $\varphi = 0$. The r_{63} constant is associated with a field E_3 and electric displacement D_3 along the $z = c$ axis, as can be seen from Eq. (7.38). We have also that $\beta_{11} = \beta_{22}$. Hence from (7.42) and (7.45) the birefringence is given by

$$(n_2 - n_1)l = \Delta B = \frac{n^3}{2} l \left(\frac{2\beta_{12}}{\beta_0} \right) = n^3 l r_{63} D_3 = n^3 z_{63} E_3 l \qquad (7.46)$$

From Eq. (7.41) with $\theta = \varphi = 0$; $\beta_{11} = \beta_{22}$; β_{33}; β_{12}; β_{13}; $\beta_{23} \neq 0$, the values of ψ are $\pm 45°$. Hence the directions $45°$ from the x and y crystallographic axes are the directions of the principal axes. In making measurements the polarizer is set so the plane of polarization is along the x or y axis and the analyzer is set at right angles to the polarizer. The polarized ray is broken up into the slow and fast waves, which have their electric displacements in directions $45°$ from the x and y directions respectively. As the voltage is increased, one wave is retarded with respect to the other; and when this retardation has reached a half wavelength, the direction of the slow wave has reversed its electric displacement direction with respect to the fast wave. The addition of the two waves produces a plane wave with its direction of polarization parallel to the analyzer direction, and the light is a maximum. For static voltages for ADP this requires a voltage of 9.6 kilovolts for a wavelength of 5,461 Å $= 5.461 \times 10^{-7}$ meter. It is seen from Eq. (7.46) that the birefringence is independent of the light-path length since the field E_3 is the applied voltage divided by the length. Hence the constant z_{63}^T, measured at constant stress, is given by

$$\Delta B = \frac{\lambda}{2} = 2.73 \times 10^{-7} \text{ meter} = (1.4798)^3 \times z_{63}^T \times 9,600 \text{ volts}$$

or

$$z_{63}^T = 8.8 \times 10^{-12} \text{ meter/volt} \qquad (7.47)$$

In this equation the ordinary index of refraction n_0 is used, since the light is sent along the optic axis.

As the frequency is increased, then as shown by Fig. 7.6, the light for a given voltage increases to a maximum and drops to about 60 per cent of its value for statically applied voltages. Hence it requires about 16 kilovolts to produce a half-wavelength retardation. With this value the electro-optic constant at constant strain becomes

$$z_{63}^{S} = 5.4 \times 10^{-12} \text{ meter/volt} \qquad (7.48)$$

This value has been shown to be constant up to 500 megacycles per second and should not change up to frequencies in the kilomegacycle range.

It is seen from Eq. (7.35) that the difference between the constant-stress and constant-strain electro-optic constants should be given by

$$z_{63}^{T} - z_{63}^{S} = m_{66} \, d_{36} \qquad (7.49)$$

where m_{66} is the piezo-optic constant connected with the shearing strain S_6 that is produced by a voltage acting along the $z = x_3$ direction through the piezoelectric constant d_{36}. Separate measurements[8] of these two constants give

$$m_{66} = 0.076; \qquad d_{36} = 48.6 \times 10^{-12} \text{ meter/volt} \qquad (7.50)$$

where m_{66} is a nondimensional constant since it is the ratio of the retardation in meters to the length in meters, and the strain is defined as the change in length per unit length. As discussed in Section 7.7, for photoelastic work another unit for the piezo-optic constant is in general use. This is the "brewster," which is defined as the retardation in Angstrom units—that is, 10^{-10} meter—for a path length measured in millimeters when the stress in bars (10^6 dynes/cm² or 10^5 newtons/meter²) is applied to the material. This constant is related to the π_{ijkl} piezo-optic constants discussed in Eq. (7.36). Using the values of m_{66} and d_{36} given by Eq. (7.50) the difference between z_{63}^{T} and z_{63}^{S} is calculated to be 3.7×10^{-12} meter/volt, compared to the optically measured value of 3.4×10^{-12} meter/volt.

KDP is a more advantageous electro-optic crystal than ADP, since it has a half-wave retardation at 5471 Å for static and high-frequency voltages of 7.5 kilovolts and 8.2 kilovolts respectively. These values correspond to the values of

$$z_{63}^{T} = 11.6 \times 10^{-12} \text{ meter/volt}; \qquad z_{63}^{S} = 10.6 \times 10^{-12} \text{ meter/volt} \quad (7.51)$$

A still larger effect is obtained[9] by replacing the hydrogens in KDP (KH_2PO_4) by deuterium atoms. For this case

$$z_{63}^{S} = 20 \times 10^{-12} \text{ meter/volt} \qquad (7.52)$$

[8] See *American Institute of Physics Handbook* [12].
[9] See Zwicker and Scherrer [13].

Table 20

ELECTRO-OPTIC CONSTANTS FOR THE MOST USED CRYSTALS

Crystal	$z_{ij}^T \times 10^{12}$ meters/volt	K_{ij}^T	r_{ij}^T meters/coulomb	$z_{ij}^S \times 10^{12}$ meters/volt	K_{ij}^S	r_{ij}^S meters/coulomb	n_o
NH$_4$H$_2$PO$_4$ (ADP)	$z_{63}^T = 8.8$	$K_{33}^T = 15.8$	$r_{63}^T = 0.0636$	$z_{63}^S = 5.4$	$K_{33}^S = 14.0$	$r_{63}^S = 0.0435$	1.4798
KH$_2$PO$_4$ (KDP)	$z_{63}^T = 11.6$	$K_{33}^T = 22.2$	$r_{63}^T = 0.059$	$z_{63}^S = 10.4$	$K_{33}^S = 21.8$	$r_{63}^S = 0.054$	1.4684
KD$_2$PO$_4$	$z_{63}^T = 20.0$	$K_{33}^T = 90.0$	$r_{63}^T = 0.025$	—	—	—	—
CuCl	$z_{41}^T = 6.15$	$K_{11}^T = 8.5$	$r_{41}^T = 0.083$	$z_{41}^S = 6.0$	$K_{11}^S = 8.4$	$r_{41}^S = 0.083$	1.93
ZnS	$z_{41}^T = 2.4$	$K_{11}^T = 10.25$	$r_{41}^T = 0.0264$	—	—	—	2.368

Actually, this increase is due to the nearer approach to the Curie temperature of KD_2PO_4, which occurs at a temperature of $212°$ K compared to $123°$ K for KH_2PO_4. It is interesting to note, as shown in Table 20, that the r_{ij} electro-optic constants do not differ much from crystal to crystal, and it is found that they are practically independent of the temperature even for the ferroelectric crystals KH_2PO_4 and KD_2PO_4.

The cubic crystals ZnS and CuCl have properties that make them more suitable for modulating light, since a modulation can be obtained for a field applied perpendicular to the light direction. Both of these crystals have the symmetry $\bar{4}3m$. The electro-optic tensor is a third-rank tensor, and from Table B.3 it has the components

$$z_{41} = z_{52} = z_{63} \tag{7.53}$$

One can obtain a longitudinal effect, as in ADP, by sending light along a crystallographic axis and applying the voltage in the same direction. In terms of the applied voltage, the birefringence becomes

$$\Delta B = n^3 z_{63} V \tag{7.54}$$

An even more interesting case is the transverse electro-optic effect, which occurs when the applied voltage direction is perpendicular to the light transmission direction. For this case the birefringence can be made large by making the light-transmission path long and the thickness in the direction of the applied voltage small. To obtain this effect the light-transmission direction is taken as the $z = \langle 001 \rangle$ axis while the voltage has to be applied along the $\langle 1\bar{1}0 \rangle$ direction. For this case

$$\beta_{23} = \beta_0\Delta_4 = \beta_0 z_{41}\frac{E}{\sqrt{2}} ; \qquad \beta_{13} = \beta_0\Delta_5 = \beta_0 z_{41}\frac{E}{\sqrt{2}} ; \qquad \beta_{12} = \beta_0\Delta_6 = 0 \tag{7.55}$$

Inserting these values, with $\theta = 0°$ and $\varphi = 45°$—that is, taking x' along the voltage direction—we have from (7.42)

$$\beta'_{11} = \beta_1 \pm \beta_0 z_{41} E \tag{7.56}$$

The directions of the slow and fast axes lie along the crystallographic x and y axes.

Table 20 shows the values of the electro-optic constants for the four most used crystals. A number of other measurements have been made and are listed in the *American Institute of Physics Handbook*,[10] p. 6–188. A number of piezo-optic constants are also listed. The constants r_{ij}, which relate the birefringence to the electric displacement, are also listed. These values vary much less from crystal to crystal and are practically independent of the temperature.

[10] Reference [12].

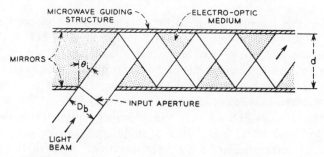

FIG. 7.7 Basic zigzag modulator configuration (after Rigrod and Kaminow).

When a crystal is used as an electro-optic modulator, a light bias is usually given to the crystal so that a positive voltage increases the light and the negative voltage decreases the light. The simplest way to do this is to introduce a quarter-wave plate between the polarizer and analyzer, since half the retardation is accomplished by the quarter-wave plate. The voltage-induced birefringence then adds or subtracts from this. Only half the voltage is required to give the total modulation.

Traveling-wave light values have also been used to modulate light waves[11]—as from a laser—at very high radio frequencies. For this purpose the light waves and the radio waves have to propagate with the same velocity. Since the light waves are usually faster, one method for keeping them in phase is shown by Fig. 7.7. The angle is adjusted so that the two velocities are equal, and modulation occurs throughout the total length. Cubic crystals are advantageous since the voltage can be applied at right angles to the optic and radio-wave propagation. By using the quadratic electro-optic ("Kerr") effect discussed in the next section, a larger sensitivity to the radio waves can be obtained.

7.6 Quadratic Electro-optic ("Kerr") Effect in Cubic Crystals

In addition to the linear electro-optic effect discussed in Section 7.5, there has been considerable recent interest in an electro-optic effect that depends on the square of the applied electric displacement. In fact, the original work on the electro-optic effect in liquids, as carried out by Kerr,[12] depended on the square of the applied voltage or electric displacement. The liquid nitrobenzene was found to have one of the largest Kerr-effect constants, and Kerr cells using this liquid have been rather widely used for optical research and for optical instruments.

[11] See Kaminow [3], Rigrod and Kaminow [4], Didomenico and Anderson [5], and Sterzer *et al.* [6].
[12] See Kerr [14].

Recently ferroelectric crystals of the barium titanate type—perovskite lattice with symmetry $m3m$—have been shown to have large quadratic electro-optic effects in the paraelectric region—that is, in the temperature range above the Curie temperature. Since the effect depends on the square of the applied voltage or electric displacement, large linear effects can be obtained by applying a large static electric field and a small alternating field that is effective into the microwave frequency region. It is found that the f_{ijkl} constants of Eq. (7.20), which describe this effect, are much more constant over the temperature range than constants defining the effect in terms of applied electric fields. To use this effect with good efficiency requires holding the temperature constant a few degrees above the Curie temperature. One crystal that seems likely[13] to receive considerable application is $KTa_xNb_{(1-x)}O_3$, where $x = 0.65$, for room-temperature applications. Using a 0.4 millimeter cube, a dc bias of 330 volts, and a peak ac drive voltage of 16 volts, 100 per cent modulation has been obtained with a sample dissipation of 15 milliwatts. This crystal has been given the designation of KTN. Such a modulator can operate over the very wide frequency band of 0 to 250 megacycles per second.

In terms of the four-index symbols of Eq. (7.20), the impermeability constants can be written in the form

$$\beta_{mn}^S(D) - \beta_{mn}^S(0) = \beta_0 \frac{f_{mnop}}{2} D_o{}^0 D_p{}^0 \tag{7.57}$$

If we designate the combination mn by the single letter r—which goes from 1 to 6—and the combination op by the letter s, the equation can be written in terms of the two-index symbols

$$\beta_r{}^S(D) - \beta_r{}^S(0) = \beta_0 f_{rs} D_o{}^0 D_p{}^0 \tag{7.58}$$

where s has the values from 1 to 6 for the combination of o and p shown by Eq. (7.24). In both of these equations the impermeability constants are measured with the applied electric displacements $\beta_r{}^S(D)$ and with no applied electric displacement $\beta_r{}^S(0)$.

From Table B.4 we see that for the symmetry $m3m$ there are three quadratic electro-optic constants

$$f_{11}, \quad f_{12}, \quad f_{44} \tag{7.59}$$

when expressed in terms of the two-index symbols. The expressions for Δ_1 to Δ_6 equivalent to (7.38) become

$$\Delta_1 = f_{11}D_1{}^2 + f_{12}(D_2{}^2 + D_3{}^2); \quad \Delta_2 = f_{12}(D_1{}^2 + D_3{}^2) + f_{11}D_2{}^2 ;$$
$$\Delta_3 = f_{12}(D_1{}^2 + D_2{}^2) + f_{11}D_3{}^2 ; \quad \Delta_4 = f_{44}D_2 D_3 ; \tag{7.60}$$
$$\Delta_5 = f_{44}D_1 D_3 ; \quad \Delta_6 = f_{44}D_1 D_2$$

[13] See Geusic et al. [15].

The transverse effect can be obtained by transmitting light along the z axis and applying an electric displacement along the x axis—that is, D_1. Then $\theta = \varphi = 0$ and

$$\frac{\beta'_{11}}{\beta_0} = \frac{\beta_1}{\beta_0} + \frac{(f_{11} + f_{12})D_1{}^2}{2} \pm \frac{(f_{11} - f_{12})D_1{}^2}{2} \tag{7.61}$$

Hence

$$\frac{1}{n_1{}^2} = \frac{1}{n_o{}^2} + f_{11}D_1{}^2 \, ; \qquad \frac{1}{n_2{}^2} = \frac{1}{n_o{}^2} + f_{12}D_1{}^2 \tag{7.62}$$

and

$$\frac{1}{n_1{}^2} - \frac{1}{n_2{}^2} = \frac{(n_2 - n_1)(n_2 + n_1)}{n_1{}^2 n_2{}^2} \doteq \frac{2(n_2 - n_1)}{n_o{}^3} = (f_{11} - f_{12})D_1{}^2 \tag{7.63}$$

The birefringence along a crystal of length l in the direction of the light propagation is given by

$$\Delta B = (n_2 - n_1)l = \frac{n_o{}^3}{2}(f_{11} - f_{12})lD_1{}^2 \tag{7.64}$$

In terms of an applied voltage V acting along the thickness l_t,

$$D_1 = \varepsilon_{11}E_1 = \frac{\varepsilon_{11}V}{l_t}$$

and the birefringence is given by the equation

$$\Delta B = \frac{n_o{}^3(f_{11} - f_{12})l^2 E^2}{2l_t{}^2} \tag{7.65}$$

The f_{44} constant can also be used to generate a transverse electro-optic effect. From (7.60) it is evident that a displacement $45°$ from the x and y axes will generate a birefringence along z. Introducing the values $D_1 = D_2 = D/\sqrt{2}$, it is readily shown that

$$\Delta B = n_o{}^3 l f_{44} D^2 = \frac{n_o{}^3 l f_{44}(\varepsilon_{11}V)^2}{l_t{}^2} \tag{7.66}$$

The two constants f_{11} and f_{12} can be separated[14] by deflecting a light beam by a prism of the material. By putting on an applied field of 11,500 volts per centimeter the index of refraction can be changed by as much as 0.005, and this is sufficient to cause a measurable light deflection. When the light is polarized in the field direction—that is, along x—the change in the index, as seen from (7.62), is

$$\Delta n = -\frac{n_o{}^3 f_{11}V^2 \varepsilon^2}{2l_t{}^2} \tag{7.67}$$

[14] See Geusic et al. [15].

whereas if the light is polarized in the perpendicular direction

$$\Delta n = - \frac{n_0^{\,3} f_{12} V^2 \varepsilon^2}{2 l_t^{\,2}} \tag{7.68}$$

Using such techniques the quadratic "Kerr" electro-optic constants have been measured for a number of ferroelectric crystals. Table 21 gives some of these measurements.[15] Over the temperature range shown the values were constant to the order of accuracy ± 10 per cent. The Curie temperatures

Table 21

VALUES FOR KERR ELECTRO-OPTIC CONSTANTS FOR SEVERAL CRYSTALS

Crystal	$f_{11} - f_{12}$ m⁴/coul²	f_{44} m⁴/coul²	f_{11} m⁴/coul²	f_{12} m⁴/coul²	n_0	Temp range
KTaO₃	+0.16	+0.12			2.216	2° K to 77° K
KTa₀.₆₅Nb₀.₃₅O₃ (KTN)	+0.174	+0.147	+0.136	−0.038	2.287	285° K to 310° K
SrTiO₃	+0.14				2.380	4.2° K to 300° K
BaTiO₃	+0.13				2.4	135° C to 160° C

for the various materials are $KTaO_3$, $\sim 1°$ K; KTN, $\sim 283°$ K; $SrTiO_3$, — ; $BaTiO_3$, 401° K.

Such modulators are used near the Curie temperature since the dielectric constant becomes very large and from (7.65) the voltage becomes small for a constant value of $f_{11} - f_{12}$. Furthermore, since the result depends on the square of the electric displacement, a linear effect can be obtained by having a small alternating value of electric displacement superposed on a large constant value D_0—that is,

$$\Delta B = \frac{n_p^{\,3} l}{2}(f_{11} - f_{12})(D_0 + D_1 e^{i\omega t})^2$$

$$= \frac{n_o^{\,3} l}{2}(f_{11} - f_{12})D_0^{\,2} + n_0^{\,3} l (f_{11} - f_{12}) D_0 D_1 e^{i\omega t} \tag{7.69}$$

The constant value of ΔB can be used to provide an optical bias or can be adjusted by the addition of a compensation plate. The details of such a light value are given in reference [15]. In general the ac voltage sensitivity is about two orders of magnitude higher than in the linear electro-optic crystals KDP or CuCl. This sensitivity requires the use of a temperature control, since the dielectric constant varies so rapidly with temperature.

[15] *Ibid.*

7.7　Piezo-optic Effect in Isotropic Materials

The piezo-optic effect (usually called the photoelastic effect) in isotropic solids has been used extensively[16] in studying the stresses set up in machine parts and in irregular samples for which stresses cannot be easily calculated. For this purpose a plastic model cut in the shape of the original is used; it is loaded in a manner similar to that of the machine part to be studied. Since stresses are applied, the π_{ij} piezo-optic constants are most useful. If we look along the z axis, it is readily shown from Eq. (7.42) that the birefringence is equal to

$$\Delta B = \frac{n^3 l}{2} \sqrt{(\beta_1 + \beta_0 \Delta_1 - \beta_2 - \beta_0 \Delta_2)^2 + 4\beta_0^2 (\Delta_6)^2} \tag{7.70}$$

Since, for an isotropic substance, $\beta_1 = \beta_2$, we have Eq. (7.71), after substituting the values of Δ_1 and Δ_2, with the appropriate piezo-optic constants from Table B.4 (last tensor):

$$\Delta B = \frac{n^3 l}{2} (\pi_{11} - \pi_{12}) \sqrt{(T_1 - T_2)^2 + 4T_6^2} \tag{7.71}$$

If we transform to axes rotated by an angle θ about z, the values of T'_{11} and T'_{22} are given by

$$T'_{11} = \cos^2 \theta \, T_1 + 2 \sin \theta \cos \theta \, T_6 + \sin^2 \theta \, T_2$$
$$T'_{22} = \sin^2 \theta \, T_1 - 2 \sin \theta \cos \theta \, T_6 + \cos^2 \theta \, T_2 \tag{7.72}$$

If, now, we choose the angle θ so that T'_{11} is a maximum, we find

$$\tan 2\theta = \frac{+2T_6}{T_1 - T_2} \tag{7.73}$$

Inserting this value of $\tan 2\theta$ in (7.72) we find

$$T_1' = \frac{T_1 + T_2}{2} + \frac{1}{2} \sqrt{(T_1 - T_2)^2 + 4T_6^2}$$
$$T_2' = \frac{T_1 + T_2}{2} - \frac{1}{2} \sqrt{(T_1 - T_2)^2 + 4T_6^2} \tag{7.74}$$

and, hence,

$$T_1' - T_2' = \sqrt{(T_1 - T_2)^2 + 4T_6^2} \tag{7.75}$$

Hence the birefringence obtained by stressing a material is proportional to the difference in the principal stresses. Methods are available for determining the stresses in the model by observing both the isochromatic lines (the ones

[16] See Coker and Filon [7] and Frocht [8].

obtained with quarter-wave plates inserted in the system) and the isoclinic lines (those obtained with only crossed polaroids).

Figure 7.8 shows a photograph[17] of the isochromatic lines for the photo-elastic sample shown by Fig. 7.5. This sample was used in determining the stresses induced by winding copper wires under tension around the photo-elastic material and was used in a study of the stresses occurring in the

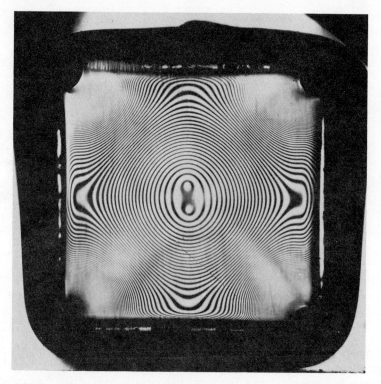

FIG. 7.8 Photoelastic picture of a square model wrapped with 19 turns of 0.050 inch copper wire with a constant load of 28 pounds.

solderless wrapped connection.[18] This is a connection formed by wrapping copper wire under tension around a nickel silver terminal. Figure 7.9 shows such a connection and the photoelastic sample used to simulate the connection. Such connections are now used to the extent of about a half billion per year and are one of the principal connections of the Bell System. The high concentration of lines near the corners in Fig. 7.8 shows that the compressional stress is high enough to cause some plastic flow in the wires

[17] Taken by T. F. Osmer.
[18] See Mason and Osmer [16].

and terminal, which produces a mating surface between them. By counting the number of fringes from the "eyes"—which are so-called isotropic points where the principal stress differences are zero—and knowing the stress optical constant, the difference in the principal stresses can be calculated at any point.

FIG. 7.9   Photograph of photoelastic model and solderless wrapped connection.

Instead of using the constants $(\pi_{11} - \pi_{12})$ it is customary to use a single constant C determined by the retardation per unit length, r_0. The constant is defined by

$$\Delta B_o = r_o = n_e - n_o = C(T'_{11} - T'_{22}) \qquad (7.76)$$

where the constant C is called the relative stress optical constant. The dimension of C is the reciprocal of a stress and is measured in meter^2 per newton. A convenient unit for most purposes is one of 10^{-13} centimeter^2 per dyne or 10^{-12} meter^2 per newton. If this is used, the stress optical coefficients of most glasses are from 1 to 10 and those of most plastics from 10 to 100. This unit so defined is called the "brewster." In terms of the brewster, with l the length of the optical path in millimeters and the stress difference expressed in bars (10^6 dynes/cm^2) then R as given by the formula is expressed in Angstrom units.

The process for stress analysis makes use of the isochromatic lines and the isoclinic lines, and requires in addition the equations of equilibrium. In the form for plane stress, these are

$$\frac{\partial T_1}{\partial x_1} + \frac{\partial T_6}{\partial x_2} = 0; \qquad \frac{\partial T_2}{\partial x_2} + \frac{\partial T_6}{\partial x_1} = 0 \qquad (7.77)$$

Since this process is described extensively in other references[19] it will not be discussed further here.

RESUME CHAPTER 7

1. General Features

In general a light wave entering a crystalline medium is broken up into two waves called the ordinary wave and the extraordinary wave. As proved in Appendix A, from Maxwell's equations, these waves have polarization vectors at right angles to each other and in general their ray paths (direction of energy transmission) are separated in direction from each other. These properties make possible the separation of the two rays, as in the Nicol prism, and the production of a polarization of the emergent light beam, since the other beam can be absorbed. A more common method for polarizing waves is by means of long fibers of crystals lined up in one direction. This is the basis for polaroid materials. Light having a polarization direction along the fiber axis is readily transmitted, while that at right angles is highly attenuated. Such sheets can be obtained in large sizes and are the most widely used polarizing device.

In all crystals there are directions, called optic-axis directions, for which the two rays have the same velocities and the same paths. For cubic crystals all directions are optic axes. For trigonal, hexagonal, and tetragonal crystals, the optic axis lies along the unique c axis. Orthorhombic, monoclinic, and triclinic crystals have two optic axes.

2. Indicatrix Ellipsoid

Light transmission in any direction is determined by the indicatrix ellipsoid, which is defined in terms of the principal impermeability axes by the equation

$$\frac{x_1^2}{n_1^2} + \frac{x_2^2}{n_2^2} + \frac{x_3^2}{n_3^2} = 1 = B_1 x_1^2 + B_2 x_2^2 + B_3 x_3^2$$

Here n_1, n_2, and n_3 are the indices of refraction defined as the ratio of the velocity of light in a vacuum to the actual velocity along the x_1, x_2, and x_3 axes respectively. The inverse of the square of the ratio was shown to be equal to the relative impermeability constant, measured at optical frequencies, for the respective axes. The elliptical cross section parallel to the wave-normal fronts includes the two dielectric displacement values of the ordinary and extraordinary rays. The extraordinary wave has its dielectric displacement vector in the plane of the principal axis x_2 and the wave direction OP. The ordinary wave has its direction of polarization along D_1 perpendicular to the plane OP, θx_2.

For transmission of light along an x_1 axis, $l_1 = 1$, $l_2 = l_3 = 0$, and the velocities are determined by the two relative impermeability constants B_2 and B_3. These velocities are designated by

$$v^2 = b^2 = B_2 V^2 \; ; \qquad n_2^2 = \frac{V^2}{b^2} \; ; \qquad v^2 = c^2 = B_3 V^2 \; ; \qquad n_3^2 = \frac{V^2}{c^2}$$

where v is the actual velocity and V the velocity of light in a vacuum. The index of refraction in each case is defined as the ratio of the free-space velocity V to the actual velocity

[19] See Coker and Filon [7] and Frocht [8].

v. For transmission along an x_2 axis, the two velocities and the corresponding indices of refraction are

$$v^2 = a^2 = B_1 V^2 ; \qquad n_1^2 = \frac{V^2}{a^2} ; \qquad v^2 = c^2 = B_3 V^2 ; \qquad n_3^2 = \frac{V^2}{c^2}$$

For the x_3 direction the corresponding velocities are $a^2 = B_1 V^2$ and $b^2 = B_2 V^2$.

It is shown in Appendix A that these velocities satisfy the equation

$$\frac{l_1^2}{a^2 - v^2} + \frac{l_2^2}{b^2 - v^2} + \frac{l_3^2}{c^2 - v^2} = 0$$

It is usual to take $a > b > c$ to determine the three axes of the ellipsoid. In the above equations l_1, l_2, and l_3 are the direction cosines for the light-propagation direction with respect to the principal axes. By introducing the direction cosines from Fig. A.2 of Appendix A, a quadratic, Eq. (A.23b) of Appendix A, is obtained for the velocities. By solving this equation it is shown that there are two directions for which the velocities of the ordinary and extraordinary rays are equal. These directions, known as optic axes, lie in a plane perpendicular to the intermediate velocity b.

Every direction is an optic axis for a cubic crystal. For trigonal, hexagonal, and tetragonal crystals, there is one optic axis, which lies along the unique c axis. For these crystals the velocity of the ordinary ray is independent of the direction of propagation and hence lies on a spherical surface. The velocity of the extraordinary ray lies on an ellipsoid of revolution. If the ellipsoid lies inside the spherical surface, the crystal is called optically positive, whereas if it lies outside, the crystal is optically negative.

Orthorhombic, monoclinic, and triclinic crystals have two optic axes for which the ordinary and extraordinary waves have equal velocity. The plane containing the optic axes, called the optic plane, is perpendicular to the intermediate-velocity axis b. The angle θ between the bisectrix of the two axes and either axis is given by

$$\tan \theta = \pm \sqrt{\frac{a^2 - b^2}{b^2 - c^2}}$$

If this angle is less than 45 degrees, the crystal is called positive, if greater negative.

3. Direction of Ray Path

In general the direction of the ray path does not coincide with the direction of the wave normal, since the ray path is perpendicular to the electric and magnetic fields, while the wave-normal path is perpendicular to the dielectric displacement and the magnetic field. For a uniaxial crystal the energy direction always coincides with the wave-normal direction for the ordinary ray but lies at an angle α from the wave-normal direction for the extraordinary wave. The angle α is given by

$$\tan \alpha = \frac{(B_1 - B_3) \sin 2\theta}{2(B_1 \cos^2 \theta + B_3 \sin^2 \theta)}$$

where θ is the angle of the light beam from the optic axis. α will be a maximum α_m, when

$$\tan \theta = \sqrt{\frac{B_1}{B_3}} = \frac{n_e}{n_o} ; \qquad \tan \alpha_m = \frac{B_1 - B_3}{2\sqrt{B_1 B_3}} = \frac{n_e^2 - n_o^2}{2 n_o n_i}$$

where n_e is the index of refraction of the extraordinary ray in the α direction perpendicular to the optic axis and n_o the index of refraction of the ordinary ray in any direction. For calcite α_m is $6°\ 16'$. This displacement of the extraordinary ray from the ordinary ray has been made the basis for a system for optically storing information. Other uses for birefringent material are in producing polarizers, such as the Nicol prism, and for producing optical quarter-wave plates, optical ring sights, and so on.

4. Effects of Electric Displacements and Elastic Strains on Light Propagation

The effect of electric or mechanical variables applied to a crystal is to produce a slight displacement of the optic axis. If light was originally propagating along on optic axis, the material becomes birefringent—that is, the light wave is broken up into an ordinary and an extraordinary ray, which propagate with slightly different velocities. This effect is made use of in producing very fast acting light valves for modulating a light wave or a maser beam. A mechanical stress on an isotropic material produces a birefringence that is proportional to the difference between the principal stresses. This effect, called the photoelastic effect, has been used to analyze stress patterns in complicated shapes.

All of these effects can be investigated by obtaining higher-order derivatives of the internal energy U. The relevant derivatives for the linear electro-optic, the quadratic electro-optic and the piezo-optic effects can be written in the form

$$E_m = -h_{mij}S_{ij} + \beta^S_{mn}D_n + \beta_0 m_{ijmn}S_{ij}D_n + \beta_0 r_{mno}D_nD_0 + \beta_0 f_{mnop}D_nD_oD_p \cdots$$

$$T_{kl} = c^D_{ijkl}S_{ij} - h_{nkl}D_n + \frac{\beta_0 m_{klno}D_nD_o}{2}$$

where $\beta_0 m_{ijmn}$, $\beta_0 r_{mno}$, and $\beta_0 f_{mnop}$ are third and fourth derivatives of the internal energy U. All other derivatives are neglected since they do not contribute to these effects. These effects are the piezo-optic effect, the linear electro-optic effect, and the quadratic electro-optic effect. The piezo-optic tensor also appears as an electrostrictive term in the last equation, and it appears that there is a relation between these two effects.

Only the first equation is required for optical effects. In general the electric fields and displacements have low-frequency components and the high optical-frequency components are at least 1,000 times as large as the highest applied modulating frequencies. Hence the equation can be separated into low-frequency applied dielectric displacements and stresses and the high-frequency (denoted by ω) optical frequencies. Then the optical effects are given by the equation

$$E_m e^{j\omega t} = D_n e^{j\omega t}\left[\beta^S_{mn} + \beta_0\left(m_{ijmn}S_{ij} + r^S_{mno}D_o{}^0 + \frac{f_{mnop}}{2}D_o{}^0 D_p{}^0\right)\right]$$

where the superscript 0 denotes low frequency dielectric displacements. The four- and three-index symbols can be replaced by the two-index terms

$$m_{ijmn} = m_{rs} \neq m_{sr}\ ;\qquad \frac{f_{mnop}}{2}(D_o{}^0 D_p{}^0 + D_p{}^0 D_o{}^0) = f_{rs}D_p{}^0 D_o{}^0\ ;\qquad r^S_{mno} = r^S_{qo}$$

These effects add terms to the dielectric impermeability constants that cause a shift in the optic axes and produce a birefringence along the light-transmission direction. The effect of these terms on light transmission in a general direction has been worked out and is given in Eq. (7.42). For all the applications, light is transmitted along the $c = x_3$ optic axis, and for this case the birefringence ΔB, defined as $(n_2 - n_1)l$—where n_2 and n_1 are the indices of refraction for the two waves—becomes

$$\left(\frac{1}{n_1{}^2} - \frac{1}{n_2{}^2}\right)l = \sqrt{\frac{(\beta_{11} - \beta_{22})^2 + (2\beta_{12})^2}{\beta_0{}^2}}\, l$$

or

$$\Delta B = (n_2 - n_1)l = \frac{n^3 l}{2} \sqrt{(\Delta_1 - \Delta_2)^2 + (2\Delta_6)^2}$$

Two tetragonal crystals, ADP and KDP, have been used to produce light valves through the constant r_{63} or the equivalent voltage constant $z_{63} = r_{63}\varepsilon_{33}^T$. At low frequencies the piezo-optic constant m_{66} produces an effect since the piezoelectric constant d_{36} produces a strain. At high frequencies—that is, above 100 kilocycles—this effect drops out and the "true" electro-optic effect produces a modulation of light into high gigacycle ranges. By arranging that the light velocity is equal to the electromagnetic wave velocity in a wave guide, as shown by Fig. 7.7, modulation can be made to occur over the whole path length; such devices are being used to modulate lasers. Cubic crystals are more advantageous than tetragonal crystals since they can have a transverse modulation as well as a longitudinal modulation.

The quadratic ("Kerr") electro-optic effect, which involves the constants f_{rs}, can be used to produce a greater sensitivity to an applied ac voltage since a high applied dc field or displacement can be used to bias the effect. This effect is applied in a ferroelectric crystal such as KTN ($KTa_xNb_{(1-x)}O_3$ with $x = 0.65$), which has a Curie temperature slightly below room temperature. By holding the temperature constant near the Curie temperature and applying a high dc field, the sensitivity can be made about two orders higher than that for a linear electro-optic crystal. This crystal can be used to produce either a longitudinal or transverse effect.

The principal use of the piezo-optic (photoelastic) effect is in determining the stresses set up in an irregular sample by applied external loads. For plane stresses, the birefringence is equal to

$$\Delta B = \frac{n^3 l}{2} \sqrt{(\Delta_1 - \Delta_2)^2 + (2\Delta_6)^2} = \frac{n^3 l}{2}(\pi_{11} - \pi_{12}) \sqrt{(T_1 - T_2)^2 + 4T_6{}^2}$$

where π_{11} and π_{12} are the two stress photoelastic constants possible for an isotropic material. By changing to other axes it is shown that the expression under the square-root sign reduces to the difference between the principal stresses $(T_1' - T_2')$. When the polarizer with quarter-wave plates is used, a simple count from some known point such as the "eyes" of Fig. 7.8—which are regions for which $T_1' - T_2' = 0$—will establish the difference between the principal stresses. By using a polarizer without quarter-wave plates, the axes of the principal stress configuration can be determined. The complete solution for the three stresses T_1, T_2, and T_6 requires starting from some point of known stress and integrating the equations of equilibrium (7.77).

PROBLEMS CHAPTER 7

1. What are the ordinary and extraordinary velocities for waves transmitted along the bisectrix direction of a biaxial crystal?

2. Prove that the energy-transmission direction and the wave-normal direction are the same for waves transmitted along the bisectrix direction of a biaxial crystal.

3. Using Eq. (7.42) and dividing through by $\beta_0 V^2$ so that β''_{11} can be replaced by v^2, β_{11} by a^2, β_{22} by b^2, and β_{33} by c^2 show that the birefringence along an optic axis of a biaxial crystal is zero.

4. If an isotropic material is replaced by a cubic crystal, what stresses determine the birefringence along a crystallographic axis?

5. For a trigonal crystal such as quartz (32), what electro-optic constants are involved in a transverse electro-optic effect for light transmitted down the z = optic axis direction? If the field is applied along the $x_2 = y$ axis, is it possible to generate a birefringence?

REFERENCES CHAPTER 7

1. B. H. Billings, *J. Opt. Soc. Am.* **39**, 797 (1949); **39**, 802 (1949).
2. R. O. B. Carpenter, *J. Opt. Soc. Am.* **40**, 225 (1950).
3. I. P. Kaminow, *Phys. Rev. Letters* **6**, 528 (1961).
4. W. W. Rigrod and I. P. Kaminow, *Proc. IEEE* **51**, 137 (1963).
5. M. Didomenico Jr. and L. K. Anderson, *Bell System Tech. J.* **42**, 2621 (1963).
6. F. Sterzer, D. Blattner, and S. Minter, *J. Opt. Soc. Am.* **54**, 62 (1964).
7. E. G. Coker and L. N. G. Filon, "Treatise on Photoelastisity." Cambridge Univ. Press, London and New York, 1931.
8. M. Frocht, "Photoelasticity." Wiley, New York, 1941.
9. T. J. Nelson, *Bell System Tech. J.* **43**, 821 (1964).
10. M. Born and E. Wolf, "Principles of Optics." Macmillan (Pergamon), New York, 1959.
11. Elizabeth A. Wood, "Crystals and Light," Plate VII. Nostrand, Princeton, New Jersey, 1964.
12. "American Institute of Physics Handbook," pp. 6–194 and 9–100. McGraw-Hill, New York, 1963.
13. B. Zwicker and P. Scherrer, *Helv. Phys. Acta.* **17**, 346 (1944).
14. J. Kerr, *Phil. Mag.* **1**, 337 (1875). Also papers in subsequent issues.
15. J. E. Geusic, S. K. Kuntz, L. G. Von Uitert, and S. H. Wemple, *J. Appl. Phys. Letters* **4**, 141 (1964).
16. W. P. Mason and T. F. Osmer, *Bell System Tech. J.* **32**, 558 (1953).

CHAPTER 8 ● Optical Activity and the Faraday Effect

8.1 Introduction

When polarized light is transmitted along an optic axis of a crystal it is sometimes found that the plane of polarization is rotated as the light is propagated through the crystal or material. This effect is discussed mathematically in Appendix A, Section A.3, and it is shown that it is due to the fact that the effective electric field acting on individual molecules is the sum of the applied field plus an average field due to neighboring molecules, which may not act in the same direction as the applied field. Phenomenologically, the effect can be explained as being due to the fact that a plane polarized wave entering the medium is broken up into two circularly polarized waves that are transmitted with different velocities. When these two waves combine to produce a wave outside the medium, a plane wave is obtained, but with the plane of polarization rotated with respect to the original plane, the amount of rotation being proportional to the length of the light path.

One of these waves has a right-handed circular polarization, as shown by Fig. (8.1a), while the other wave has a left-handed polarization. A right circular polarization is defined in terms of the observer looking back in the direction from which the light comes, and it appears to rotate counter-clockwise when viewed by an observer looking along the direction of light propagation. The mathematical expressions for right- and left-handed circular waves are given by Eq. (A.59) and it is shown that since the velocities are slightly different they are equivalent to a plane wave whose plane of polarization changes in proportion to the distance of the propagation path. When the light leaves the material it is rotated by an angle φ, as shown by Fig. (8.1b). The sense of the rotation is the same as the sense of the faster of the two circularly polarized rays. In Fig. 8.1 this is the right-handed component.

188

In Appendix A, Section A.3, the rotation is expressed in terms of the refractive indices n_r and n_l of the two components. It is shown that the rotation of the plane of polarization, expressed in radians, is

$$\Phi = \frac{\pi l}{\lambda}(n_l - n_r) \qquad (8.1)$$

where λ is the wavelength in a vacuum. The rotation ρ per unit length is known as the rotary power.

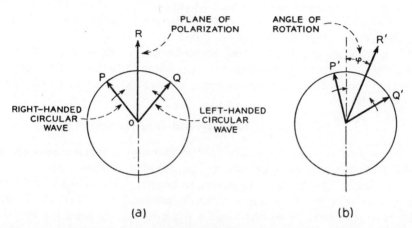

FIG. 8.1 Optical activity in a crystal produced by two circularly polarized waves of slightly different velocities. (a) Entrance of plane wave and break up into two circularly polarized waves. (b) Exit point where the two circularly polarized waves combine to produce a plane polarized wave, with its plane of polarization at an angle φ with respect to the entering wave.

In practice the difference $(n_l - n_r)$ is very small compared to unity. For example, in quartz for sodium light (near $\lambda = 5893$ Å), the rotation is 21,700 degrees or 379 radians per meter. Hence the difference between the index of refraction for left- or right-handed quartz is 7.11×10^{-5}, and the rotation term of (A.57) is 2.3×10^{-5} times as large as the index-of-refraction term.

A rotation of the plane of polarization of a light wave can also occur when a magnetic field is applied in the direction of light propagation. This effect is known as the Faraday effect. As shown in Section A.4 of Appendix A, it is related to the Zeeman effect, which splits the resonant frequency, causing the optical dispersion, into two frequencies whose separation is proportional to the magnetic field. This produces two circularly polarized waves with slightly different velocities, and hence, a rotation of the plane of polarization. Since a magnetic field is an axial vector, the rotation does not reverse sign when the light is reflected back through the same path; instead, the polarization angle is doubled. This effect has been used[1] to provide a one-way transmission system for an optical maser. For this system polarized light is sent through the system made from an optical glass, and a field is introduced in a solenoid surrounding the glass of such a value that a 45° rotation occurs in the forward direction. Any light reflected from the surface has an additional 45° rotation in traversing the return path and hence it is cancelled out by the original polarizer. Similar effects are obtained at microwave frequencies by putting a ferrite material inside a microwave transmission system. A solenoid outside the wave-transmitting system provides the necessary magnetic field. Such systems are rather widely used to provide freedom from reflections in microwave systems.

Finally, when the light is sent perpendicular to the magnetic field, a birefringence occurs in the medium. This effect is known as the Cotton-Mouton effect. The birefringence is proportional to the square of the magnetic field.

8.2 Optical Activity and Birefringence

When light is transmitted in a general direction z' making an angle θ with the optic axis of a uniaxial crystal, both birefringence and optical activity occur. As shown by Appendix A, Section A.33, the light wave will in general be broken up into two elliptically polarized waves that travel with different velocities. When the angle θ is zero—that is, when the light is transmitted along the optic axis—these two waves become circular and a rotation of the plane of polarization without birefringence results.

The amount of rotation per unit length—the rotary power—depends on a *gyration tensor* g_{ij}. As seen from Eq. (A.50), this tensor can be defined in the two equivalent ways

$$g_{ij} = \frac{\varepsilon_0\mu_0 E_i}{\frac{\partial}{\partial t}B_j} = -\frac{\mu_0\varepsilon_0 H_j}{\frac{\partial}{\partial t}(D_i)} \tag{8.2}$$

Since the tensor is the ratio of a polar vector to an axial vector it does not obey the same symmetry relations that the ratios of two polar vectors or two axial vectors do—that is, those given in Table B.2. The components of this gyration tensor were first worked out by Nye,[2] and the results are given by Table B.6. In particular, for the symmetry of quartz (class 32) the only components are

$$g_{11} = g_{22} ; \qquad g_{33} \tag{8.3}$$

---

[1] See Geusic and Scovil [1] and Aplet and Carson [2].
[2] See Nye [3].

The values found for α quartz[3] are in the units used in Eq. (A.62)

$$g_{11} = \mp 1.56 \times 10^{-29}; \qquad g_{33} = \pm 3.47 \pm 10^{-29} \left(\frac{\text{radians} \times \text{sec}}{\text{meter}} \right) \qquad (8.4)$$

With a right-handed choice of axes, the upper sign refers to right-handed and the lower sign to left-handed quartz. These units have the advantage that the g tensor does not vary with wavelength as long as the square law holds. If we shift to the formula for g'_{11} of Eq. (A.64) then g'_{ij} is independent of the wavelength for all useful wavelengths. This makes less than a 4 per cent increase for the constants of Eq. (8.4).

When the direction of wave propagation is at an angle to the optic axis both birefringence and optical activity can occur. The development of Section A.33 of Appendix A shows that a plane wave is broken up into two elliptically polarized light waves of equal ellipticity, which propagate with slightly different velocities through the crystal. Since quartz is the only crystal for which both optical rotation and birefringence have been investigated, discussion will be limited to this crystal.

If n_l and n_r are the refractive indices of these two waves, we have from the equations of propagation

$$D_1 = D_{1_0} e^{j\omega(t - n_l x_3/v)} + D_{2_0} e^{j\omega(t - n_r x_3/v)} \qquad (8.5)$$

that the relative phases of the two waves are

$$\Delta = \frac{\omega}{v}(n_l - n_r) = \frac{2\pi}{\lambda}(n_l - n_r) \qquad (8.6)$$

Inserting the values from Eq. (A.70), noting that

$$B'_{11} = \frac{1}{n_1^2}; \qquad B'_{22} = \frac{1}{n_2^2} \qquad \text{and setting} \quad n_1 n_2 \doteq n_o^2$$

we have, to the first order,

$$\Delta = \frac{2\pi}{\lambda} \left[\sqrt{(n_2 - n_1)^2 + \frac{\omega^2 n_o^2}{\varepsilon_0 \mu_0}(n_2^2 g'_{22} + n_1^2 g'_{11})(g'_{11} + g'_{22})} \right] \qquad (8.7)$$

If the gyration tensor is zero the phase difference becomes

$$\delta = \frac{2\pi}{\lambda}(n_2 - n_1) \qquad (8.8)$$

If the birefringence is zero, as it is along an optic axis, $n_1 = n_2$ and $g_{11} = g_{22} = g$ for crystal quartz. The rotary power ρ is half the phase difference, so that

$$\rho = \frac{\pi}{\lambda} \left[\frac{4\pi n_o^2 f}{\sqrt{\varepsilon_0 \mu_0}} g \right] = \left[\frac{2\pi}{\lambda} n_0 V \right]^2 g \qquad (8.9)$$

[3] See Szivessy [4] and Szivessy and Münster [5].

in agreement with Eq. (A.62). When $n_1 \neq n_2$ and $g'_{11} \neq g'_{22}$, the expression becomes more complicated, as shown by Eq. (8.7).

When both terms exist there is a superposition for the two effects, which results in the equation for the phase difference

$$\Delta^2 = \delta^2 + (2\rho)^2 \tag{8.10}$$

The wave surfaces for α quartz are shown by Fig. 8.2. The solid lines show the wave surfaces without optical activity—that is, for plane polarized light.

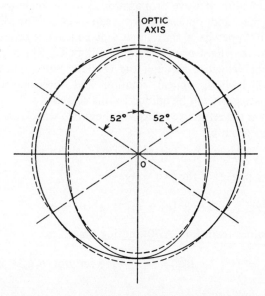

FIG. 8.2   Wave surfaces of α-quartz. Solid lines represent ordinary and extraordinary ray surfaces; dashed lines show in exaggerated form the effect of the optical rotation.

The dashed lines show in an exaggerated fashion the wave surfaces for the elliptically polarized ordinary and extraordinary rays. Except in the region between 51° 50′ and 52° 5′ the optical rotation always increases the birefringence.

The ellipticities for the ordinary and extraordinary rays are given by Eq. (A.72), the faster wave being the ordinary wave and the slower wave the extraordinary wave. The values start at unity (circularly polarized wave) along the optic axis, decrease to zero in the region of 52°, and become negative when $\theta = 90°$. When the values are zero, plane polarized waves are propagated and no rotation occurs. Figure 8.3 shows graphically the directions of rotation of the two waves.

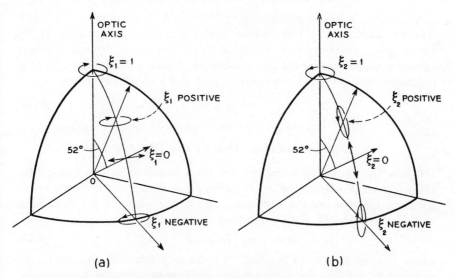

(a)   (b)

FIG. 8.3 (a) Ellipticity ξ of the ordinary ray as a function of orientation. Arrows show the direction of rotation of the elliptical wave. (b) Ellipticity for the extraordinary wave.

The additivity rule of Eq. (8.10) determines an angle β, as shown by Fig. 8.4, whose sine and cosine values are

$$\sin \beta = \frac{2\rho}{\Delta} = \frac{2\rho}{\sqrt{\delta^2 + (2\rho)^2}} \; ; \quad \cos \beta = \frac{\delta}{\sqrt{\delta^2 + (2\rho)^2}} \qquad (8.11)$$

The ellipticity values for the fast and slow waves, as shown by Eqs. (A.72) and (A.73) are related to this angle. If we divide both top and bottom through by $B'_{11}B'_{22}$ we have

$$\xi_1 = \frac{2\rho}{\delta + \sqrt{\delta^2 + (2\rho)^2}} = \frac{\sin \beta}{1 + \cos \beta} = \tan \frac{\beta}{2}$$

$$\xi_2 = \frac{\delta - \sqrt{\delta^2 + (2\rho)^2}}{2\rho} = \frac{1 - \cos \beta}{\sin \beta} = \tan \frac{\beta}{2} \qquad (8.12)$$

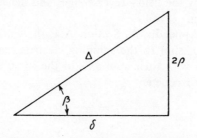

FIG. 8.4 Triangle illustrating the principle of superposition for optical activity and birefringence.

This *principle of superposition* for optical activity and birefringence, which was first discussed by Nye,[4] shows that the state of polarization of the emerging wave is the same as though the light had passed through a doubly refracting section giving the phase difference δ plus a section giving a pure optical rotation ρ. δ and ρ are related to Δ and ξ by the equations

$$2\rho = \Delta \sin \beta; \qquad \delta = \Delta \cos \beta; \qquad \xi = \tan \frac{\beta}{2} \qquad (8.13)$$

8.3 Optical Rotation Due to the Faraday Effect

It was first shown by Faraday[5] that when polarized light is transmitted along a glass or crystal rod in the presence of a magnetic field applied in the same direction as the light propagation, the plane of polarization is rotated. The rotation ρ is expressed by the formula

$$\rho = \bar{V}Hl \cos \theta \qquad (8.14)$$

where $\bar{V}$ is called Verdet's constant, H is the magnetic field strength, l is the path length, and θ is the angle between the light-propagation direction and the magnetic-field direction. The sense of the rotation when viewed by an observer looking at the source is the same as the current direction that produces the field in the surrounding solenoid. The rotation is the same for either direction of the propagation of light, and hence the rotation of the plane of polarization for light reflected from one end is twice that occurring in one direction of transmission. The rotation scalar ρ is a true scalar since it does not change sign when the direction of light transmission is reversed, which is equivalent to changing from a right-handed system to a left-handed system. Optical activity in a crystal does reverse the sign of rotation when the direction of light transmission is reversed, and hence the rotation is represented by a pseudo-scalar that has a plus sign for one system and a minus sign when the handedness is reversed. The origin of this difference in the character of the rotation is that the gyration tensor is proportional to the applied magnetic field for the Faraday rotation. Therefore, when the handedness of the system reverses, both numerator and denominator of Eq. (8.2)—divided by the field—reverse sign, and hence the gyration tensor does not change sign.

As discussed in Appendix A, Section A.4, the Faraday rotation results from the addition of spin to the spinning electron, caused by the applied field. For electrons with their spin axis in the same direction as the field this adds an angular frequency

$$\omega_L = \frac{e}{2mc} H \qquad (8.15)$$

---

[4] See Nye [3].
[5] See Faraday [6].

to the spin frequency, whereas for electrons with their spin axis in the direction opposite to the field, the same angular frequency is subtracted from ω_o. This formula was first determined by Larmor and is known as the Larmor precession.[6] This classical formula is in fair agreement with the

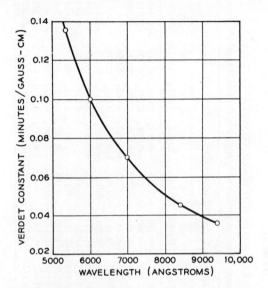

FIG. 8.5   Verdet constant for Corning No. 8363 glass as a function of the wavelength in angstroms (after Aplet and Carson).

Zeeman effect—that is, the separation of ω into the two frequencies— although quantum effects not discussed here make some difference in the quantitative results. Qualitatively,

$$\omega = \omega_0 \pm \omega_L \tag{8.16}$$

This separation produces the two circularly polarized waves when the light wave is transmitted in the direction of the magnetic field.

   The Faraday effect exists for gases, liquids, and solids but is principally of interest for glasses. Table 22 gives some recent measurements[7] of the Verdet constant for a number of glasses. These glasses were formed by fusing a high percentage of the rare-earth materials listed with the glass-forming material B_2O_3. Some of these compositions show very large Verdet constants. This constant is usually expressed in minutes of arc per centimeter times oersteds. Figure 8.5 shows the Verdet constant as a function of

[6] See Larmor [7].
[7] See Rubinstein et al. [8].

Table 22

Faraday Rotation of Rare-Earth (III) Borate Glasses

Measured Verdet constants of rare-earth (III) borate glasses at various wavelengths in the visible spectral region

Verdet Constant, $\bar{V}$ (min of arc/cm · Oe)

Wavelength (mμ)	1[b] La	2 Pr-La	3 Nd-La	4 Sm-La	5 Eu-La	6 Gd-La	7 Tb-La	8 Dy-La	9 Ho-La	10 Er-La	11 Tm-La	12 Vb-La	13 Tb-Pr	14 Dy-Pr	15 Pr
405	0.043_8	-0.380	-0.180	0.032_5	-0.081_5	0.032_5	-0.512_5	-0.436	-0.269	-0.093_7	0.060_6	0.115	-0.940	-0.850	-0.843_6
420	0.038_3	-0.336	-0.163	0.031_8	-0.069_3	0.030_1	-0.458	-0.398	-0.228	-0.088_6	0.051_2	0.104	-0.836	-0.768	-0.727
436	0.036_4	-0.307	-0.147	0.030_2	-0.060_2	0.026_8	-0.419	-0.361_5	-0.252	-0.078_5	0.046_4	0.094_9	-0.786	—	-0.646
450	0.034_0	$—$[a]	-0.139	0.031_7	-0.051_5	0.026_7	-0.382	-0.345	-0.240	-0.087_8	0.042_5	0.086_5	—	—	-0.507
465	0.031_5	—	-0.128	0.027_6	-0.046_4	0.025_5	-0.347	-0.317	-0.101_5	-0.077_2	0.041_5	0.079_6	—	—	-0.482
480	0.029_5	-0.230	-0.120	0.025_0	-0.038_3	0.024_5	-0.319_5	-0.299_5	-0.123	-0.068_9	0.039_6	0.073_3	-0.560	-0.497	-0.471
500	0.026_7	-0.220	-0.111	0.024_5	-0.033_3	0.022_3	-0.288	-0.273	-0.131	-0.082_1	0.034_9	0.066_4	-0.536	-0.465	-0.480
520	0.023_8	-0.201	-0.096	0.022_2	-0.029_6	0.021_1	-0.262	-0.246	-0.112	—	0.031	0.060_4	-0.489	-0.413	-0.432
546	0.022_6	-0.178	-0.094	0.019_8	-0.024_2	0.020_6	-0.234	-0.220_5	-0.128	-0.045_7	0.026_9	0.054_0	-0.436	-0.358	-0.390
578	0.019_8	-0.153	-0.100	0.017_0	-0.019_2	0.018_0	-0.205	-0.193	-0.104	-0.042_9	0.023_4	0.046_9	-0.380	-0.332	-0.334
600	0.018_2	-0.146	-0.059	0.016_1	-0.016_9	0.017_2	-0.186	-0.177_5	-0.096	-0.040_9	0.021_0	0.043_0	-0.348	-0.290	-0.317
635	0.016_3	-0.128	-0.056	0.014_3	-0.014_3	0.015_1	-0.167	-0.159	—	-0.035_9	0.018_5	0.037_5	-0.306	-0.252	-0.271
670	0.014_7	-0.110	-0.046	0.010_5	-0.012_9	0.013_2	-0.142	-0.138	-0.074	-0.034_6	0.016_5	0.033_3	-0.265	—	-0.243

[a] Blank spaces occur in regions of intra-f-orbital transitions.
[b] Numbers at heads of columns identify samples.

wavelength for a Corning No. 8363 glass.[8] This glass has been used to provide a nonreflecting transmission of a laser light wave of 6943 Å. For a length of 14 centimeters the flux required for a 45° rotation was 2700 gauss or 0.27 weber per meter². An analyzer set at 45° from the polarizer lets most of the light through, but any reflected from the surface will be rotated by 90° with respect to the polarizer and hence will not be transmitted toward the light source.

By using a 45° rotation with polaroids rotated 45° from each other on the two ends, a one-way transmission system—equivalent to the balanced gyrator of Chapter 11—is obtained.

8.4   Microwave Faraday Effect

It was first shown by Polder[9] that a ferromagnetic substance should show an appreciable Faraday rotation at microwave frequencies. The analysis is discussed in Appendix A, Section A.4. The result is to generate two circularly polarized microwaves that produce a rotation of the plane of polarization that is proportional to the square of the frequency for frequencies much smaller than the natural spin frequency ω_0. For frequencies much greater than ω_0, the rotation becomes independent of the frequency. Hence the microwave Faraday effect gyrator is used for frequencies above ω_0. However, the rotation is not strictly independent of the frequency, a fact that limits the possible bandwidth.

The microwave gyrator has been used[10] in producing one-way transmission systems, microwave circulators, mircowave switches, and electrically controlled variable attenuators and modulators. Figure 8.6a shows one method used for obtaining essentially a one-way transmission system. A plane polarized wave $TE_{1,0}$, as shown by Fig. 8.6b, is transmitted in the rectangular wave guide. Since this guide cannot transmit a circularly polarized wave, the guide is tapered until it reaches a circular section. The electric field strength will now be in the form of a $TE_{1,1}$ mode, which has the form of Fig. 8.6c. In the center is inserted a ferrite rod of the proper length to cause a rotation of the plane of polarization of 45° for the $TE_{1,1}$ mode; the rod is magnetized by the dc power source to give the required rotation. On the output side the chamber is tapered down to fit the wave guide, which is usually then given a 45° rotation so that the output orientation is the same as the input. A wave traveling in the reverse direction is rotated through 45° in the same direction as before, and since it will be below the cutoff frequency for which waves polarized along the width can be transmitted, no transmission

---

[8] See Geusic and Scovil [1] and Aplet and Carson [2].
[9] See Polder [9].
[10] See Hogan [10], Fox et al. [11], and Katz [12].

occurs in the backward direction. The absorbing vane D attenuates the reflected wave sufficiently so that a negligible portion emerges from the output side. Although the principles are the same as for a Faraday rotator, consideration has to be given to loss mechanisms, reflections in coupling members, and so on, before practical designs can be obtained. The reader is referred to a number of sources[11] for these details. Commercially available

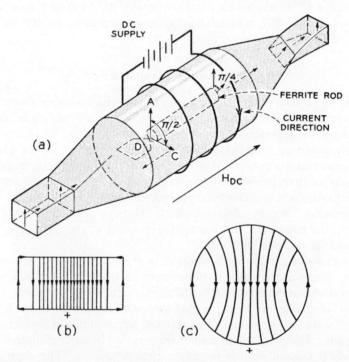

FIG. 8.6—(a) Microwave isolator utilizing Faraday rotation in a ferrite rod. (b) Direction of electric wave vector for a $TE_{1,0}$ wave. Separation of lines indicates relative strength. (c) Electric vector for a $TE_{1,1}$ wave.

devices at X band—8 to 12 gigacycles/sec—have typical characteristics of 0.1 decibel or less forward loss and a minimum backward loss of 15 decibels.

8.5 Cotton-Mouton Effect

When the magnetic field is applied perpendicularly to the light direction, a birefringence is observed in the transmitted light that is proportional to the

[11] See Hogan [10], Fox *et al.* [11], and Katz [12].

square of the magnetic flux density. This effect is known as the Cotton-Mouton[12] effect after its discoverers. For liquids and glasses, for flux densities of 10^4 gauss or one weber per meter^2, the Cotton-Mouton effect is only about 10^{-3} of the Faraday effect. At microwave frequencies, however, the effect is larger[13] and has been used to produce birefringence and circularly polarized waves in a wave guide.

Phenomenologically, the effect is similar to the quadratic electro-optic effect and can be obtained by expanding the electric field in a series of derivatives. Since the magnetic flux does not directly affect the electric field, the only derivatives of interest are

$$E_m = \frac{\partial E_m}{\partial D_n} D_n + \frac{1}{3!} \frac{\partial^3 E_m D_n B_o B_p}{\partial D_n \, \partial B_o \, \partial B_p} + \cdots \tag{8.17}$$

Denoting the partial derivatives by the definitions

$$\frac{\partial E_m}{\partial D_n} = \beta_{mn}; \qquad \frac{1}{3} \frac{\partial^2 E_m}{\partial D_n \, \partial B_o \, \partial B_p} = \beta_0 l_{mnop} \tag{8.18}$$

and noting that since the indices m and n can be interchanged as can also the indices o and p, Eq. (8.17) can be written

$$E_m = D_n[\beta_{mn} + \beta_0 l_{qr} B_o B_p] \tag{8.19}$$

where q and r have the values 1 to 6 according to the usual relations (see Eq. [3.87]). For an isotropic material, which is normally employed, the three possible constants are, from Table B4,

$$l_{11}, \quad l_{12}, \quad \text{and} \quad l_{44} = l_{11} - l_{12} \tag{8.20}$$

We see from Eq. (7.60) for the quadratic electro-optic effect that the expressions for the changes in the impermeability constants at optical or microwave frequencies become

$$\Delta_1 = l_{11}B_1{}^2 + l_{12}(B_2{}^2 + B_3{}^2) \qquad \Delta_2 = l_{12}(B_1{}^2 + B_3{}^2) + l_{11}B_2{}^2$$
$$\Delta_3 = l_{12}(B_1{}^2 + B_2{}^2) + l_{11}B_3{}^2 \qquad \Delta_4 = (l_{11} - l_{12})B_2B_3 \tag{8.21}$$
$$\Delta_5 = (l_{11} - l_{12})B_1B_3 \qquad \Delta_6 = (l_{11} - l_{12})B_1B_2$$

If, for example, the magnetic flux is applied perpendicular to the light-propagation direction, $\theta = 0$ in Eq. (7.42) and

$$B_1 = B_0 \cos \varphi; \qquad B_2 = B_0 \sin \varphi; \qquad B_3 = 0 \tag{8.22}$$

From Eq. (7.42) the two extreme values of β'_{11} are

$$\frac{\beta'_{11}}{\beta_0} = \frac{\beta_1}{\beta_0} + \left(\frac{l_{11} + l_{12}}{2}\right)B_0{}^2 \pm \tfrac{1}{2}(l_{11} - l_{12})B_0{}^2 \tag{8.23}$$

---

[12] See Cotton and Mouton [13].
[13] See Fox et al. [11].

Hence the two indices of refraction are given by

$$\frac{1}{n_1^2} = \frac{1}{n_o^2} + l_{11}B_0^2 \; ; \qquad \frac{1}{n_2^2} = \frac{1}{n_o^2} + l_{12}B_0^2 \qquad (8.24)$$

The birefringence along the direction of light propagation is

$$\Delta B = (n_2 - n_1)l \doteq \frac{n_o^3}{2}(l_{11} - l_{12})lB_0^2 \qquad (8.25)$$

It is readily shown from Eqs. (8.21) and (7.42) that the effect of putting the flux at an angle γ from the direction of wave propagation is to vary the birefringence according to the formula

$$\Delta B = \frac{n_o^3}{2}(l_{11} - l_{12})lB_0^2 \sin^2 \gamma \qquad (8.26)$$

A secondary effect of the field, as seen from Eq. (8.23) is to reduce the average value of the index of refraction—that is, to increase the velocity. For a field perpendicular to the light direction, the change is

$$n_o' = n_o - \frac{(l_{11} + l_{12})B_0^2 n_0^3}{4} \qquad (8.27)$$

When the flux is in the same direction as the light-wave propagation, the birefringence is zero and the decrease in the index of refraction is

$$n_o' = n_o - \frac{l_{12}B_0^2 n_0^3}{2} \qquad (8.28)$$

The formulation discussed here is general enough to apply to an anisotropic medium. However, the only device employed so far using this effect is one for which birefringence is induced in a circular wave guide by employing[14] four rods of ferrite with their axes along the direction of polarization of microwaves. By applying appropriate static magnetic fields, birefringence can be generated and plane polarized waves can be turned into circularly polarized waves.

RESUME CHAPTER 8

I. Description of Optical Activity

When polarized light is transmitted along the optic axes of crystals, it is found that for certain crystal classes the plane of polarization is rotated as the light is transmitted through the crystal. The amount of rotation is proportional to the length of the light path, and the

---

[14] See Fox et al. [11].

rotation reverses direction when the direction of the light path is reversed. The cause of this rotation is that the effective electric field acting on individual molecules is the sum of the applied field plus an average field due to neighboring molecules. As shown in Section A.3 of Appendix A, the electric and magnetic fields are given by the equations

$$E_j = \beta_{ij}D_i + \frac{g_{ij}}{\varepsilon_0\mu_0}\frac{\partial B_i}{\partial t}\ ; \qquad H_i = \frac{B_i}{\mu_0} - \frac{g_{ij}}{\varepsilon_0\mu_0}\frac{\partial D_j}{\partial t}$$

where g_{ij} is a new gyration tensor that determines the rotation of the plane of polarization of the light. Since it is the ratio of a polar vector to an axial vector, it is an axial second-rank tensor, whose components are given by Table B.6 of Appendix B. For any crystal with a center of symmetry, this tensor disappears. Hence calcite, which has a symmetry $\bar{3}m$, has no rotation of its plane of polarization. However, quartz with a symmetry 32 has a rotation of 21,700 degrees/meter or 379 radians/meter.

The effect of the relation given above is to cause a plane polarized light to be broken up into right- and left-handed circularly polarized waves that travel with slightly different velocities. When the two circularly polarized waves are joined together at the output, this is equivalent to a rotation of the plane of polarization of the original plane wave. The rotation in radians is

$$\Phi = \frac{\pi l}{\lambda}(n_l - n_r)$$

where n_l and n_r are the indices of refraction of the left and right polarized waves.

2. Transmission of Light at an Angle to the Optic Axis

When light is transmitted in a direction making an angle θ with the optic axis, both birefringence and a rotation of the plane of polarization occur. For the general direction, it is shown that the light is broken up into two elliptically polarized waves that travel with different velocities. These velocities differ not only on account of the rotation of the plane of polarization but also because of the birefringence of the crystal. It is shown in Appendix A that there is a superposition formula of the type

$$\Delta^2 = \delta^2 + (2\rho)^2$$

where Δ is the phase difference between the two waves, δ the phase difference due to the natural birefringence, and ρ the rotation of the plane of polarization due to the gyration tensor g_{ij} .

Apparently the only crystal for which both birefringence and optical rotation have both been measured is quartz. This has a gyration tensor having the components

$$g_{11} = g_{22}\ ; \qquad g_{33}$$

Using the evaluated components, the velocity surfaces for quartz are shown by Fig. 8.2. The solid lines show the velocities of the ordinary and extraordinary rays when rotation is neglected, while the dashed lines show, in exaggerated form, the effect of rotation on the velocities for the elliptically polarized waves. Figure 8.3 shows the ellipticities and the directions of rotation or polarization for the ordinary (a) and the extraordinary (b) waves. At an angle of about 52 degrees, the two rays become plane polarized. For angles greater than 52 degrees, the direction of rotation reverses for both waves. If an angle β is determined, as shown by Fig. 8.4, by the relation between the rotation ρ and the birefringence δ, the ellipticity for both waves is determined by the relation

$$\xi = \tan\frac{\beta}{2}$$

3. Optical Rotation Due to the Faraday Effect

When light is transmitted in a material in the same direction as that for which a magnetic field H is applied, there is a rotation of the plane of polarization of a plane polarized light with a value

$$\rho = VHl \cos \theta$$

where V is called the Verdet constant, H is the magnetic field, l is the length of the path, and θ the angle between the magnetic field and the light-path direction. This constant is usually expressed in minutes of arc per centimeter when H is expressed in the cgs unit, the oersted. When the light is reflected back over the same path, the rotation continues in the same sense. This makes possible transmission of light in one direction with no light out of the system in the other direction. As shown in Appendix A, the Verdet constant is given by the equation

$$V = \frac{Bf^2}{(f_1{}^2 - f^2)^2}$$

where f is the frequency of the light wave, B a constant, and f_1 the frequency of absorption that causes the dispersion of the index of refraction n shown by Eq. (A.93). The optical Faraday effect is caused by the Zeeman effect which splits the absorption frequency into two frequencies

$$(f_1 \pm FH)$$

in the presence of the magnetic field H. This causes the light wave to be broken up into right and left circularly polarized light waves that propagate with slightly different velocities. When these combine, the plane of polarization has been rotated by an angle proportional to the length of path and the magnetic field.

4. Faraday Effect in Microwave Systems

A similar rotation of the plane of polarization occurs in a microwave radio system when a magnetized ferrite specimen is introduced into a wave guide section that can support circularly polarized microwaves. As shown in Appendix A, Section A.4, this effect is caused by the alignment of the spin axis of the ferromagnetic material by the applied magnetic field. For fields less than that required to produce saturation, these spinning particles will react to a perturbing force supplied by the radio wave at right angles to their spin by precessing about the spin axis. This precession causes the radio wave to break up into two circularly polarized waves that propagate with slightly different velocities, and when they are combined again at the output, a rotation of the plane of polarization occurs. The nearer the frequency of the radio wave is to the precessing frequency, the larger is the rotation of the plane of polarization.

When the radio-wave frequency is higher than the precessing frequency, the rotation becomes approximately independent of the frequency. All applications occur in this frequency range. This microwave "gyrator" is capable of producing a one-way transmission system. An example is shown by Fig. 8.6. Commercially available devices working in the range from 8 to 12 gigacycles have typical forward losses of 0.1 decibel or less and a minimum backward loss of 15 decibels.

5. Cotton-Mouton Effect

When a magnetic flux is applied perpendicular to the light direction, a birefringence is produced in the transmitted light that is proportional to the square of the magnetic flux density. For liquids or glasses this effect for fields in the order of a weber per meter squared is only about $\frac{1}{1000}$ as large as the Faraday effect. However, for microwave systems it can produce appreciable results and has been used to produce circularly polarized radio waves from plane polarized waves.

Phenomenologically the effect can be obtained by expanding the electric field in a series of derivatives involving the optical electrical displacement and the applied magnetic flux. The relation obtained is

$$E_m = D_n[\beta_{mn} + \beta_0 l_{mnop} B_o B_p]$$

where the electrical variables apply to optical frequencies while the magnetic fluxes are statically applied values. For an isotropic material there are two independent constants

$$l_{11}, \quad l_{12} \quad \text{and} \quad l_{44} = (l_{11} - l_{12})$$

PROBLEMS CHAPTER 8

1. Using the general formula (8.7) determine the ellipticity for a light wave traversing a quartz crystal at an angle of 10 degrees from the optic axis. The relative indices of refraction can be determined from the values of n_o and n_l given by Table 10 of Chapter 7.

2. Show that the rotation for a cubic crystal of class 432 or 23 is independent of the direction of light transmission.

3. For the tetragonal crystals ADP or KDP (symmetry $\bar{4}2m$) show that there is no rotation when light is sent along the optic axis (x_3) but that the velocity may be increased by a very small amount due to the gyration term g_{12}.

4. For a cubic crystal, assuming $l_{44} > (l_{11} - l_{12})$, what direction for the applied magnetic field will produce the largest birefringence?

REFERENCES CHAPTER 8

1. J. E. Geusic and H. E. D. Scovil, *Bell System Tech. J.* **41**, 1371 (1962).
2. L. J. Aplet and J. W. Carson, *Appl. Opt.* **3**, 544 (1964).
3. J. F. Nye, "Crystal Optics," Chap. XIV. Oxford Univ. Press (Clarendon), London and New York, 1957.
4. G. Szivessy, *Handbuch der Physik* **20**, 804 (1928).
5. G. Szivessy and C. Münster, *Ann. Physik* **20**, 703 (1934).
6. M. Faraday, *Phil. Trans. Roy. Soc. London, Ser. A.* **136**, 1 (1846).
7. J. Larmor, "Aether and Matter," p. 341. Cambridge Univ. Press, London and New York, 1900.
8. C. B. Rubinstein, S. B. Berger, L. G. Van Uitert, and W. A. Bonner, *J. Appl. Phys.* **35**, 2338 (1964).
9. D. Polder, *Phil. Mag.* **40**, 99 (1949).
10. C. L. Hogan, *Bell System Tech. J.* **31**, 1 (1952).
11. A. G. Fox, S. E. Miller, and M. T. Weiss, *Bell System Tech. J.* **32**, 1358 (1953).
12. H. W. Katz, *in* "Solid State Magnetic and Dielectric Devices" (H. W. Katz, ed.), Chapter VII. Wiley, New York, 1959.
13. A. Cotton and H. Mouton, *Compt. Rend.* **145**, 229 (1907).

Part III

Transport Crystalline Properties That Depend on Irreversible Thermodynamics

CHAPTER 9 • Thermal and Electrical Conductivities

9.1 Introduction: Nonreversible Thermodynamics

All the crystalline properties discussed in the first eight chapters are consistent with reversible thermodynamics, sometimes called thermostatics. The symmetry of the tensors involved followed directly from the fact that the energy of the crystal can be considered as a definite function of the mechanical, electrical, magnetic, or thermal variables. The crystal properties were described by reference to equilibrium states. For example, the elastic energy of a strained crystal can be described in terms of an initial state and a strained state. In straining the crystal, it is possible to apply the strain and remove it as slowly as desired, and in so doing the energy goes from one equilibrium state to another. In the language of thermodynamics, the changes are reversible.

There exist, however, a group of phenomena for which this process does not apply. Most of these phenomena can be described as transport properties in which, for example, electrical charge or thermal energy is transported from one position in the crystal to another under the influence of an electrical potential gradient or a thermal temperature gradient. Neither type of conduction can be described in terms of equilibrium states. There is a continuous flow of charge or thermal energy, which may reach a steady state—that is, its state may not change with time—but it cannot be considered in equilibrium because the sources or sinks of charge or thermal energy are undergoing continuous changes.

Before the introduction of the Onsager[1] irreversible thermodynamic relations, there were a large number of phenomenological laws describing irreversible properties in the form of proportionalities. Examples are Fourier's law between heat flow and temperature gradients, Ohm's law

[1] See Onsager [1].

between electric current and potential gradients, as well as the thermo-electric relations between temperature and electric potential gradients and thermal and electric current flow. The last phenomenon represents an interaction between two processes. The cross effects are mathematically described by the addition of new terms to the phenomenological laws mentioned above. All these relations are phenomenological in the sense of being experimentally verified laws that are not part of a comprehensive theory of irreversible processes. No general rules are evident between cross phenomena arising from mutual interference between two irreversible effects.

Such a principle was supplied by the Onsager *thermodynamics of irreversible processes*, published in 1931 and later refined by Casimir and others.[2] A number of causes can give rise to irreversible phenomena. Examples are temperature and potential gradients, magnetic fields, concentration gradients and so on. These quantities are usually called forces in the thermodynamics of irreversible processes and are designated by X_k $(k = 1, \ldots, n)$. These forces cause certain irreversible phenomena such as current flow, heat flow, diffusion flow, and so on. There are called the fluxes or sometimes the flows, and are designated by the letters $J_i (i = 1, \ldots, n)$.

In general, any force can give rise to any flow if there is a cross-coupling coefficient L_{ik}, and we can write

$$J_i = \sum_{k=1}^{n} L_{ik} X_k \tag{9.1}$$

The coefficients L_{ii} are, for example, the heat conductivity or the electrical conductivity. The coefficients L_{ik} can be such effects as the electric current caused by a thermal gradient or a heat flow caused by a potential gradient. Onsager's principal states that if the flows and forces are chosen correctly

$$L_{ij} = L_{ji} \tag{9.2}$$

In case a magnetic field is involved, or any axial vector, this rule has to be generalized to

$$L_{ij}(H) = L_{ji}(-H) \tag{9.3}$$

as was first shown by Meixner.[3] This means that the coefficient of L_{ij} is the same function of H that L_{ji} is of $-H$. This has physical meaning even for $i = j$—that is, the diagonal elements of the matrix are even functions of the magnetic field strength.

[2] See Casimir [2].
[3] See Meixner [3].

A complete discussion of the proper choices for the "forces" X_i and the flows J_i has been given by DeGroot.[4] It is shown that the proper choice is related to the time rate of change of the entropy by the equation

$$\delta\dot\sigma \propto \sum J_i X_i \tag{9.4}$$

For the only two systems considered in the present book—that is, electrical and thermal conductivities and their interactions in the presence of stresses and magnetic fields—the two forces are

$$E_i = -\frac{\partial\Phi}{\partial x_i} \quad \text{and} \quad -\frac{\partial\Theta/\partial x_i}{\Theta} \tag{9.5}$$

while the flows are the current densities I_j and the flow of heat per unit area h_j. In Eq. (9.5) Φ is the electric potential, E_i the electric fields, and Θ the absolute temperature.

A critical discussion of Onsager's equation was given by Casimir.[5] He showed, for example, that the three heat flows discussed in the next section are not directly determinable, because strictly it is only the divergence of the heat flow that can be measured. That is, only the difference between the heat entering and the heat leaving the crystal can be determined. Casimir shows that this allows an antisymmetric tensor k_{ij}^A to be added to the symmetric tensor k_{ij}^S as long as

$$k_{ij}^A = -k_{ji}^A \quad \text{and the divergence} \quad \frac{\partial k_{ij}^A}{\partial x_i} = 0 \tag{9.6}$$

Hence Onsager's theory correctly applied gives

$$\frac{\partial(k_{ij} - k_{ji})}{\partial x_i} = 0 \quad \text{rather than} \quad k_{ij} - k_{ji} = 0 \tag{9.7}$$

However, the antisymmetric part of k_{ij} gives no observable physical consequences as long as it satisfies (9.6). Since this part is not observable there is no reason we should not put it equal to zero and obtain

$$k_{ij} = k_{ji} \tag{9.8}$$

One of the consequences of not accepting (9.8) is that it is necessary to assume that the conductivity of a vacuum is not zero. For example, if we take a flat surface of the crystal perpendicular to x_3, then if (9.8) is not accepted,

$$k_{31} - k_{13} \neq 0 \tag{9.9}$$

By Eq. (9.7), Onsager's relation gives

$$\frac{\partial(k_{31} - k_{13})}{\partial x_3} = 0 \tag{9.10}$$

[4] See DeGroot [4].
[5] See Casimir [2].

which holds for the crystal boundary between the crystal and the vacuum. Therefore (9.9) still holds in the vacuum, and hence k_{31} and k_{13} cannot both be zero. Conversely, if k_{ij} is assumed symmetrical in the vacuum it must also be symmetrical in the crystal. It follows that if k_{ij} is assumed zero in the vacuum it must be symmetrical in the crystal. Hence, Onsager's relation (9.2) appears to be the general relation for nonreversible thermodynamics.

The proof given for Onsager's relations[6] depends upon statistical mechanics and is beyond the scope of the book. Briefly, it depends upon the principle of microscopic reversibility, which states that if the velocities of all the particles are reversed simultaneously, the particles will retrace their former paths, reversing the entire series of configurations. Alternatively, we can say that the mechanical equations of motion of individual particles are symmetrical with respect to the past and the future; that is to say, they are invariant to the transformation $t = -t$. This is not true when a magnetic field is present unless we reverse the field as well as the velocity. This leads to the relation of Eq. (9.3). A consequence of this is that any symmetric tensor depends on only even powers of H, whereas any antisymmetric tensor depends on odd powers of H. These relations have an important consequence for the Hall effect and magnetoresistance effect discussed in Chapter 11. These effects are expressed by the equation

$$E_i = \rho_{ij}(B)I_j + \varepsilon_{ijk}I_jR_k(B) \tag{9.11}$$

where the resistivities ρ_{ij} are even functions of the magnetic field, and the vector R_k, determining the Hall effect—which is an antisymmetric, second-rank tensor by the relations of Eqs. (2.20)—is an odd function of the magnetic flux.

9.2 Thermal Conductivities and Resistivities

9.21 Thermal Conductivities

When a difference of temperature occurs between two parts of a solid, there is a flow of heat throughout the solid. h_i $(i = 1, 2, 3)$ are the components of the heat flow along the x_1, x_2, x_3 axes. The MKS unit is the watt per square meter (the watt is 1 joule per second). From Fourier's law for an isotropic solid

$$h_i = -k \frac{\partial \Theta}{\partial x_i} \tag{9.12}$$

where k is the coefficient of thermal conductivity. The units are watts per square meter divided by °K per meter. Hence k is in watts per meter × °K.

---

[6] See Onsager [1], Casimir [2], and DeGroot [4].

In a crystal h_i is in general not parallel to the temperature gradient, and (9.12) is replaced by

$$h_i = -k_{ij} \frac{\partial \Theta}{\partial x_j} \qquad (9.13)$$

Each component now depends on all three components of the temperature gradient. Since the thermal-conductivity coefficients k_{ij} relate two polar vectors, it is a second-rank tensor. By Onsager's principle, discussed in

Table 23
THERMAL CONDUCTIVITIES OF CRYSTALS[a]

Crystal	Chemical Formula	System	k_1	k_2	k_3
Diamond	C	cubic	550.0	550.0	550.0
Silicon	Si	cubic	175.0	175.0	175.0
Germanium	Ge	cubic	65.0	65.0	65.0
Lithium fluoride	LiF	cubic	16.0	16.0	16.0
Sodium chloride	NaCl	cubic	6.5	6.5	6.5
Potassium chloride	KCl	cubic	6.4	6.4	6.4
Calcite	$CaCO_3$	trigonal	4.18	4.18	4.98
Quartz	SiO_2	trigonal	6.5	6.5	11.3
Sapphire	Al_2O_3	trigonal	22.5	22.5	25.1
Rutile	TiO_2	tetragonal	9.3	9.3	12.85
ADP	$NH_4H_2PO_4$	tetragonal	1.32	1.32	0.71
KDP	KH_2PO_4	tetragonal	1.42	1.42	1.22
Rochelle salt	$NaKC_4H_4O_6 \cdot 4H_2O$	orthorhombic	0.46	0.61	0.56
Aluminum	Al	cubic	235.0	235.0	235.0
Copper	Cu	cubic	400.0	400.0	400.0
Gold	Au	cubic	311.0	311.0	311.0
Lead	Pb	cubic	35.0	35.0	35.0
Silver	Ag	cubic	418.0	418.0	418.0

[a] In watts/m · °K at 300° K.

Section 9.1, it is seen that the tensor is symmetrical and the components for various crystal symmetries are similar to the tensor components of Table B.2.

If the components of the tensor are referred to the principal axes—which are also the crystal axes, except for monoclinic and triclinic crystals—the equations can be more simply written as

$$h_1 = -k_1 \frac{\partial \Theta}{\partial x_1}; \qquad h_2 = -k_2 \frac{\partial \Theta}{\partial x_2}; \qquad h_3 = -k_3 \frac{\partial \Theta}{x_3} \qquad (9.14)$$

As with other second-rank tensors the representation quadric is always of the form of Eq. (2.35) and, referred to the principal axes, takes the form of (2.36). Since the principal thermal conductivities k_1, k_2, and k_3 are always positive, the quadric is an ellipsoid. Table 23 shows a few numerical values for crystals of interest.

9.22 Thermal Resistivities

If the three equations represented by (9.13) are solved for the temperature gradients $\partial \Theta / \partial x_j$, we can express them in terms of the three components of heat flow by the equation

$$\frac{\partial \Theta}{\partial x_i} = -r_{ij} h_j \tag{9.15}$$

where

$$r_{ij} = \frac{(-1)^{i+j} \Delta_{ij}}{|k_{ij}|}$$

Here Δ_{ij} is the determinant obtained from the matrix (k_{ij}) by suppressing the ith row and jth column. In terms of the principal axes

$$r_1 = \frac{1}{k_1} \; ; \qquad r_2 = \frac{1}{k_2} \; ; \qquad r_3 = \frac{1}{k_3} \tag{9.16}$$

As discussed in the next section, thermal conductivities are measured when the dimension along the heat path is small compared to the cross-sectional dimensions, whereas resistivities are measured when the heat-flow direction is large compared to the cross-sectional dimensions.

9.23 Configurations Used to Measure Thermal and Electrical Conductivities

It is a matter of some interest to determine what properties are measured when the heat-flow direction does not coincide with the thermal-gradient direction. The same considerations also apply to electrical conductivities since, as shown by Section 9.3, they satisfy a similar set of equations found for heat flow. The two methods usually used involve a thin plate or a long bar.

9.231 HEAT FLOW ACROSS A FLAT PLATE. For the flat crystal plate shown by Fig. 9.1, the two surfaces are in contact with two good thermal conductors that maintain the surfaces at different temperatures. The other surfaces are usually well insulated to cut down any thermal losses through these surfaces. Temperature differences can be measured by accurate thermocouples, and the heat flow can be measured by taking the difference between the electrical energies required to keep the two end plates at temperatures Θ_2 and Θ_1 respectively. This assumes perfect insulation on all surfaces except those between the end plates and the crystal. If this is not so, corrections have to be made for the heat losses through these surfaces.

For a perfectly insulated crystal, the isothermal surfaces must run parallel to the crystal except at the edge. This follows because at the edge the heat flow can only be down the axis x_3, and hence from (9.15) there must be other components of the temperature gradient. However, this effect extends in for only a region of the same order as the thickness, as shown in the dashed lines, and if the cross-sectional dimensions are much larger than the thickness, this end effect can be neglected. In general the heat flow h_i is not parallel to the gradient but is at an angle, as shown. It may be remarked that for all crystal systems except the monoclinic and triclinic, crystals cut normal to the crystallographic axes will have their heat flow in the same direction as the gradient, and hence no conductivities other than k_1, k_2, and k_3 are measured. If this is not the case and the temperature gradient is along the x_3 axis, it is obvious from the discussion of the parallel-plate

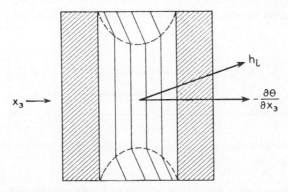

FIG. 9.1   Heat flow across a flat crystalline plate. Solid lines show isothermal lines. Dashed lines indicate regions of end effects. Arrows show thermal-gradient and heat-flow directions.

condenser of Section 3.11 that the thermal flow is determined by the thermal conductivity k_{33}. This is obvious also because the coefficients k_{13} and k_{23} determine the sidewise heat flow, and since no heat flows across the boundaries the integrated sidewise flow must be zero.

To determine the cross components it is usually necessary to cut crystals with their thermal-axis directions at an angle from crystallographic axes and depend on transformation equations of the type

$$k_{33}' = \frac{\partial x_3'}{\partial x_i} \frac{\partial x_3'}{\partial x_j} k_{ij} \tag{9.17}$$

to relate the measured value k_{33}' to the values pertaining to the crystal axes— that is, to the constants k_{ij}. For example, if we cut a crystal with its thermal axis at 45° between the x_1 and x_3 axes, the direction cosines are

$$\frac{\partial x_3'}{\partial x_1} = \frac{1}{\sqrt{2}} ; \qquad \frac{\partial x_3'}{\partial x_2} = 0; \qquad \frac{\partial x_3'}{\partial x_3} = \frac{1}{\sqrt{2}} \tag{9.18}$$

and the measured thermal conductivity k'_{33} is

$$k'_{33} = \frac{k_{11} + 2k_{13} + k_{33}}{2} \tag{9.19}$$

Hence, three measurements along the crystallographic axes and three measurements for crystals with their thermal axes 45° between two pair of the crystallographic axes will determine all the constants.

9.232   HEAT FLOW IN A LONG ROD. The other crystal configuration used for the measurement of thermal conductivity is a rod with its length large compared to the cross sectional dimensions, as shown by Fig. 9.2.

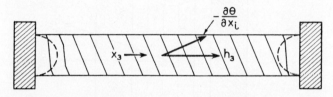

FIG. 9.2   Heat flow down a crystalline rod. Direction of heat flow and thermal gradient are shown by arrows.

When a temperature difference is maintained between the two ends and the crystal is thermally insulated around its periphery, all the heat flows down the rod and hence h_3 is a constant. Except for end effects the thermal gradient will be at an angle to the heat flow, the angle being determined by the resistivity equations

$$\frac{\partial \Theta}{\partial x_1} = -r_{13}h_3 \; ; \qquad \frac{\partial \Theta}{\partial x_2} = -r_{23}h_3 \; ; \qquad \frac{\partial \Theta}{\partial x_3} = -r_{33}h_3 \tag{9.20}$$

Since $\partial\Theta/\partial x_3$ is directly measured in the temperature differences of the two ends, this measurement gives directly r_{33}. To obtain the other constants requires oriented rods with the same thermal-path directions as for the flat plate.

9.24   Poisson's Equation and Determination of General Heat Flow

For a more general shape than a plate or rod we require an extended analysis that leads to Poisson's equation for an isotropic material and a generalized Poisson's equation for an anisotropic material. Let us suppose that there are sources and sinks of heat distributed uniformly throughout the crystal so that the net rate of production of heat per unit volume of the crystal is $\dot{Q}$. This may be either positive or negative. For a steady state the difference between the heat entering and leaving must be equal to the heat generated or lost from the volume. We have, therefore, the conservation equation

$$\frac{\partial h_i}{\partial x_i} = \dot{Q} \tag{9.21}$$

Substituting h_i from (9.13) we have

$$k_{ij}\frac{\partial^2\Theta}{\partial x_i\,\partial x_j} = -\dot{Q} \tag{9.22}$$

This equation gives the temperature Θ in terms of the thermal conductivities and the heat sources. For an isotropic material it reduces to the equation

$$k\frac{\partial^2\Theta}{\partial x_i{}^2} = -\dot{Q} \qquad (i = 1, 2, 3) \tag{9.23}$$

which is the equivalent of the Poisson's potential theory equation

$$\varepsilon\frac{\partial^2\Phi}{\partial x_i{}^2} = -\rho \tag{9.24}$$

where Φ is the electric potential function, ε the dielectric constant of the material, and ρ the charge density of the medium. Poisson's equation follows directly from the third of Maxwell's equations in (3.39) when the substitution is made

$$D_i = \varepsilon E_i = -\varepsilon\frac{\partial\Phi}{\partial x_i} \quad\text{in}\quad \frac{\partial D_i}{\partial x_i} = \rho \quad\text{or}\quad \varepsilon\frac{\partial^2\Phi}{\partial x_i{}^2} = -\rho \tag{9.25}$$

Since many solutions of Poisson's equations are known it is desirable to transform Eq. (9.22) into the form of (9.23). As a first step we can express (9.22) in terms of the principal axes. This results in

$$k_1\frac{\partial^2\Theta}{\partial x_1{}^2} + k_2\frac{\partial^2\Theta}{\partial x_2{}^2} + k_3\frac{\partial^2\Theta}{\partial x_3{}^2} = -\dot{Q} \tag{9.26}$$

Nye[7] has shown that by making the substitution

$$x_1 = \frac{k_1^{1/2}X_1}{(k_1k_2k_3)^{1/6}}\;; \qquad x_2 = \frac{k_2^{1/2}X_2}{(k_1k_2k_3)^{1/6}}\;; \qquad x_3 = \frac{k_3^{1/2}X_3}{(k_1k_2k_3)^{1/6}} \tag{9.27}$$

the desired transformation can be made. The volume element remains fixed during this transformation since

$$dX_1\,dX_2\,dX_3 = dx_1\,dx_2\,dx_3 \tag{9.28}$$

and hence the source strength remains unchanged. Introducing the transformation Eqs. (9.27) into the differential Eq. (9.26), there results

$$(k_1k_2k_3)^{1/3}\left(\frac{\partial^2\Theta}{\partial X_1{}^2} + \frac{\partial^2\Theta}{\partial X_2{}^2} + \frac{\partial^2\Theta}{\partial X_3{}^2}\right) = -\dot{Q} \tag{9.29}$$

[7] See Nye [5].

This, then, is the equation for the temperature distribution produced by a source $\dot{Q}$ in an isotropic medium of conductivity $(k_1 k_2 k_3)^{1/3}$.

Nye[8] gives a description of how to apply this process and an example of a heat flow from a point source.

9.3 Electrical Conductivity

The analysis of electrical conductivity in a crystal is formally equivalent to that for thermal conductivity. Ohm's law is generalized to the form

$$I_i = \sigma_{ij}E_j = -\sigma_{ij}\frac{\partial \Phi}{\partial x_j} \tag{9.30}$$

where I_i are the current densities, σ_{ij} the electrical conductivity tensor, E_j the electric fields, and Φ the electric potential. Alternatively, the electric-field components can be written in the form

$$E_i = \rho_{ij}I_j \tag{9.31}$$

where ρ_{ij} is the resistivity tensor, which is related to the conductivity tensor by the usual relation

$$\rho_{ij} = \frac{(-1)^{i+j}\Delta\sigma_{ij}}{|\sigma_{ij}|} \tag{9.32}$$

Numerical values of the principal resistivities are given by Table 24 for a number of crystals.

Table 24

ELECTRICAL RESISTIVITIES OF METALLIC CRYSTALS[a]

Crystal	System	ρ_1, ρ_2	ρ_3
Aluminum (Al)	cubic	2.72	2.72
Copper (Cu)	cubic	1.68	1.68
Gold (Au)	cubic	2.21	2.21
Lead (Pb)	cubic	20.7	20.7
Mercury (Hg)	cubic	95.4	95.4
Platinum (Pt)	cubic	10.5	10.5
Silver (Ag)	cubic	1.58	1.58
Tin (Sn)	tetragonal	9.9	14.3
Bismuth (Bi)	trigonal	109.0	138.0
Cadmium (Cd)	hexagonal	6.8	8.3
Zinc (Zn)	hexagonal	5.91	6.13

[a] Values of principal resistivities at 20° C (unit: 10^{-8} ohm-meter).

[8] *Ibid.*

The rate of Joule heating is expressed by the scalar product of the current densities by the fields—that is,

$$I_i E_i = \rho_{ik} I_i I_k = \rho I^2 \tag{9.33}$$

where I is the magnitude of the current density and ρ the resistivity in the direction of the current flow. Since the Joule heating divided by the absolute temperature Θ is the time rate of increase of entropy, it is evident that I_i and E_i satisfy the criterion of (9.4) for the generalized forces and flows.

RESUME CHAPTER 9

I. Nonreversible Thermodynamics

When a thermal gradient or an electric field is applied to a crystal or an electrical conductor, thermal energy or electric charges are transported from one position in the crystal to another. These transport properties cannot be described in terms of equilibrium thermodynamics. Steady-state conditions, for which the flows do not vary with time, can be realized, but the sources and sinks are undergoing continuous changes.

Although a number of phenomenological postulates were used in explaining these results, it was not until the Onsager theory of the thermodynamics of irreversible processes was proposed that a firm basis for these processes was obtained. This principle is expressed in terms of a series of forces such as temperature gradients and electric fields and corresponding flows such as heat flows and current flows. In general any force can give rise to any flow through a cross coupling coefficient L_{ik}. Onsager's principle states that

$$L_{ij} = L_{ji}$$

In case a magnetic field is involved, or any axial vector, the rule has to be generalized to

$$L_{ij}(H) = L_{ji}(-H)$$

This relation, which means that L_{ij} is the same function of H that L_{ji} is of $-H$, has physical meaning even when $i = j$, for then the diagonal terms are even functions of the magnetic flux whereas the nondiagonal terms are odd functions of the flux.

The necessary conditions for a series of forces and fluxes to satisfy Onsager's principle is that they determine the time rate of change of entropy $\delta\dot{\sigma}$ according to the relation

$$\delta\dot{\sigma} \propto \sum J_i X_i$$

For the only two systems considered in the present book—that is, electrical and thermal conductivities and their interactions—the two forces are

$$E_i = -\frac{\partial\Phi}{\partial x_i} \quad \text{and} \quad -\frac{\partial\Theta/\partial x_i}{\Theta}$$

while the fluxes are

$$I_i \quad \text{and} \quad h_i$$

where Φ is the electrical potential, Θ the absolute temperature, I_i the electrical current density, and h_i the heat flow.

The proof of Onsager's principle depends on statistical mechanics and results from the principle of microscopic reversibility, which states that if the velocities of all the particles are reversed simultaneously, the particles will retrace their former paths, reversing the entire series of configurations. When a magnetic field or flux or any axial vector is applied, it is necessary to reverse the sign of this vector as well as the sign of the velocity.

2. Thermal Conductivities and Resistivities

A generalized representation of Fourier's law of heat conduction, applicable to crystals, is

$$h_i = -k_{ij} \frac{\partial \Theta}{\partial x_j}$$

where k_{ij} are the terms of the thermal-conductivity tensor. Since h_i and $(\partial \Theta / \partial x_i)$ satisfy the conditions for fluxes and forces it follows that

$$k_{ij} = k_{ji}$$

Thermal resistivities r_{ij} are determined from the equation

$$\frac{\partial \Theta}{\partial x_i} = -r_{ij} h_j$$

where

$$r_{ij} = (-1)^{i+j} \Delta_{ij} / |k_{ij}|$$

where Δ_{ij} is the determinant obtained from the matrix $|k_{ij}|$ by suppressing the ith row and jth column.

Thermal conductivities are usually measured by determining the temperature differences across thin flat plates when known amounts of heat are flowing through the crystal. Thermal resistances are measured by using long, thin rods.

Poisson's equation, applicable to isotropic substances, can be generalized by using a set of coordinates X_1, X_2, X_3 defined in terms of the spatial coordinates by the equations

$$X_1 = \frac{(k_1 k_2 k_3)^{1/6}}{k_1^{1/2}} x_1 ; \qquad X_2 = \frac{(k_1 k_2 k_3)^{1/6}}{k_2^{1/2}} x_2 ; \qquad X_3 = \frac{(k_1 k_2 k_3)^{1/6}}{k_3^{1/2}}$$

These substitutions lead to a generalized Poisson's equation

$$(k_1 k_2 k_3)^{1/3} \left(\frac{\partial^2 \Theta}{\partial X_1^2} + \frac{\partial^2 \Theta}{\partial X_2^2} + \frac{\partial^2 \Theta}{\partial X_3^2} \right) = -\dot{Q}$$

which can be used to investigate heat flow in more complicated shapes than flat plates and long rods. k_1, k_2, and k_3 are the principal thermal conductivities of the conductivity ellipsoid.

3. Electrical Conductivities

The analysis of electrical conductivity in a crystal is similar to that for thermal conductivity. Ohm's law can be written as

$$I_i = \sigma_{ij} E_j = -\sigma_{ij} \frac{\partial \Phi}{\partial x_j}$$

where I_i are the electrical current densities, σ_{ij} the electrical conductivity tensor, and E_j the electric fields, which are the negative of the space gradients of the electrical potential Φ.

Alternatively, the electric field components can be written

$$E_i = \rho_{ij} I_j$$

where ρ_{ij} is the resistivity tensor related to the conductivity tensor by the usual relations

$$\rho_{ij} = \frac{(-1)^{i+j} \Delta \sigma_{ij}}{|\sigma_{ij}|}$$

where $\Delta \sigma_{ij}$ is the determinant obtained from the matrix σ_{ij} by suppressing the ith row and jth column. Both σ_{ij} and ρ_{ij} are symmetrical tensors by virtue of Onsager's principle.

The rate of Joule heating is the scalar product

$$I_i E_i = \rho_{ik} I_i I_k = \rho I^2$$

where I is the magnitude of the current density and ρ the resistivity in the direction of the current flow.

PROBLEMS CHAPTER 9

1. If the components of a conductivity tensor along three axes x_1, x_2, x_3 are

$$\sigma_{ij} = \begin{vmatrix} 20 \times 10^7 & 0 & 0 \\ 0 & 5 \times 10^7 & -2 \times 10^7 \\ 0 & -2 \times 10^7 & 4 \times 10^7 \end{vmatrix}$$

determine the components for a set of axes given by the direction cosines

	x_1	x_2	x_3
x_1'	$a_{11} = \cos \theta \cos \varphi;$	$a_{12} = \cos \theta \sin \varphi;$	$a_{13} = -\sin \theta$
x_2'	$a_{21} = -\sin \varphi;$	$a_{22} = \cos \varphi;$	$a_{23} = 0$
x_3'	$a_{31} = \cos \varphi \sin \theta;$	$a_{32} = \sin \varphi \sin \theta;$	$a_{33} = \cos \theta$

2. What values do the resistivity tensor ρ_{ij} have for the x_1, x_2, x_3 axes and the x_1', x_2', x_3' axes?

3. If an electric field of one volt per meter is established in the direction OP, having the direction cosines of problem 1, calculate the components of the field along the x_1, x_2, x_3 axes and hence determine the current densities I_1, I_2, and I_3.

4. The solution for heat flow in an arbitrarily shaped specimen for an isotropic medium is obtained from Laplace's equation

$$k \left(\frac{\partial^2 \Theta}{\partial x_1^2} + \frac{\partial^2 \Theta}{\partial x_2^2} + \frac{\partial^2 \Theta}{\partial x_3^2} \right) = -\dot{Q}$$

For a point source of heat flow, the solution is

$$\Theta = \frac{A}{r} + \Theta_\infty$$

where

$$r = \sqrt{x_1^2 + x_2^2 + x_3^2}$$

What is the temperature distribution for an anisotropic medium with the principal heat conductivities k_1, k_2, and k_3?

REFERENCES CHAPTER 9

1. L. Onsager, *Phys. Rev.* **37,** 405 (1931); *ibid.* **38,** 2265 (1931).
2. H. B. G. Casimir, *Rev. Mod. Phys.* **17,** 343 (1945).
3. J. Meixner, *Ann. Physik* **40,** 165 (1941).
4. S. R. DeGroot, "Thermodynamics of Irreversible Processes." North-Holland Publ., Amsterdam, 1951.
5. J. F. Nye, "Physical Properties of Crystals," pp. 200–204. Oxford Univ. Press (Clarendon), London and New York, 1957.

CHAPTER 10 • Piezoresistivity

10.1 Introduction: Energy Surfaces and Electron Flows

One of the most important problems in mechanics is the measurement of stress and strain. A photoelastic method that depends on the birefringence generated in an isotropic material by a strain has been discussed in Chapter 7. However, the most widely applied methods result in the conversion of a mechanical signal into an electric signal—that is, they employ some type of electromechanical transducer. For an alternating strain or a very short strain pulse, piezoelectric crystals and ceramics are often used. However, these will not work for static or slowly varying stresses, and some form of gage whose resistance varies as a function of strain is normally applied.

The metallic-wire strain gage has been employed for many years. The first publication that showed that metallic conductors exhibit a change in resistance when subjected to a strain is due to Lord Kelvin and dates back to 1856. Part of this effect is due to the change of dimensions, but another nearly equal part in most metal wires is caused by a change in the resistivity. This effect is called piezoresistivity. Studies and evaluations of the piezoresistive effects in metals and crystals, including silicon for hydrostatic pressures, were first reported by Bridgman[1] beginning in 1917. His extensive studies in this field span 35 years and provide an important contribution related to the development of the wire strain gage.

A much larger piezoresistive effect connected with linear longitudinal and shearing strains has been found in the semiconductors silicon and germanium. This increases the gage factor—that is, the ratio of the change in resistance to the original resistance times the strain—as shown in Eq. (10.1)

$$G = \frac{\Delta R}{R_0 S} \tag{10.1}$$

[1] See Bridgman [1].

from a value of 2 to 4 for a wire strain gage to a value of 175 at room temperature for a lightly doped p-silicon sample.

The effect was first discovered by C. S. Smith[2] when he was verifying the form of the energy surfaces for the semiconductors silicon and germanium. For n-type material the energy surfaces are of the multivalley type, for which electrons congregate around certain momentum values in certain directions

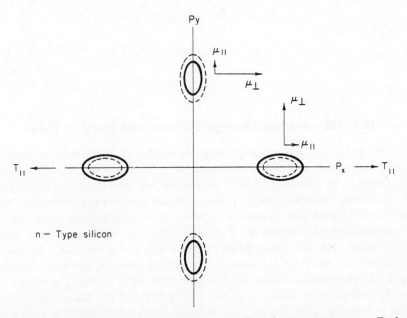

FIG. 10.1 Multivalley energy surfaces for n-type silicon in momentum space. Dashed lines show effect of a stress T_{11} on the relative positions of the energy surfaces. $\mu_\parallel$ and $\mu_\perp$ are the relative mobilities parallel and perpendicular to a valley.

in the crystal. For n-silicon the surfaces are as shown by Fig. 10.1. These surfaces, which lie along the cube axes, are elliptical in shape. As discussed in other references[3], this results in mobilities—that is, drift velocities divided by the applied electrical fields—that are larger in a perpendicular direction, $\mu_\perp$, than they are for the parallel direction, $\mu_\parallel$. The effect of a stress applied along a crystal axis is to raise the energy levels of the two valleys parallel to the direction of the stress $T_\parallel$ and to lower the energy levels at right angles to the direction of the applied stress. The dashed lines show the positions and sizes of the valleys for equal energies. Since the valleys along the stress have a higher energy, there is a flow of electrons from the

[2] See Smith [2].
[3] See Mason [3].

high energy levels to the low energy levels, which for equilibrium conditions results in more electrons in the low energy surfaces and less in the high energy surfaces. For an electric field applied in the same direction as the stress, this results in a decrease in resistivity over the nonstressed case, since the number of electrons increases for the valleys for which the mobility is larger in the applied-field direction. The time required to obtain equilibrium conditions has been shown[4] to be of the order of 10^{-12} seconds, so that such devices can be used for dynamic as well as static applications.

When the electric field is applied perpendicular to the directions of the applied stress it is readily seen that the resistivity increases rather than decreases, and calculations indicate that the effect is half as large as that parallel to the stress. A shearing stress in the plane of the crystallographic axes affects all the valleys in the same way and hence does not produce a change in resistance. From a phenomenological point of view, as discussed in the next section, these relations can be expressed in the form

$$\pi_{12} = \frac{-\pi_{11}}{2}; \qquad \pi_{44} = 0 \qquad (10.2)$$

For n-germanium the energy valleys lie along the $\langle 111 \rangle$ crystallographic direction. A longitudinal stress along a crystallographic axis affects all the wells equally, and hence

$$\pi_{11} = \pi_{12} = 0 \qquad (10.3)$$

A shearing strain in the plane of the crystallographic axes does separate these valleys, and hence π_{44} is large. As shown in the next section the largest effect for this type of symmetry is found for a longitudinal strain along a $\langle 111 \rangle$ axis with the electric field applied in the same direction.

In n-type material, the conduction is by electrons, which are supplied by the impurity atoms introduced into the material. These have to have a valence of $+5$ and are such materials as arsenic, antimony, and phosphorus. When the semiconductors are doped with materials such as boron, gallium, aluminum, or indium—all of which have a valence of $+3$—conduction occurs by an electron jumping from a complete bond to an uncompleted bond, resulting in a motion of the uncompleted bond, which is called a "hole." The hole acts like an electron having a positive charge. The energy surfaces for p-type material are much more complicated[5] and will not be discussed here. The resulting symmetry of the effect, however, is the same as that for n-type germanium, and the phenomenological constants are given by

$$\pi_{11} = \pi_{12} = 0; \qquad \pi_{44} \qquad (10.4)$$

[4] See Mason and Bateman [4].
[5] See Mason [3].

p-type silicon has the best sensitivity and the greatest mechanical strength, and this material is largely used in semiconductor strain gages.

When the *p*-type doping, usually boron, is less than 10^{23} atoms per meter³—which corresponds to one impurity atom in 500,000 silicon atoms—the equilibrium populations are determined by using Boltzmann statistics. In this region the gage factor varies inversely as the absolute temperature, as indicated by the top curve of Fig. 10.2, which is for boron-doped silicon.

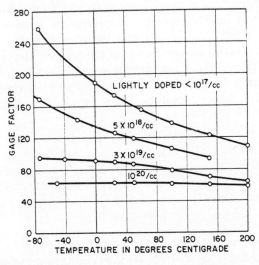

FIG. 10.2   Effect of doping levels on the value and temperature variation of the gage factor of *p*-type silicon.

As the doping increases, the holes get closer together and the type of statistics changes from the classical Boltzmann statistics to the quantum mechanical Fermi-Dirac statistics. One result of this is that the gage factor no longer varies inversely as the temperature but varies less rapidly. For a doping of 10^{26} atoms per meter³—that is, one impurity atom for 500 silicon atoms—the so-called degeneracy temperature Θ_D, which separates the region of quantum statistics from the region of Boltzmann statistics, has risen to approximately 900° K. Over the temperature range from 0° K to 500° K, the gage factor remains constant to about ±3 per cent. This is obtained by sacrificing some sensitivity, as shown in Fig. 10.2. The resistivities for the most highly doped samples, plotted in ohm-centimeter—which is 100 times the mks unit the ohm-meter—are shown by Fig. 10.3. Since the gage factor is decreasing slightly while the resistivity is increasing slightly, the change in resistance ΔR—which determines the voltage unbalance for the Wheatstone bridge usually used in such measurements—is much more independent of the temperature than is either the resistivity or the gage factor.

Another advantage of the heavily doped p-silicon strain gage is that the response remains linear for much higher values of strain than it does in the

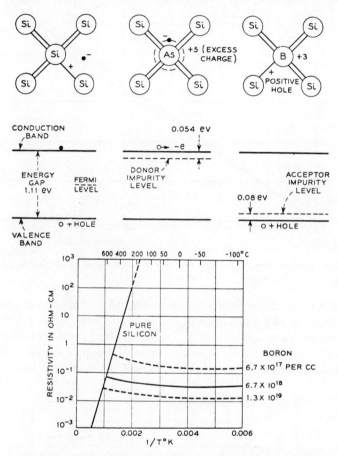

FIG. 10.3 Resistivity as a function of temperature for three different dopings of p-type silicon.

lightly doped specimen. Figure 10.4 shows the change in resistance divided by the resistance, plotted as a function of the strain in microinches per inch for three temperatures. The response remains linear out to a strain of 2,500 microinches—that is, a strain of 2.5×10^{-3}—before it starts to become nonlinear.

The maximum strain that a gage will take depends on the breaking strength of the silicon crystal. This strength has been investigated by Pearson, Read,

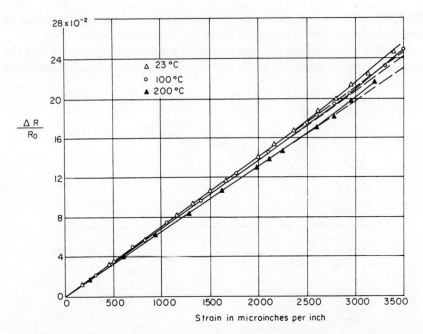

FIG. 10.4   Change in resistance as a function of strain for p-type silicon doped with 10^{26} boron atoms per cubic meter.

and Feldman[6] with the results shown in Fig. 10.5. These results hold both for naturally grown whiskers and for rods cut from larger crystals. For square rods 6 mil in. square and larger, the breaking stress is quite uniformly 3.5×10^8 newtons/meter^2. Since the value of Young's modulus along the $\langle 111 \rangle$ direction is 1.87×10^{11} newtons/meter^2, the maximum strain possible

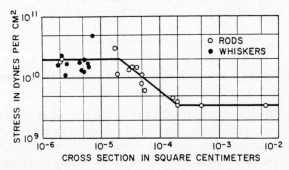

FIG. 10.5   Effect of size on room-temperature fracture stress for silicon rods (after Pearson, Read, and Feldman).

[6] See Pearson et al. [5].

is 1.87×10^{-3} or 1,870 microin./in. If, however, the dimension is reduced to 2 mil in. or less, the average strength increases to 2×10^9 newtons/meter², and the minimum strength to 10^9 newtons/meter². It is found that if one dimension satisfies this criterion, the breaking strength still holds. Hence, most semiconductor strain gages are on the order of 1 mil in. thick and may be 10 to 15 mil in. wide, which, for the highest doping value, gives a resistance around 15 ohms. The maximum strain that they will stand is at least 5,000 microin./in., with an average value of around 10,000 microin.—that is, a strain of 1 per cent. Since the small dimension allows a considerable bending around a small radius before strains of this magnitude are obtained, the gage is very flexible and can be used on curved surfaces.

10.2 Phenomenological Development of Piezoresistivity

In order to determine how strain gages respond under a wide variety of strain conditions, it is convenient to develop phenomenological relations between the strains and the changes in resistivities. Such a development has been given in a previous publication,[7] and is reproduced here. It is assumed that the electric-field components are functions of the current-density components I_j and the stress components T_{kl}. When E_i is developed in a Maclaurin series about the state of zero current and stress, there results

$$dE_i = \frac{\partial E_i}{\partial I_j} dI_j + \frac{\partial E_i}{\partial T_{kl}} dT_{kl}$$
$$+ \frac{1}{2!}\left[\frac{\partial^2 E_i}{\partial I_j \, \partial I_m} dI_j \, dI_m + \frac{\partial^2 E_i}{\partial T_{kl} \, \partial T_{no}} dT_{kl} \, dT_{no} + 2 \frac{\partial^2 E_i}{\partial I_j \, \partial T_{kl}} dI_j \, dT_{kl} \right]$$

$$(10.5)$$

For all crystal classes having a center of symmetry—which covers the case of the most used semiconductors silicon and germanium—all odd-rank tensors disappear. This eliminates the second, third, and fourth terms. For crystal classes without a center of symmetry the odd-rank tensors partake of the nature of piezoelectricity—that is, the second and fourth terms— whereas the third term indicates a correction to the resistivity at high current densities. Hence, as far as the piezoresistance effect is concerned we can define the remaining derivatives as

$$\frac{\partial E_i}{\partial I_j} = \rho_{ij} = \text{resistivity components}$$

$$(10.6)$$

$$\frac{\partial^2 E_i}{\partial I_j \, \partial T_{kl}} = \rho_0 \pi_{ijkl} = \text{piezoresistance components}$$

[7] See Mason and Thurston [6].

Replacing the differential components by the variables themselves, Eq. (10.5) becomes

$$E_i = \rho_{ij}I_j + \rho_0\pi_{ijkl}I_jT_{kl} \tag{10.7}$$

For the piezoresistive effect, the subscripts i and j are interchangeable, as are also k and l. However, i and j are not interchangeable with k and l, and hence the symmetry is type M in Table B.4.

As seen from Tables B.2 and B.4 for the symmetry of silicon and germanium ($m3m$), the resistivity ρ_0 is independent of direction while the π_{ijkl} constants expressed in terms of two-index symbols, are

$$\pi_{11}, \qquad \pi_{12}, \qquad \pi_{44} \tag{10.8}$$

Hence the form of the piezoresistive equations is given by

$$\frac{E_1}{\rho_0} = I_1[1 + \pi_{11}T_1 + \pi_{12}(T_2 + T_3)] + \pi_{44}(I_2T_6 + I_3T_5)$$

$$\frac{E_2}{\rho_0} = I_2[1 + \pi_{11}T_2 + \pi_{12}(T_1 + T_3)] + \pi_{44}(I_1T_6 + I_3T_4) \tag{10.9}$$

$$\frac{E_3}{\rho_0} = I_3[1 + \pi_{11}T_3 + \pi_{12}(T_1 + T_2)] + \pi_{44}(I_1T_5 + I_2T_4)$$

The effects of various types of stresses can be calculated directly from these equations.

The effect of a pure hydrostatic pressure is obtained by setting

$$T_1 = T_2 = T_3 = -p \tag{10.10}$$

For this case the Equations. (10.9) reduce to the vector equations

$$\frac{E_i}{\rho_0} = [1 - (\pi_{11} + 2\pi_{12})p]I_i \tag{10.11}$$

Since it was shown that $\pi_{12} = -\pi_{11}/2$ for the multivalley model (10.2), the effect of a hydrostatic pressure cancels out for linear terms. From the model it has been shown[8] that a hydrostatic pressure affects the resistivity only in proportion to the square of the pressure.

To determine the effect of a stress applied in an arbitrary direction, it is necessary to transform Eqs. (10.7) or (10.9) to a new set of axes x', y', z' related to the original axes x, y, and z by the direction cosines

$$
\begin{array}{c|ccc}
 & x & y & z \\
\hline
x' & l_1 & m_1 & n_1 \\
y' & l_2 & m_2 & n_2 \\
z' & l_3 & m_3 & n_3 \\
\end{array}
\tag{10.12}
$$

[8] See Mason [3].

For the usual case, for which longitudinal stress T'_{33} is applied along the length of the strain gage with the current flowing in the same direction, we need to consider the direction cosines of the new set of axes with respect to the old set. Using the angles θ and φ of Fig. 10.6, which determine

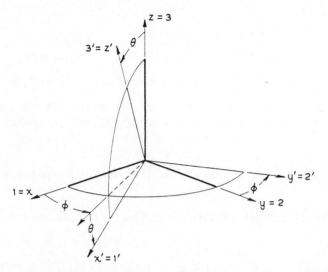

FIG. 10.6 Axes for relating a rotated crystal with respect to the crystallographic axes.

respectively the angle of z' from z and the angle of the z', z plane from the x, z plane, the direction cosines l_1 to n_3 are

	x	y	z
x'	$l_1 = \cos\theta\cos\varphi;$	$m_1 = \cos\theta\sin\varphi;$	$n_1 = -\sin\theta$
y'	$l_2 = -\sin\varphi;$	$m_2 = \cos\varphi;$	$n_2 = 0$
z'	$l_3 = \cos\varphi\sin\theta;$	$m_3 = \sin\varphi\sin\theta;$	$n_3 = \cos\theta$

$$(10.13)$$

For calibration purposes the stress T'_{33} is usually obtained by putting a weight on the end of the gage and determining the change in resistance ΔR. Hence T'_{33} is the only stress in the rotated system of axes and I_3' is the only current density. The stresses in the old system in terms of T'_{33} and the currents in terms of I_3' are given by the transformation equations (2.14) and (2.15) for first- and second-rank tensors respectively. Hence

$$I_i = \frac{\partial x_i}{\partial x_3'} I_3'; \qquad T_{kl} = \frac{\partial x_k}{\partial x_3'}\frac{\partial x_l}{\partial x_3'} T'_{33} \qquad (10.14)$$

Expanding these equations noting that

$$\frac{\partial x_1}{\partial x_3'} = l_3 = \cos \varphi \sin \theta; \qquad \frac{\partial x_2}{\partial x_3'} = m_3 = \sin \varphi \sin \theta; \qquad \frac{\partial x_3}{\partial x_3'} = n_3 = \cos \theta$$

$$(10.15)$$

we have

$$I_1 = l_3 I_3'; \qquad I_2 = m_3 I_3'; \qquad I_3 = n_3 I_3'$$

$$T_{11} = l_3{}^2 T_{33}'; \qquad T_{22} = m_3{}^2 T_{33}'; \qquad T_{33} = n_3{}^2 T_{33}' \qquad (10.16)$$

$$T_{12} = T_6 = l_3 m_3 T_{33}'; \qquad T_{13} = T_5 = l_3 n_3 T_{33}'; \qquad T_{23} = T_4 = m_3 n_3 T_{33}'$$

Inserting these values back in Eq. (10.9), the three fields in the old system of axes in terms of the new current density I_3' and the applied stress T_{33}' become

$$\frac{E_1}{\rho_0} = l_3 I_3'[1 + [\pi_{11} l_3{}^2 + \pi_{12}(m_3{}^2 + n_3{}^2)]T_{33}'] + I_3' T_{33}'[m_3{}^2 l_3 + n_3{}^2 l_3]\pi_{44}$$

$$\frac{E_2}{\rho_0} = m_3 I_3'[1 + [\pi_{11} m_3{}^2 + \pi_{12}(l_3{}^2 + n_3{}^2)]T_{33}'] + I_3' T_{33}'[l_3{}^2 m_3 + n_3{}^2 l_3]\pi_{44}$$

$$(10.17)$$

$$\frac{E_3}{\rho_0} = n_3 I_3'[1 + [\pi_{11} n_3{}^2 + \pi_{12}(l_3{}^2 + m_3{}^2)]T_{33}'] + I_3' T_{33}'[l_3{}^2 n_3 + m_3{}^2 n_3]\pi_{44}$$

Now from the equation

$$E_3' = \frac{\partial x_3'}{\partial x_i} E_i = [l_3 E_1 + m_3 E_2 + n_3 E_3] \qquad (10.18)$$

and noting that $(l_3{}^2 + m_3{}^2 + n_3{}^2) = 1$, we have

$$\frac{E_3'}{\rho I_3'} = 1 + \pi_{11}(l_3{}^4 + m_3{}^4 + n_3{}^4)T_{33}' + 2(\pi_{12} + \pi_{44})(l_3{}^2 m_3{}^2 + l_3{}^2 n_3{}^2 + m_3{}^2 n_3{}^2)T_{33}'$$

$$= 1 + T_{33}'[\pi_{11} + 2(\pi_{44} + \pi_{12} - \pi_{11})(l_3{}^2 m_3{}^2 + l_3{}^2 n_3{}^2 + m_3{}^2 n_3{}^2)] \quad (10.19)$$

$$= 1 + \pi_l T_{33}'$$

where π_l is the longitudinal piezoresistive constant valid for any direction. Inserting the values of the direction cosines from (10.15) into (10.19), the value of π_l becomes

$$\pi_l = \pi_{11} + 2(\pi_{44} + \pi_{12} - \pi_{11})(\sin^2 \theta \cos^2 \theta + \sin^4 \theta \sin^2 \varphi \cos^2 \varphi) \quad (10.20)$$

Analysis shows that this expression for the cosines is a maximum when $\cos^2 \varphi = \sin^2 \varphi = \frac{1}{2}$; $\sin^2 \theta = \frac{2}{3}$, $\cos^2 \theta = \frac{1}{3}$. This corresponds to a value of $\theta = 54° 40'$, $\varphi = 45°$, which is the direction of a $\langle 111 \rangle$ axis; the longitudinal components π_l are respectively

$$\pi_l = \pi_{11} ; \quad \langle 100 \rangle; \qquad \pi_l = \frac{\pi_{11} + \pi_{12} + \pi_{44}}{2} ; \quad \langle 110 \rangle;$$

$$\pi_l = \frac{\pi_{11} + 2(\pi_{12} + \pi_{44})}{3} ; \quad \langle 111 \rangle \tag{10.21}$$

Longitudinal piezoresistive measurements can determine directly π_{11} and the sum $(\pi_{12} + \pi_{44})$. Hence, unless the assumption is made that $\pi_{12} = -\pi_{11}/2$—which is approximately true—then one measurement has to be made for which the voltage measured is at right angles to the current. From Eq. (10.9) the constant π_{12} can be directly measured when the voltage E_2 and current I_2 are sent at right angles to the direction of the stress T_{11}. Table 25[9] gives values of the piezoresistive constants at $300°$ K ($27°$ C), measured in terms of 10^{-11} newtons/meter² for a number of semiconductor crystals. Of these silicon and germanium are the only two that have received application for strain transducers. For hydrostatic pressures, however, such crystals as indium antimonide (InSb) have interesting values. The values quoted are the three primary constants π_{11}, π_{12}, and π_{44}, the hydrostatic pressure constant $\pi_p = -(\pi_{11} + 2\pi_{12})$, and the value of π_l, which determines in Eq. (10.19) the longitudinal response for any direction. In determining the gage factor and the measured strain, it is of course the response to strain that is of interest. For a long, thin strain gage, such as used in practice, the strain is the stress divided by the Young's modulus Y_0. The column headed Y_m gives the value of the Young's modulus in 10^{11} newtons/meter² in the direction of the largest piezoresistive constant. Since the stress is Y_m times the strain, the response is determined by the product of π_l times Y_m. The resulting value m_l is shown by the next-to-last column. The value m_l determines the resistivity component of the gage factor G.

10.3 Best Orientations for Measuring Strains

When a semiconductor or wire strain gage is cemented to a sample whose strain is to be measured, it is found that the length direction has the same strain as in the sample. The width dimension, being much smaller than the length, will usually have a sidewise strain that is somewhat smaller than the strain in the sample. This follows, since the stiffness of the cement is usually smaller than that of the wire or semiconductor and hence the cement does not communicate the strain of the specimen to the strain gage. Some sidewise strain may be communicated to the sample, however, and hence it

[9] Taken from Thurston [7].

Table 25

PIEZORESISTANCE, VOLUME COMPLIANCE, AND MAXIMUM YOUNG'S MODULUS FOR SEMICONDUCTORS[a]

Material	ρ(ohm-m)	Concentration atoms/m³	π_{11}	π_{12}	π_{44}	π_p	π_i	$1/B$	Y_m	m_i	m_p
n-Ge	990.0	1.5×10^{20}	−4.7	−5.0	−137.9	+14.7	−96.7	1.333	1.540	−149.0 ⟨111⟩	+11.0
n-Ge	150.0	10^{21}	−2.3	−3.2	−138.1	+8.7	−95.0	1.333	1.540	−146.0 ⟨111⟩	+6.5
p-Ge	110.0	3×10^{21}	−3.7	+3.2	+96.7	−2.7	+65.4	1.333	1.540	+110.0 ⟨111⟩	−2.0
n-Si	1170.0	5×10^{20}	−102.2	+53.4	−13.6	−4.6	−102.0	1.02	1.30	−135.0 ⟨100⟩	−4.5
p-Si	780.0	2×10^{21}	+6.6	−1.1	+138.3	−4.4	+93.5	1.02	1.875	+175 ⟨111⟩	−4.3
p-Si	2.2	5×10^{24}	—	—	—	—	+66.6	1.02	1.875	+125 ⟨111⟩	—
p-Si	0.1	10^{26}	—	—	—	—	+33.4	1.02	1.875	+62.5 ⟨111⟩	—
n-PbTe	—	1 to 3×10^{24}	+20.0	+25.0	−107.0	−70.0	−48.0	2.40	0.356	−17.0 ⟨111⟩	−29.0
p-PbTe	—	$\sim 3 \times 10^{24}$	+24.0	+15.0	+215.0	−54.0	+161.0	2.40	0.356	+57.5 ⟨111⟩	−22.5
p-PbTe	—	1 to 3×10^{24}	+35.0	+40.0	+185.0	−115.0	+162.0	2.40	0.356	+57.5 ⟨111⟩	−48.0
p-PbSe	—	1 to 3×10^{24}	+24.0	+19.0	+57.0	−62.0	+58.5	—	—	—	—
n-PbS	—	1 to 3×10^{24}	+11.6	+6.6	−11.2	−24.7	+2.4	1.645	0.99	+2.4 ⟨100⟩	−15.0
n-PbS	—	6 to 9×10^{24}	+6.9	+2.9	−1.2	−12.8	+3.4	1.645	0.99	+3.4 ⟨100⟩	−7.8
n-InSb	—	—	+81.6	+114.2	83.0	+310.0	−81.6	2.13	0.414	−33.8 ⟨100⟩	+145.0
p-InSb	54.0	3×10^{21}	−70.0	−115.0	−10.0	+300.0	−70.0	2.13	0.414	−29.0 ⟨100⟩	+141.0

[a] B = Bulk modulus; Y_m = value of Young's modulus along most sensitive piezoresistive axis. Units of piezoresistance and volume compliance are in 10^{-11} m²/N; Young's modulus is in 10^{11} N/m².

is desirable to determine the effect of this strain in addition to the effect of the lengthwise strain. This requires writing the piezoresistive equations in terms of the strains rather than the stresses.

Since the stresses can be written in the form

$$T_{kl} = c_{mnkl}S_{mn} \tag{10.22}$$

we can replace the stresses T_{kl} in (10.7) and obtain

$$E_i = \rho_{ij}I_j + (\pi_{ijkl}c_{mnkl})I_jS_{mn} \tag{10.23}$$

The term in the parenthesis is an eight-rank tensor reduced twice and hence is a fourth-rank tensor

$$m_{ijmn} = \pi_{ijkl}c_{mnkl} \tag{10.24}$$

For the cubic symmetry of silicon and germanium and in terms of two-index symbols, Eq. (10.24) becomes

$$\begin{aligned}
m_{11} &= \pi_{11}c_{11} + 2\pi_{12}c_{12}\ ; \\
m_{12} &= \pi_{11}c_{12} + \pi_{12}(c_{11} + c_{12}); \\
m_{44} &= \pi_{44}c_{44}
\end{aligned} \tag{10.25}$$

where $m_{44} = m_{2323}/2$. Since the engineering shear strains are twice the tensor shear strains, the complete voltage-current strain relations can be written in the form for cubic crystals

$$\frac{E_1}{\rho_0} = I_1[1 + m_{11}S_1 + m_{12}(S_2 + S_3)] + m_{44}(I_2S_6 + I_3S_5)$$

$$\frac{E_2}{\rho_0} = I_2[1 + m_{11}S_2 + m_{12}(S_1 + S_3)] + m_{44}(I_1S_6 + I_3S_4) \tag{10.26}$$

$$\frac{E_3}{\rho_0} = I_3[1 + m_{11}S_3 + m_{12}(S_1 + S_2)] + m_{44}(I_2S_4 + I_1S_5)$$

For a hydrostatic pressure, since the volume change Δ is

$$\Delta = S_1 + S_2 + S_3 = -\frac{p}{B} \tag{10.27}$$

where B is the bulk modulus, Eq. (10.11) can be written in the form

$$\frac{E_i}{\rho_0} = [1 + m_p\Delta]I_i \tag{10.28}$$

where $m_p = (\pi_{11} + 2\pi_{12})/B$. The last column gives the values for the semiconductors listed, and it is seen that indium antimonide (InSb) is the most sensitive. However, the values vary rapidly with temperature and concentration.[10]

[10] See Tuzzolino [8].

When a semiconductor strain gage is rigidly attached to a sample whose strain is to be measured, it is necessary to give consideration to the effect of the strains at right angles to the crystal length as well as the shearing strain transmitted to the crystal. For the most sensitive orientations—the $\langle 100 \rangle$ direction for n-type silicon and the $\langle 111 \rangle$ direction for n- and p-type germanium and p-type silicon—the effect of sidewise strains can be appreciable. This is seen directly for n-silicon from Eqs. (10.26). For example, if the gage lies along the $3 = \langle 001 \rangle$ axis, it responds to a strain along this axis in proportion to the constant m_{11}. For strains at right angles, the response is proportional to m_{12}. Since m_{12} is about half as large as m_{11} and of opposite sign, an appreciable effect may result. There is no response for a shearing strain S_5 in the x, z plane.

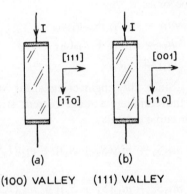

(a)   (b)

(100) VALLEY   (111) VALLEY

FIG. 10.7   Stress gages insensitive to transverse strain (after Thurston and Pfann).

There exist,[11] however, orientations for which sidewise strains and shearing strains have no effect on the resistance of the gage. The orientations for the $\langle 100 \rangle$ valley system—n-type silicon—and the $\langle 111 \rangle$ valley system—n- and p-germanium and p-silicon—are shown by Fig. 10.7.

To show that this result is correct it is necessary to transform the Eqs. (10.26) to the system of axes shown by Fig. A.2 of Appendix A. The z' axis represents the length of the crystal and the direction of current flow. The face of the specimen is taken as the x', z' plane, so that the face of the crystal with respect to the crystallographic axes is determined by x' and z'. The transformation equations for the current are given by the first of Eqs. (10.16). Since the strains are the engineering strains, account has to be taken of the fact that the shearing strains are twice the tensor shearing strains. If we

11 See Pfann and Thurston [9].

retain the strains S_1', S_3', and S_5' in the rotated system, the strains for the crystallographic directions can be written in the form

$$
\begin{aligned}
S_1 &= l_1^2 S_1' + l_3^2 S_3' + l_1 l_3 S_5' \\
S_2 &= m_1^2 S_1' + m_3^2 S_3' + m_1 m_3 S_5' \\
S_3 &= n_1^2 S_1' + n_3^2 S_3' + n_1 n_2 S_5' \\
S_4 &= 2 m_1 n_1 S_1' + 2 m_3 n_3 S_3' + (m_1 n_3 + n_1 m_3) S_5' \\
S_5 &= 2 l_1 n_1 S_1' + 2 l_3 n_3 S_3' + (l_1 m_3 + m_1 l_3) S_5' \\
S_6 &= 2 l_1 m_1 S_1' + 2 l_3 m_3 S_3' + (l_1 m_3 + m_1 l_3) S_5'
\end{aligned}
\tag{10.29}
$$

If we introduce these equations in (10.26) and combine the three fields in (10.18) we have, after some reduction and combining of terms,

$$
\begin{aligned}
\frac{E_3'}{\rho_o I_3'} = 1 &+ [m_{12} + (m_{11} - m_{12} - 2 m_{44})(l_1^2 l_3^2 + m_1^2 m_3^2 + n_1^2 n_3^2) \\
&+ 2 m_{44}(l_1 l_3 + m_1 m_3 + n_1 n_3)^2] S_1' \\
&+ [m_{11} + 2(-m_{11} + m_{12} + 2 m_{44})(l_3^2 m_3^2 + l_3^2 n_3^2 + m_3^2 n_3^2)] S_3' \\
&+ [(m_{11} - m_{12} - 2 m_{44})(l_1 l_3^3 + m_1 m_3^3 + n_1 n_3^3) \\
&+ (m_{12} + 2 m_{44})(l_1 l_3 + m_1 m_3 + n_1 n_3)] S_5'
\end{aligned}
\tag{10.30}
$$

For a strain along the length of the sample—that is, S_3'—the response will be a maximum for n-silicon when the axis lies along one of the crystallographic axes and will be equal to m_{11}. The sidewise stress S_1' produces the component m_{12}, while a shearing strain produces no effect.

In p- and n-germanium and p-silicon, the maximum effect is produced along the $\langle 111 \rangle$ axis, for which $l_3 = m_3 = n_3 = 1/\sqrt{3}$. The response to the strain S_3' is

$$
\frac{m_{11} + 2 m_{12}}{3} + \frac{4}{3} m_{44}
\tag{10.31}
$$

For a multivalley energy surface $m_{11} + 2 m_{12} = 0$ and in actual measurements the result is very small, so that the response is given by $(4/3) m_{44}$. In a direction perpendicular to the length the last term in the expression multiplying S_1' in (10.30) vanishes. This follows from Eq. (2.56), second line, which is the product of the direction cosines. Since the direction cosines $l_3 = m_3 = n_3 = 1/\sqrt{3}$, it is readily seen that the first term in the shear response can be written

$$
\tfrac{1}{3}(l_1 l_3 + m_1 m_3 + n_1 n_3)
\tag{10.32}
$$

and hence the response to a shearing stress vanishes. The complete relation for this orientation is then

$$\frac{E_3'}{\rho_0 I_3'} = 1 + \left(\frac{m_{11} + 2m_{12}}{3} - \frac{2}{3}m_{44}\right)S_1' + \left(\frac{m_{11} + 2m_{12}}{3} + \frac{4}{3}m_{44}\right)S_3' \quad (10.33)$$

Hence the sidewise response is about half as large as the lengthwise response, provided that the strain in the specimen is transmitted to the strain gage.

For the two orientations of Fig. 10.7 it is readily seen from the multivalley energy surface of Fig. 10.1 that an extensional strain along a $\langle 111 \rangle$ axis or a shear strain in the $\langle 100 \rangle$, $\langle 111 \rangle$ plane will not produce any response. The same is true for the energy minimum lying along $\langle 111 \rangle$ directions when a longitudinal strain is applied along the $\langle 001 \rangle$ direction and the shearing strain is applied in the $\langle 110 \rangle$, $\langle 001 \rangle$ plane. Phenomenologically, these two cases can be obtained by putting the two sets of direction cosines given by (10.34) into (10.30).

$$\langle 100 \rangle \text{ valley;} \quad l_1 = m_1 = n_1 = \frac{1}{\sqrt{3}}; \quad l_3 = \frac{1}{\sqrt{2}}; \quad m_3 = -\frac{1}{\sqrt{2}}; \quad n_3 = 0$$

$$\quad (10.34)$$

$$\langle 111 \rangle \text{ valley;} \quad l_1 = 0; \quad m_1 = 0; \quad n_1 = 1.0; \quad l_3 = m_3 = \frac{1}{\sqrt{2}}; \quad n_3 = 0$$

These two orientations result in the response equation given by Eqs. (10.35)

$$\langle 100 \rangle; \quad \frac{E_3'}{\rho_0 I_3'} = 1 + \left[\frac{m_{11} + 2m_{12} - 2m_{44}}{4}\right]S_1' + \left[\frac{m_{11} + m_{12} + m_{44}}{2}\right]S_3'$$

$$\quad (10.35)$$

$$\langle 111 \rangle; \quad \frac{E_3'}{\rho_0 I_3'} = 1 + m_{12}S_1' + \left(\frac{m_{11} + m_{12} + m_{44}}{2}\right)S_3'$$

Since, on the multivalley model, the relations are

$$\langle 100 \rangle; \quad m_{11} + 2m_{12} = m_{44} = 0; \quad \langle 111 \rangle; \quad m_{11} = m_{12} = 0 \quad (10.36)$$

it is seen that the cross strain S_1' produces no response. Actually the measurements of Table 25 show that the sidewise response is small. The $\langle 111 \rangle$ valley gage is much more advantageous, since the response to the strain S_3' drops from $(4/3)m_{44}$ to m_{44}, whereas for the $\langle 100 \rangle$ valley gage, the response drops from m_{11} to $m_{11}/4$.

Much more complicated arrangements can also be found,[12] such as gages that respond to the sum of the principal stresses. Also, the transverse effect and the shearing effect can be used. In the transverse effect the resistivity

[12] See Pfann and Thurston [9].

at right angles to the applied stress can be used in strain measurements. This effect can be seen from Eqs. (10.26), for if we measure the resistivity E_1/I_1, it changes in proportion to the constant m_{12} when a strain is employed that is at right angles to the gage length. The constant m_{44} can be employed in measuring shearing strains. As shown by Mason and Thurston,[13] a gage for measuring torque can be obtained by cutting a cylinder of n- or p-germanium or p-silicon along the crystal axis. By sending reverse currents along the edges of the crystal, a torque produces a voltage at right angles to the current direction. By combining two longitudinal piezoresistive measurements and a transverse measurement, it has been shown by Pfann and Thurston[14] that all the principal strains can be measured by the use of a single crystal. These types have not come into general use, however, and the reader is referred to the original articles[15] for the details.

RESUME CHAPTER 10

I. Introduction

One of the most important problems in mechanics is the measurement of stress and strain. For static or slowly varying stresses, wire strain gages have been most widely used. These suffer from the defect of being insensitive, thus requiring an amplifier to measure medium strains. Recently, certain semiconductors, namely silicon and germanium, have been shown to have a much larger change in resistance with strain than do metallic wires—larger by a factor of nearly 100. By using such materials, it is possible to measure medium strains without the use of amplifiers, or much smaller strains if amplifiers are used.

The large increase in resistance is due to the piezoresistance effect, which represents a change of resistance of the semiconductor. This effect can be discussed on the basis of semiconductor theory or on the basis of a phenomenological development.

2. Semiconductor Theory

Silicon is a covalent crystal of valence four with each silicon being bound to four other silicons by exchange of electrons with them. The covalent bond is a strong bond, and it requires an energy of about 1.11 eV to break this bond. When a bond is broken, an electron is freed to go into the conduction band and the broken bond—known as a hole—is free to move around the crystal. The hole has a positive charge and moves in a direction opposite to the electron when an electric field is put on the crystal. If all the holes and electrons are generated by thermal agitation, the resistivity follows an exponential law when plotted against $1/T$, as shown by the line of Fig. 10.3 labelled "pure silicon." The conductivity is known as intrinsic.

When atoms such as arsenic, with a valence of $+5$, are incorporated in the crystal, all the bonds are satisfied and there is one electron left over. At low temperatures the extra electron is more stable if it follows a rather large orbit around the arsenic atom that

---

[13] See Mason and Thurston [6].
[14] See Pfann and Thurston [9].
[15] See Mason and Thurston [6], Thurston [7], and Pfann and Thurston [9].

substitutionally replaces the silicon atom. This is indicated by the impurity donor level 0.054 eV below the conduction band. As the temperature increases, more electrons are freed and can traverse the conduction bands. If the number of arsenic atoms becomes large, they get close enough together so that the electron cannot follow any specific path around the atom and the impurity level raises to the conduction band. Under these conditions the resistivity becomes nearly independent of the temperature, as shown for boron by the lower curves of Fig. 10.3.

If an element such as boron with a valence of +3 is used as the doping agent, three bonds are satisfied with neighboring silicon atoms, but the fourth is not completed. This fourth bond acts as a positive electron—that is, a hole—and an electric field can cause a conduction current. Here again the hole is more stable if it traverses an orbit around the boron atom. This is indicated by the acceptor level of 0.08 eV above the valence band. High doping results in a resistivity nearly independent of the temperature until thermal agitation cuts down the mean scattering time between collisions.

The conductivity of such a semiconductor satisfies an equation of the type

$$\sigma = N_e e \mu_e + N_h e \mu_h$$

where N_e is the number of electrons per cubic meter, e the electronic charge, and μ_e the mobility in meters^2/volt-sec. The mobility is the average velocity the electron acquires when an electric field of one volt per meter is applied to the specimen. N_h is the number of holes, e the positive charge of the hole, and μ_h the mobility of the hole. Since both the hole charge and the hole mobility are of opposite sign to the electronic charge and mobility, the two currents add in the conductivity equation. The mobility is determined by the product of the ratio of charge to mass times the average time τ between collisions of the electron or hole with thermal waves (phonons) or impurity atoms. For a lightly doped germanium, $\tau = 2.5 \times 10^{-13}$ seconds so that the mobility

$$\mu = \frac{e}{m^*}\tau = \frac{1.6 \times 10^{-19}}{1.1 \times 10^{-31}} \times 2.5 \times 10^{-13} \doteq 0.365 \text{ meters}^2/\text{volt-sec}.$$

m^* is the effective mass, which for germanium is 0.12 times the electron mass, 9.1×10^{-31} kgm. For silicon the effective mass is $0.265 m_0$ and $\mu \doteq 0.16$ meters^2/volt-sec. For certain III-V semiconductors, such as indium antimonide, the effective mass is only $0.006 m_0$ and the mobility is as high as 75 meters^2/volt-sec.

The effective mass for n-type materials depends on the curvature of the conduction energy band near the bottom of the band. It has been shown that the effective mass is

$$m^* = \frac{\hbar^2}{d^2 W/dk^2}$$

where $\hbar$ is Planck's constant divided by 2π and the denominator is the second derivative of the energy surface by the wave vector $k = (\omega/V)$.

The lines of Fig. 10.3 indicate that the energy surfaces are spherical, in which case effective mass is equal to the electron mass m_0. Actually, however, the surface for n-type silicon has minima along the six cube axes. The solid lines of Fig. 10.1 show contours of constant energy plotted against the momenta, which are proportional to the value of the wave vector k. Since the effective mass is inversely proportional to the curvature of the energy surface, it is readily seen that the effective mass will be smaller for a direction perpendicular to the "valley" since the curvature is larger. The result is that the mobility $\mu_\perp$ is about five times larger than $\mu_\parallel$ for silicon. For germanium the "valleys" lie along $\langle 111 \rangle$ directions and the relative mobilities are $\mu_\perp = 19\mu_\parallel$.

The origin of the piezoresistive effect is that a stress parallel to a valley—T_{11} for n-silicon —raises the energy minima along the direction of the stress and lowers the energy of the four valleys in a direction perpendicular to the stress. The relative energy contours are shown by the dashed lines. Since the four perpendicular "valleys" have the lower energy, electrons tend to flow from the higher energy levels to the lower and the relative populations. as given by Boltzmann's principle, are

$$\frac{N_{010}}{N_{(100)}} = \exp\frac{(\alpha_1 + \alpha_2)T_{11}}{RT}$$

where $\alpha_1 T_{11}$ is the amount the parallel values are raised in energy and $\alpha_2 T_{11}$ is the amount the perpendicular "valleys" are lowered in energy. Since the perpendicular valleys have the larger electron populations, a field applied in the direction of the stress will experience a higher mobility for these electrons and hence the conductivity will increase—and the resistivity decrease—in proportion to the applied stress. For n-germanium, and also for p-type silicon and germanium, a stress along a $\langle 111 \rangle$ direction will produce a similar effect. For p-type material the resistivity increases with applied stress. Figure 10.2, lightly doped sample, shows the gage factor, defined as

$$G = \frac{\Delta R}{R_o S_{111}}$$

that is, as the change in resistance divided by the initial resistance times the resulting strain along the $\langle 111 \rangle$ direction as a function of the temperature. The gage factor varies inversely as the absolute temperature. By going to higher doping, Boltzmann's statistics are replaced by Fermi-Dirac statistics, and a consequence is that the gage factor becomes small but is independent of the temperature. Thin specimens—that is, specimens with one dimension less than 0.002 in.—become stronger, and as shown by Fig. 10.4 the change in resistance is linear with strain up to strains of 2,500 microin./in., and is practically independent of the temperature for high doping levels.

3. Phenomenological Development

By differentiating the electric field by the current density and the mechanical stress, and making use of the fact that the semiconductors silicon and germanium have a center of symmetry so that the odd-rank tensors vanish, the fields can be written in the form of the equations

$$E_i = \rho_{ij}I_j + \pi_{ijkl}I_jT_{kl}$$

Only one resistivity ρ_o occurs for a cubic crystal, while there are three piezoresistive components which in reduced tensor notation are

$$\pi_{11}, \quad \pi_{12}, \quad \pi_{44}$$

The relations between the fields, currents, and applied stresses are given by equation (10.9) for the cubic axes.

By applying the usual tensor transformation equations, it is readily shown that the resistivity of a long, thin specimen such as used in a strain gage is given by

$$\frac{E_3'}{I_3'} = \rho = \rho_0[1 + \pi_l T_{33}']$$

where

$$\pi_l = \pi_{11} + 2(\pi_{44} + \pi_{12} - \pi_{11})(\sin^2\theta\cos^2\varphi + \sin^4\theta\sin^2\varphi\cos^2\varphi)$$

Here θ and φ are the angles indicated by Fig. 10.6. For a $\langle 100 \rangle$ valley semiconductor such as shown by Fig. 10.1, π_l is a maximum along $\langle 100 \rangle$ and is equal to π_{11}. For such a system $\pi_{11} + 2\pi_{12} = 0$; $\pi_{44} = 0$. Hence, if the resistivity is measured in a direction perpendicular to the direction of the applied stress, the change is half as much as π_{11} and in the opposite direction. A shearing stress produces a zero result.

For the $\langle 111 \rangle$ valley system—n-germanium and p-type silicon and germanium—the maximum is obtained for $\varphi = 45°$, $\theta = 54° \, 40'$, which is the direction of the $\langle 111 \rangle$ axis

$$\left(\frac{\pi_{11} + 2\pi_{12}}{3} \right) + \frac{2}{3} \pi_{44} = \frac{2}{3} \pi_{44}$$

since the first term is zero. Since the strain along the $\langle 111 \rangle$ direction is

$$S'_{33} = \frac{T'_{33}}{Y_0}$$

the gage factor is

$$\frac{\rho - \rho_o}{\rho_o S_{111}} = \left(\frac{2}{3} \pi_{44} Y_0 \right)$$

Table 25 shows the value of the gage factor (m_l) for a number of materials. Special orientations, such as shown by Fig. 10.7, are also possible for measuring certain strains in the presence of other strains or for measuring all types of strains with a single gage.

PROBLEMS CHAPTER 10

1. For certain types of cubic semiconductors—such as indium antimonide, which does not have a center of symmetry—what properties do the second and third terms of equation (10.5) represent?

2. Semiconductor gages can be obtained by using the transverse effect—that is, by measuring the ratio of the voltage to the current in a direction perpendicular to the plane of the gage. Determine the ratio of the field to current density in terms of the stresses T_1', T_2', T_6' for any orientation of the gage. (Note that if the gage is bonded directly to the surface, the stresses T_3', T_4', and T_5' are zero.)

3. What orientation of the plane of the gage of problem 2 will cause the response to be proportional to the sum of the principal stresses?

REFERENCES CHAPTER 10

1. P. W. Bridgman, *Proc. Am. Acad. Arts Sci.* **57**, 41 (1922); *ibid.* **60**, 43 (1925); Phys. Rev. **42**, 858 (1932).
2. C. S. Smith, *Phys. Rev.* **94**, 42 (1954).
3. W. P. Mason *in* "Physical Acoustics" (Warren P. Mason, ed.), Vol. IB, Chapter X. Academic Press, New York, 1964.
4. W. P. Mason and T. B. Bateman, *Phys. Rev.* **134**, A1387 (1964).
5. G. L. Pearson, W. T. Read, and W. L. Feldman, *Acta Met.* **5**, 181 (1957).
6. W. P. Mason and R. N. Thurston, *J. Acoust. Soc. Am.* **29**, 1096 (1957).
7. R. N. Thurston *in* "Physical Acoustics" (Warren P. Mason, ed.) Vol. IB, Chapter XI. Academic Press New York, 1964.
8. A. Tuzzolino, *Phys. Rev.* **109**, 1980 (1958).
9. W. G. Pfann and R. N. Thurston, *J. Appl. Phys.* **32**, 2008 (1961).

CHAPTER 11 • Hall Effect and Magnetoresistance Effect and Their Uses in Transducers

11.1 Introduction

When a magnetic field is applied at a definite angle with respect to the current flow in a conductor, it produces two effects that are potentially interesting for measurements and for displacement transducers. The first of these effects is the Hall effect, which results in a voltage at right angles to the direction of the normal current flow. This effect is caused by the force exerted on a moving charge by the magnetic field. As shown by Eq. (3.45), this force is proportional to the product of the charge, the velocity, and the magnetic flux density. It is directed at right angles to the velocity direction and the flux direction in a right-handed system of coordinates. This force produces a displacement of charges at right angles to the normal current flow, and this result is the origin of the Hall effect. Another result of the magnetic flux is a change in the current density distribution across the width of the conductor, with a higher density in the direction of the force component. Then since the current effectively flows through a smaller area and a longer path, the resistance of the conductor increases. This phenomenon is known as the magnetoresistance effect.

The velocity entering Eq. (3.45) is the so-called drift velocity—that is, the average velocity attained by an electron (or hole) under the effect of an applied electric field. The force on a charge q produced by a field E_3 can be equated to the mass times the acceleration or

$$ma = m \frac{d^2z}{dt^2} = +qE_3 \qquad (11.1)$$

where m is the mass of the charge, z the displacement along the $z = x_3$, axis and q the charge carried. Integrating this with respect to t, the time,

241

the velocity $dz/dt = v$ is given by

$$\frac{dz}{dt} = v_3 = + \frac{qE_3 t}{m} \qquad (11.2)$$

The velocity increases in proportion to the time until the charge suffers a collision with an impurity atom, a thermal wave, or some other obstruction. It is usually assumed that the charge is scattered isotropically, so that on the average the velocity is reduced to zero and the acceleration has to start over again. If the average time between collisions is written as τ, Eq. (11.2) shows that the drift velocity is given by

$$v_3 = + \frac{qE_3 \tau}{m} ; \qquad \mu = + \frac{v_3}{E_3} = + \frac{q\tau}{m} \qquad (11.3)$$

The mobility μ, which is the drift velocity divided by the applied field, is given by the right-hand side of Eq. (11.3). The current carried per square meter—that is, the current density—is

$$I_3 = + Nqv_3 = \frac{Nq^2 \tau}{m} E_3 = \sigma E_3 \qquad (11.4)$$

where N is the number of charges per cubic meter. σ, the conductivity, can be used to determine τ provided the number of electrons N_e is known, since $q = -e$ is known and for most metals the mass is comparable to the mass of a free electron, 9.1×10^{-31} kgm. For copper the mass is about 1.35^1 times m_0, the electron mass. The number of electrons is usually taken as equal to the number of atoms, so that for copper $N_e = 8.5 \times 10^{28}$ electrons/cubic meter[3]. With a resistivity of 1.68×10^{-8} ohm-meter, the relaxation time τ is

$$\tau = \frac{1.35 \times 9.1 \times 10^{-31}}{1.68 \times 10^{-8} \times 8.5 \times 10^{28} \times (1.6 \times 10^{-19})^2} \doteq 3.4 \times 10^{-14} \text{ sec} \quad (11.5)$$

The mobility μ is then

$$\mu = \frac{1.6 \times 10^{-19} \times 3.4 \times 10^{-14}}{1.35 \times 9.1 \times 10^{-31}} = 4.4 \times 10^{-3} \frac{\text{meter}^2}{\text{volt-sec}} \qquad (11.6)$$

This is a very small mobility, and the Hall voltage and magnetoresistance effect are small for a metal. These effects can be much larger for a semiconductor, and semiconductor materials are universally used for all applications of the Hall or magnetoresistance effects. Part of the greater mobility results from the larger relaxation time associated with the smaller number of carriers, but for some semiconductors, such as the III-V types,

[1] See *American Institute of Physics Handbook* [1].

the largest effect is associated with the very small effective mass of the carrier. As discussed in other references,[2] the effective mass varies inversely as the curvature of the energy surfaces. In the multivalley energy surface of Fig. 10.1, the curvature normal to the valley is five times as large as it is parallel to the valley. Since the charge and relaxation times are approximately equal for both directions, the mobility $\mu_\perp$ is about five times the mobility $\mu_\parallel$. The ratio is larger for germanium. For some of the III-V compounds, such as indium arsenide and indium antimonide, the curvature in the conduction

Table 26

CARRIER MOBILITIES AND ENERGY GAPS FOR SEVERAL SEMICONDUCTORS

Material		μ (electrons)[a]	μ (holes)[a]	Energy gap[b]
Germanium	Ge	0.36	0.17	0.80
Silicon	Si	0.12	0.025	1.11
Indium antimonide	InSb	7.5	0.11	0.25
Indium arsenide	InAs	2.3	0.024	0.45
Gallium arsenide	GaAs	0.85	0.040	1.35
Gallium antimonide	GaSb	0.45	0.09	0.7

[a] Mobilities in m²/volt-sec.
[b] Energy gaps in electron volts.

band is so high that the effective mass may be as small as $\frac{1}{77}$ of the electron mass. Table 26 shows the mobilities and the energy gaps of several useful semiconducting materials.

The energy gap shown in the last column determines how high in temperature some of these devices are operative. For example, Fig. 11.1 shows the electrical conductivity of indium antimonide for two doping values as a function of the inverse absolute temperature. The values are given as the inverse of ohm-centimeters. In these units the conductivity is 100 times smaller than the MKS unit of (ohm-meter)⁻¹. It is seen that the conductivity due to the doping by impurity atoms—that is, extrinsic conductivity—is nearly constant up to 350° K (77° C). For temperatures higher than this, the conductivity increases in a linear fashion when plotted on a logarithmic scale as a function of $10^3/T$. The slope of the curve is determined by the energy gap and the increased conductivity is due to the electrons and holes generated by thermal agitation—that is, intrinsic conductivity. When

[2] See Mason [2] and Wert and Thomson [3].

electrons and holes are present, they both contribute to the electrical conductivity according to the equation

$$\sigma = N_e e \mu_e + N_h e \mu_h \qquad (11.7)$$

where N_h is the number of holes per meter³, e the charge carried by the holes in coulombs, and μ_h the mobility of the holes in meter²/volt-sec. A given electric field will send electrons in one direction and holes in the other, but

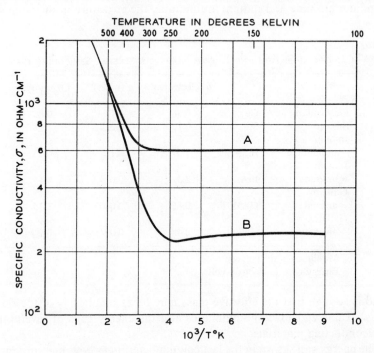

FIG. 11.1   Conductivity of InSb for two doping values (after Welker and Weiss [6]).

since their charges are of opposite sign they add in the conductivity equation. Silicon, with its larger energy gap has extrinsic conduction up to a temperature of above 1,000° K for fairly heavily doped specimens.

Conductivity measurements alone cannot determine the number of electrons, their mobility, or the sign of the charge carrier. If we combine this measurement with a Hall-voltage measurement, these quantities can be separated for the single type of charge carrier that is found in doped crystals. Using the force equation of (3.56), a simple derivation of the Hall voltage can be obtained for a single type of flow. Let us consider the conductor shown by Fig. 11.2 with an electric field applied along the x_1 direction and the magnetic flux applied in the x_3 direction. Without the flux the velocity will be only in the x_1 direction, but with the flux electrons tend to flow along both the x_1 and x_2 directions and hence the velocity will have the components v_1 and v_2. From Eq. (3.45) the force components on the moving charge q are

$$F_1 = qv_2B_3 ; \qquad F_2 = -qv_1B_3 \qquad (11.8)$$

Hence the magnetic field produces a force in the x_2 direction and drives electrons toward the left hand face and charges it negatively, while the positive x_1, x_3 face becomes positively charged. This corresponding field

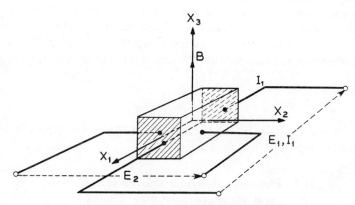

FIG. 11.2 Hall-effect plate showing direction of current flow, Hall voltage, and magnetic flux.

E_2 produces an additional force on the electrons of magnitude $-qE_2$, which is a force directed opposite to that produced by the magnetic field. At equilibrium the net force must vanish, and hence,

$$-qE_2 + qv_1B_3 = 0 \qquad (11.9)$$

Since v_1 is the drift velocity we can replace it, using the results of (11.3), and obtain

$$qE_2 = \frac{q^2\tau E_1 B_3}{m} \qquad (11.10)$$

Rewriting this expression in the form of (11.11) the Hall mobility becomes

$$\frac{E_2}{E_1 B_3} = \frac{q\tau}{m} \qquad \text{(Hall mobility)} \qquad (11.11)$$

Replacing E_1 in Eq. (11.10) by I_1/σ, where $\sigma = q^2\tau N/m$, gives

$$E_2 = \left(\frac{1}{Nq}\right)I_1 B_3 \qquad (11.12)$$

The factor of proportionality is called the Hall constant R. Then

$$R = \left(\frac{1}{Nq}\right) \tag{11.13}$$

Conduction by electrons produces a negative field E_2, so that R is negative for electrons and positive for holes. It will be noted that the product

$$|R|\, \sigma = \mu \tag{11.14}$$

where $|R|$ is the absolute value of R.

It is seen from the discussion that the measurements of the electrical conductivity and the Hall constant determine the sign of the charge carrier, the number of carriers per unit volume, and the mobility of the carriers. When both holes and electrons are present, the Hall constant is given by

$$R = \frac{1}{|q|}\left[\frac{N_h\mu_h^{\,2} - N_e\mu_e^{\,2}}{(N_h\mu_h + N_e\mu_e)^2}\right] \tag{11.15}$$

Some transducer uses for Hall effect and magnetoresistance devices are discussed in Section 11.3.

11.2 Phenomenological Derivation of Hall and Magnetoresistance Effects

When it is desired to investigate the effects of crystal symmetry on the Hall and magnetoresistance effects, it is considerably simpler to resort to phenomenological treatments than to generalize the simple treatment of Section 11.1, which strictly applies only to isotropic or cubic crystals. Such a derivation has been given previously[3] and is summarized in this section.

By employing the thermodynamics of irreversible processes and Onsager's principle of microscopic reversibility, Casimir[4] first showed that the magneto-resistance and Hall effects can be expressed by the tensor equation

$$E_i = \rho_{ij}(B)I_j + \varepsilon_{ijk}I_jR_k(B) \tag{11.16}$$

where the resistivities ρ_{ij} and the Hall-effect constant R_k are considered to be functions of the magnetic flux B applied to the sample. From the discussion of Eq. (9.11) of Chapter 9 it is found that ρ_{ij} is an even function of B, whereas R_k is an odd function of B. As a consequence, we can develop these functions in a Maclaurin series about the point of zero magnetic flux, with even-order terms occurring in ρ_{ij} and odd-order terms in R_k. The derivatives can be carried to high-order terms, and in the present case derivatives are carried

---

[3] See Mason et al. [4].
[4] See Casimir [5].

to sixth-rank tensors. These terms are sufficient to fit the magnetoresistance and Hall effect measurements so far made for germanium. Out to correction terms involving sixth-rank tensors, the results are

$$\rho_{ij}(B) = \rho_{ij_0} + \frac{1}{2!} \frac{\partial^2 \rho_{ij}}{\partial B_k \, \partial B_l} B_k B_l + \frac{1}{4!} \frac{\partial^4 \rho_{ij} B_k B_l B_m B_n}{\partial B_k \, \partial B_l \, \partial B_m \, \partial B_n} + \cdots$$

$$R_k(B) = \frac{\partial R_k}{\partial B_m} B_m + \frac{1}{3!} \frac{\partial^3 R_k}{\partial B_m \, \partial B_n \, \partial B_o} B_m B_n B_o \qquad (11.17)$$

$$+ \frac{1}{5!} \frac{\partial^5 R_k}{\partial B_m \, \partial B_n \, \partial B_o \, \partial B_p \, \partial B_q} B_m B_n B_o B_p B_q + \cdots$$

Designating the partial derivatives by the tensor letters

$$\frac{1}{2!} \frac{\partial^2 \rho_{ij}}{\partial B_k \, \partial B_l} = \xi_{ijkl} ; \qquad \frac{1}{4!} \frac{\partial^4 \rho_{ij}}{\partial B_k \, \partial B_l \, \partial B_m \, \partial B_n} = \zeta_{ijklmn}$$

$$\frac{\partial R_k}{\partial B_m} = R_{km} ; \qquad \frac{1}{3!} \frac{\partial^3 R_k}{\partial B_m \, \partial B_n \, \partial B_o} = \gamma_{kmno} ; \qquad (11.18)$$

$$\frac{1}{5!} \frac{\partial^5 R_k}{\partial B_m \, \partial B_n \, \partial B_o \, \partial B_p \, \partial B_q} = \chi_{kmnopq}$$

Eq. (11.16) can be written in the form

$$E_i = I_j[\rho_{ij_0} + \xi_{ijkl}B_k B_l + \zeta_{ijklmn}B_k B_l B_m B_n + \cdots]$$

$$+ \varepsilon_{ijk}I_j[R_{km}B_m + \gamma_{kmno}B_m B_n B_o + \chi_{kmnopq}B_m B_n B_o B_p B_q + \cdots] \qquad (11.19)$$

The only crystals for which both the Hall effect and the magnetoresistance effect have been measured are germanium ($m3m$) and some of the III-V compounds having the symmetry ($\bar{4}3m$). The terms for second- and fourth-rank polar tensors—which apply for both the magnetoresistance and Hall effects—are given by Tables B.2 and B.4. The two second-rank tensors have the single components

$$\rho_{11} = \rho_{22} = \rho_{33} = \rho_0 ; \qquad R_{11} = R_{22} = R_{33} = R_0 \qquad (11.20)$$

In the fourth-rank tensor ξ_{ijkl}, i, and j can be interchanged since they refer to the electrical variables, and k and l can be interchanged. However, i and j cannot be interchanged with k and l, and hence the symmetry is type M in Table B.4. The independent constants in terms of two-index symbols are

$$\xi_{11} , \qquad \xi_{12} , \qquad \xi_{44} \qquad (11.21)$$

with the relations between other terms shown by the table. For the other fourth-rank tensor γ_{kmno} all four subscripts can be interchanged, since three of the symbols refer to the same variable, the magnetic flux, and the fourth index is determined by the product of the other three. According to Table

B.4, symmetry K, this reduces the number of independent constants to two through the relation

$$\gamma_{12} = \gamma_{1122} = \gamma_{1212} = \gamma_{44} \tag{11.22}$$

The terms of the sixth-rank tensor ζ_{ijklmn} are determined by the fact that the first two can be interchanged, as can also the last four. As discussed previously,[5] this reduces the number of independent coefficients to six with the relationships shown by Eq. (11.23) for the symmetries $\bar{4}3m$, 432, and $m3m$. It will be noted that the terms left are not the same as those for the third-order elastic constants given by Table B.8. The relations found are for the subscripts

A $111 = 222 = 333$

B $112 = 113 = 121 = 131 = 221 = 223 = 212 = 232 = 331$

$\quad = 332 = 313 = 323 = 155 = 166 = 244 = 266 = 344 = 355$

C $122 = 133 = 211 = 233 = 311 = 322$

D $123 = 132 = 213 = 231 = 312 = 321 = 144 = 255 = 366$ $\qquad$ (11.23)

E $441 = 552 = 663 = 414 = 525 = 636 = 456 = 465 = 546$

$\qquad\qquad\qquad\qquad\qquad\qquad = 564 = 645 = 654$

F $442 = 443 = 551 = 553 = 661 = 662 = 424 = 434 = 515$

$\qquad\qquad\qquad\qquad\qquad\qquad = 535 = 616 = 626$

Using the subscripts of the first row the independent terms are

$$\zeta_{111}; \quad \zeta_{112}; \quad \zeta_{122}; \quad \zeta_{123}; \quad \zeta_{441}; \quad \zeta_{442} \tag{11.24}$$

For the tensor χ_{kmnopq} all six subscripts can be interchanged. This results in the relations

$$B = C = F; \quad D = E \tag{11.25}$$

and there are three independent constants

$$\chi_{111}, \quad \chi_{112}, \quad \text{and} \quad \chi_{123} \tag{11.26}$$

Using these relations, Eq. (11.19) can be written out with the appropriate constants and variables appearing in the equation. For example, the field equations along the x_1 axis take the form

$$\begin{aligned}
E_1 = {} & \rho_0 I_1 + \xi_{11}I_1B_1{}^2 + \xi_{12}I_1(B_2{}^2 + B_3{}^2) + 2\xi_{44}B_1(I_2B_2 + I_3B_3) + \zeta_{111}I_1B_1{}^4 \\
& + \zeta_{112}I_1B_1{}^2(B_2{}^2 + B_3{}^2) + \zeta_{122}I_1(B_2{}^4 + B_3{}^4) + 6\zeta_{123}I_1B_2{}^2B_3{}^2 \\
& + 12\zeta_{441}B_1B_2B_3(I_2B_3 + I_3B_2) \\
& + 4\zeta_{442}B_1[I_3B_3(B_1{}^2 + B_3{}^2) + I_2B_2(B_1{}^2 + B_2{}^2)] + [R_0 + 3\gamma_{12}B^2 + 5\chi_{112}B^4] \\
& \times [I_2B_3 - I_3B_2] + (\gamma_{11} - 3\gamma_{12})(I_2B_3{}^3 - I_3B_2{}^3) + (\chi_{111} - 5\chi_{112}) \\
& \times (I_2B_3{}^5 - I_3B_2{}^5) + (30\chi_{123} - 10\chi_{112})[I_2B_3B_1{}^2B_2{}^2 - I_3B_2B_1{}^2B_3{}^2] + \cdots
\end{aligned} \tag{11.27}$$

[5] See Mason *et al.* [4].

To obtain the equation pertaining to the x_2 axis, all the subscripts of E, I, and B are increased by 1 with the understanding that $B_{3+1} = B_1$, and so on. Equation (11.27) is obvious if we use all the relations of Table B.4 and Eqs. (11.23), (11.24), and (11.25). For example, for the fourth-rank tensor ξ_{ijkl}, the relations are

$$\xi_{11} = \xi_{1111} = \xi_{2222} = \xi_{3333}$$

$$\xi_{12} = \xi_{1122} = \xi_{1133} = \xi_{2233} = \xi_{2211} = \xi_{3311} = \xi_{3322} \qquad (11.28)$$

$$\xi_{44} = \xi_{1212} = \xi_{1221} = \xi_{1313} = \xi_{1331} = \xi_{2323} = \xi_{2332}$$

The first number indicates the field subscript, the second the current-density subscript, and the last two the flux-density subscripts. For example, for the tensor $\xi_{1111} = \xi_{11}$ the electric field and current density are E_1 and I_1 and the flux has to be $B_1{}^2$. For ξ_{1122} with the relations of (11.28) it is seen that the square of the fluxes are $B_2{}^2$ and $B_3{}^2$, as given in (11.27). For the $\xi_{44} = \xi_{1212} = \xi_{1313}$, and so on, the electric field is given, and hence we cannot employ such interchanges as $\xi_{1212} = \xi_{2112}$. Next the appropriate current is specified, and this determines the two fluxes. We can interchange the last two terms, since

$$\xi_{1212} = \xi_{1221} \qquad (11.29)$$

and this is the origin of the factor 2 multiplying ξ_{44}. Similar but more complicated relations are required for the sixth-rank tensor ζ_{ijklmn}. As an example, consider the term multiplying ζ_{112}. From (11.23) the terms that have unity for the first index are

$$\zeta_{112} = \zeta_{113} = \zeta_{121} = \zeta_{131} = \zeta_{155} = \zeta_{166} \qquad (11.30)$$

The first four terms obviously will give

$$2I_1 B_1{}^2 (B_2{}^2 + B_3{}^2) \qquad (11.31)$$

Since the indices 5 and 6 are shorthand for 13 and 12 in the complete expression and there are four possible combinations of the 1313 and 1212 indices, the multiplier for (11.31) has to be increased to 6. The other terms follow in a similar manner.

For the Hall-effect part, the first term becomes

$$\varepsilon_{ijk} I_j R_{km} B_m = I_2 R_{3m} B_m - I_3 R_{2m} B_m = R_0 (I_2 B_3 - I_3 B_2) \qquad (11.32)$$

Hence $R_0 = R_{11} = R_{22} = R_{33}$, and no cross-terms appear. For the second term involving the fourth-rank tensor, the form is

$$I_2 (\gamma_{3mno} B_m B_n B_o) - I_3 (\gamma_{2mno} B_m B_n B_o) \qquad (11.33)$$

Since the terms are

$$\gamma_{1111} = \gamma_{2222} = \gamma_{3333}$$

$$\gamma_{1122} = \gamma_{2211} = \gamma_{1133} = \gamma_{3311} = \gamma_{1212} = \gamma_{2112} = \gamma_{1221} ; \qquad (11.34)$$

$$\gamma_{2121} = \gamma_{1313} = \gamma_{1331} = \gamma_{3113} = \gamma_{3131} = \gamma_{2323} = \gamma_{2332} = \gamma_{3223} = \gamma_{3232}$$

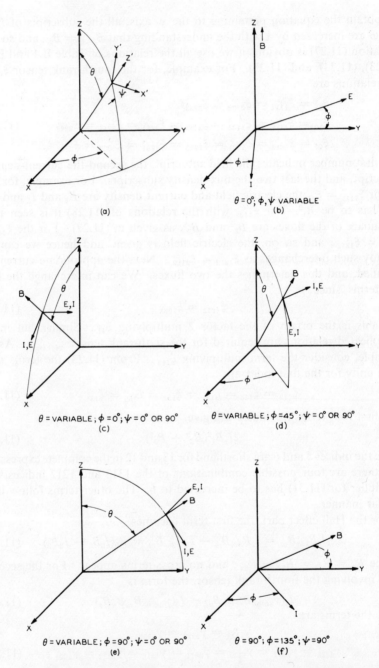

$\theta = 0°;\ \phi,\ \psi$ VARIABLE
(b)

$\theta =$ VARIABLE; $\phi = 0°$; $\psi = 0°$ OR $90°$
(c)

$\theta =$ VARIABLE; $\phi = 45°$; $\psi = 0°$ OR $90°$
(d)

$\theta =$ VARIABLE; $\phi = 90°$; $\psi = 0°$ OR $90°$
(e)

$\theta = 90°$; $\phi = 135°$; $\psi = 90°$
(f)

FIG. 11.3   Orientations for zero cross magnetoresistance effects.

it is obvious that the terms in (11.33) are

$$\gamma_{11}(I_2 B_3{}^2 - I_3 B_2{}^3) + 3\gamma_{12}[I_2 B_3(B_1{}^2 + B_2{}^2) - I_3 B_2(B_1{}^2 + B_3{}^2)] \quad (11.35)$$

By adding and subtracting

$$3\gamma_{12} I_2 B_3{}^2 - 3\gamma_{12} B_2{}^3 I_3 \quad\quad\quad (11.36)$$

the equation can be written in the form of Eq. (11.27) with $B^2 = B_1{}^2 + B_2{}^2 + B_3{}^2$. A similar procedure works for the sixth-rank tensor.

It is instructive to consider the simple case of Fig. 11.2, with the field E_1 applied along the length x_1 and the magnetic flux along B_3. Equation (11.27) simplifies to the form

$$E_1 = I_1[\rho_0 + \xi_{12} B_3{}^2 + \zeta_{122} B_3{}^4] + I_2 B_3[R_0 + \gamma_{11} B_3{}^2 + \chi_{111} B_3{}^4]$$
$$E_2 = I_2[\rho_0 + \xi_{12} B_3{}^2 + \zeta_{122} B_3{}^4] - I_1 B_3[R_0 + \gamma_{11} B_3{}^2 + \chi_{111} B_3{}^4] \quad (11.37)$$

Hence, as long as the sidewise current $I_2 = 0$, the magnetoresistance effect is determined by measuring the ratio of the applied field to the current density I_1, while the Hall-voltage gradient E_2 is determined by the product of the current density times the flux multiplied by the Hall-effect constants, which here are carried out to two correction terms.

By introducing transformation equations such as those used in Chapter 10, it is readily shown[6] that for electric fields applied along the x' direction, magnetic fluxes along z', and cross electric fields measured along y', most orientations will produce a component of E_2' due to the magnetoresistance effect, and this field will not reverse sign when the magnetic flux reverses direction. Hence, to measure the Hall voltage it is desirable to have an orientation for which this does not occur. One such orientation, as shown by Eq. (11.37), is one for which the length lies along a cube axis, the Hall voltage being measured along another cube axis with the magnetic flux being applied along the third cube axis. Other orientations are also possible, and Fig. 11.3 shows all the values. The first, Fig. 11.3a, gives the general coordinate system with the angles of rotation considered; the next five figures show particular orientations for which a cross magnetoresistance voltage disappears.

In order to evaluate the variation of the Hall effect as a function of the angle, use[7] was made of two cylinders of germanium, one having its axis along the $\langle 100 \rangle$ direction and the other with its axis along the $\langle 110 \rangle$ direction. The Hall-effect voltages were measured at right angles to the current and magnetic flux directions by using a potentiometer of high impedance so

[6] See Mason et al. [4].
[7] See Mason et al. [4].

that I_2 could be considered zero. To take account of the fact that the specimen was cylindrical, use is made of the definition of the Hall coefficient

$$R_H = \frac{V_2/D}{I_1 B_3} = \frac{V_2/D}{(4i/\pi D^2)B_3} = \frac{\pi}{4}\left(\frac{V_2 D}{i B_3}\right) \tag{11.38}$$

where D is the diameter of the cylinder, V_2 the measured voltage, and i the total current in the x_1 direction.

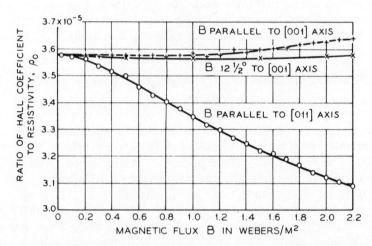

Fig. 11.4   Ratio of R_H/ρ_0 for a $\langle 100 \rangle$ cylinder for three directions of the magnetic flux.

Figure 11.4 shows the ratio of R_H to ρ_0 (the resistivity) plotted as a function of the flux density in webers/meter²—10 kilogauss equals 1 weber per meter²—for various orientations of the magnetic field, all of which are perpendicular to the crystal axis. From Fig. 11.3c all of these orientations satisfy the condition of zero cross magnetoresistance effects. The top and bottom curves can be fitted by the equations

$$\frac{R_H}{\rho_0} = 3.58 \times 10^{-5}[1 + 2.4 \times 10^{-3} B^4] \qquad\qquad B \text{ along } \langle 001 \rangle$$

$$\tag{11.39}$$

$$\frac{R_H}{\rho_0} = 3.58 \times 10^{-5}[1 - 7.9 \times 10^{-2} B^2 + 1.55 \times 10^{-2} B^4] \qquad B \text{ along } \langle 011 \rangle$$

These equations are valid up to flux densities of 1.5 webers/meter², but it requires more terms to fit the curve for higher flux densities. The intermediate curve marked 12.5° from the $\langle 001 \rangle$ axis is the orientation for which the Hall coefficient is most nearly constant for fluxes as high as 2.2 webers/meter².

Figure 11.5 shows the curves obtained for a cylinder cut along the $\langle 110 \rangle$ direction with fluxes at five directions with respect to cylinder axis. All five of these directions are normal to the length of the crystal, and they satisfy the conditions of Fig. 11.3d with θ variable, $\varphi = -45°$, $\psi = 90°$. For all values of θ, the field E_1 and current density I_1 lie along the $\langle 110 \rangle$ axis, while the magnetic flux varies from the $\langle 001 \rangle$ direction—$\theta = 0$—to the $\langle 1\bar{1}0 \rangle$

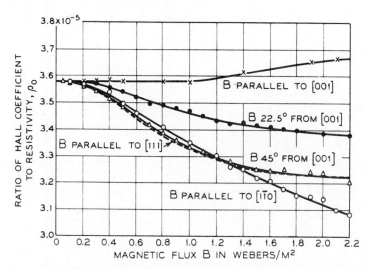

FIG. 11.5   Ratio of R_H/ρ_0 for a $\langle 110 \rangle$ cylinder for five directions of the magnetic flux.

direction for which $\theta = 90°$. The voltage gradient E_2 is at right angles to the other two vectors. All these measurements can be fitted to the equations—valid up to 1.5 webers/meter²—given by Eq. (11.40)

$$\frac{R_H}{\rho_0} = 3.58 \times 10^{-5}[1 + 2.4 \times 10^{-3}B^4]; \qquad \begin{matrix} \theta = 0; \\ B \text{ along } \langle 001 \rangle \end{matrix}$$

$$\frac{R_H}{\rho_0} = 3.58 \times 10^{-5}[1 - 4.13 \times 10^{-2}B^2 + 0.92 \times 10^{-2}B^4 + \cdots]; \quad \theta = 22.5°$$

$$\frac{R_H}{\rho_0} = 3.58 \times 10^{-5}[1 - 8.8 \times 10^{-2}B^2 + 1.96 \times 10^{-2}B^4 + \cdots]; \quad \theta = 45°$$

$$(11.40)$$

$$\frac{R_H}{\rho_0} = 3.58 \times 10^{-5}[1 - 9.8 \times 10^{-2}B^2 + 2.5 \times 10^{-2}B^4 + \cdots]; \quad \theta = 55°$$

$$\frac{R_H}{\rho_0} = 3.58 \times 10^{-5}[1 - 7.9 \times 10^{-2}B^2 + 1.55 \times 10^{-2}B^4 + \cdots]; \quad \theta = 90°;$$

$$B \text{ along } \langle 1\bar{1}0 \rangle$$

Table 27

RELATIONS BETWEEN CRYSTAL DIRECTIONS AND HALL-EFFECT EQUATIONS

Direction of I	θ	φ	ψ	Equations for the Hall effect
$\langle 100 \rangle$	0	0	0	$R_H = R_0 + \gamma_{11}B^2 + \chi_{111}B^4$
$\langle 100 \rangle$	45°	90°	−90°	$R_H = R_0 + \left(\dfrac{\gamma_{11}+3\gamma_{12}}{2}\right)B^2 + \left(\dfrac{\chi_{111}+15\chi_{112}}{4}\right)B^4$
$\langle 110 \rangle$	0	−45°	+90°	$R_H = R_0 + \gamma_{11}B^2 + \chi_{111}B^4$
$\langle 110 \rangle$	22.5°	−45°	+90°	$R_H = R_0 + (0.7393\gamma_{11} + 0.7821\gamma_{12})B^2 + (0.623\chi_{111} + 1.75\chi_{112} + 0.414\chi_{123})B^4$
$\langle 110 \rangle$	45°	−45°	+90°	$R_H = R_0 + \left(\dfrac{3}{8}\gamma_{11} + \dfrac{15}{8}\gamma_{12}\right)B^2 + \left(\dfrac{5}{32}\chi_{111} + \dfrac{105}{32}\chi_{112} + \dfrac{90}{32}\chi_{123}\right)B^4$
$\langle 110 \rangle$	55°	−45°	+90°	$R_H = R_0 + \left(\dfrac{\gamma_{11}+6\gamma_{12}}{3}\right)B^2 + \left(\dfrac{\chi_{111}+30\chi_{112}+30\chi_{123}}{9}\right)B^4$
$\langle 110 \rangle$	90°	−45°	+90°	$R_H = R_0 + \left(\dfrac{\gamma_{11}+3\gamma_{12}}{2}\right)B^2 + \left(\dfrac{\chi_{111}+15\chi_{112}}{4}\right)B^4$

Table 28

RELATIONS BETWEEN CRYSTAL DIRECTIONS AND THE RESISTIVITY EQUATIONS

Direction of I	Direction of B	θ	φ	ψ	Equation for resistivity
$\langle 100 \rangle$	$\langle 100 \rangle$	$90°$	$0°$	$0°$	$\rho = \rho_0 + \xi_{11}B^2 + \zeta_{111}B^4$
$\langle 100 \rangle$	$\langle 010 \rangle$	$0°$	$0°$	$90°$	$\rho = \rho_0 + \xi_{12}B^2 + \zeta_{122}B^4$
$\langle 110 \rangle$	$\langle 001 \rangle$	$0°$	$45°$	$0°$	$\rho = \rho_0 + \xi_{12}B^2 + \zeta_{122}B^4$
$\langle 110 \rangle$	$\langle 110 \rangle$	$90°$	$45°$	$0°$	$\rho = \rho_0 + \left(\dfrac{\xi_{11} + \xi_{12} + 2\xi_{44}}{2} \right)B^2 + \left(\dfrac{\zeta_{111} + 6\zeta_{112} + \zeta_{122} + 8\zeta_{442}}{4} \right)B^4$
$\langle 110 \rangle$	$\langle 1\bar{1}0 \rangle$	$90°$	$135°$	$-90°$	$\rho = \rho_0 + \left(\dfrac{\xi_{11} + \xi_{12} - 2\xi_{44}}{2} \right)B^2 + \left(\dfrac{\zeta_{111} + 6\zeta_{112} + 3\zeta_{122} - 8\zeta_{442}}{4} \right)B^4$

The Hall voltage for low values of B reaches a minimum along the $\langle 111 \rangle$ direction, which agrees with theory.

To evaluate the constants in the phenomenological equation, the voltage at right angles to the current and flux has to be evaluated for a rotated system of coordinates. The method for doing this is identical with the method used for the piezoresistive equations of Chapter 10 and is left as an exercise for the reader. The results of these transformations are given by Table 27. The first equation is the same as the last of Eqs. (11.37). Only the absolute value of R_H is given in the table. It is seen that the equations for the Hall voltages measured for the $\langle 100 \rangle$ and $\langle 110 \rangle$ cylinder are identical when the fluxes are along the $\langle 001 \rangle$ axes and the $\langle 110 \rangle$ directions for both cases.

Evaluating the ratio of the second- and third-order Hall constants to the first order, R_0, the experimentally determined values of Eqs. (11.39) and (11.40) are fitted best by the values

$$\frac{\gamma_{11}}{R_0} = 0; \qquad \frac{\gamma_{12}}{R_0} = -5.0 \times 10^{-2}; \qquad \frac{\chi_{111}}{R_0} = 2.4 \times 10^{-3}$$

$$\frac{\chi_{112}}{R_0} = 4.0 \times 10^{-3}; \qquad \frac{\chi_{123}}{R_0} = 3.0 \times 10^{-3}$$

The sum of the terms in B^2 agree with the measured values within 6 per cent for all cases, whereas the terms in B^4 check to within 12 per cent. Since the γ_{11} term is found to be zero, the best orientations for measuring flux densities are the first and third orientations of Table 27.

Measurements of the magnetoresistance effect have also been made[8] for germanium. From these measurements all the second-order terms, ξ_{11}, ξ_{12}, and ξ_{44}, can be evaluated, and four of the third-order terms. To evaluate the terms ξ_{11} and χ_{111} the magnetic flux has to be applied in the same direction as the current flow. From Eq. (11.27) with $B = B_1$, the first equation of Table 28 follows. The second one follows by putting $B = B_2$. The other three equations require rotated-axes expressions, which are left to the reader.

Up to 1.5 webers/meter^2, the measurements of Pearson and Suhl are best fitted by the resistivity equations of (11.41). In these equations $\Delta\rho$ is the difference between the resistivity at zero magnetic flux ρ_0 and the measurements at the flux density B—that is, $\rho(B)$.

---

[8] See Pearson and Suhl [6].

$$
\begin{array}{ccl}
I & B & \\
\langle100\rangle & \langle100\rangle & \dfrac{\Delta\rho}{\rho_0} = 1.6 \times 10^{-1}B^2 - 1.5 \times 10^{-2}B^4 \\[2ex]
\langle100\rangle & \langle010\rangle & \dfrac{\Delta\rho}{\rho_0} = 0.7 \times 10^{-1}B^2 - 0.6 \times 10^{-2}B^4 \\[2ex]
\langle110\rangle & \langle1\bar{1}0\rangle & \dfrac{\Delta\rho}{\rho_0} = 1.5 \times 10^{-1}B^2 - 2.3 \times 10^{-2}B^4 \quad (11.41) \\[2ex]
\langle110\rangle & \langle001\rangle & \dfrac{\Delta\rho}{\rho_0} = 0.7 \times 10^{-1}B^2 - 0.6 \times 10^{-2}B^4 \\[2ex]
\langle110\rangle & \langle110\rangle & \dfrac{\Delta\rho}{\rho_0} = 0.8 \times 10^{-1}B^2 - 1.7 \times 10^{-2}B^4
\end{array}
$$

From these measurements all the constants ξ_{mn}/ρ_0 and four of the six constants ζ_{mno}/ρ_0 can be evaluated. These constants are independent of the resistivity of the material for moderately doped samples and for n-type germanium have the values

$$
\frac{\xi_{11}}{\rho_0} = 1.6 \times 10^{-1}; \qquad \frac{\xi_{12}}{\rho_0} = 0.7 \times 10^{-1}; \qquad \frac{\xi_{44}}{\rho_0} = -0.35 \times 10^{-1}
$$

$$
\frac{\zeta_{111}}{\rho_0} = -1.5 \times 10^{-2}; \qquad \frac{\zeta_{122}}{\rho_0} = -0.6 \times 10^{-2}; \qquad \frac{\zeta_{112}}{\rho_0} = -0.88 \times 10^{-2}
$$

$$
\frac{\zeta_{442}}{\rho_0} = +0.075 \times 10^{-2} \qquad (11.42)
$$

The largest variation occurs when the field is in the same direction as the current; the change is a 28 per cent increase for a flux density of 1.5 webers/meter^2.

The variations for materials with higher mobilities are much larger. For spherical energy surfaces, which occur for n-type III-V semiconductors such as InSb, theory[9] shows that the increase of resistance, for a flux applied perpendicular to the current direction, is given to first powers of B by

$$
\frac{\Delta\rho}{\rho_0} = \frac{9\pi}{16}\left(\frac{4-\pi}{4}\right)(\mu B)^2 \qquad (11.43)
$$

where μ is the mobility in meters^2/volt-sec and B is the flux in webers/meter^2. For example, this formula indicates an increase of 21 times for indium antimonide for a field of 1 weber/meter^2 since the drift velocity is 7.5

[9] See Welker and Weiss [7].

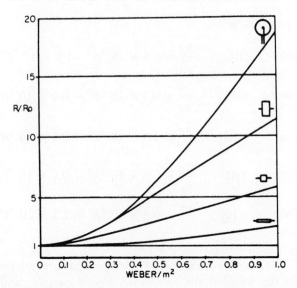

FIG. 11.6 Magnetoresistance effect for various shaped specimens of InSb (after Welker and Weiss [6]).

meters^2/volt-sec. Measurements for this material as a function of the flux are shown by Fig. 11.6. It is seen that the results depend on the shape of the sample, and the shape having the largest effect is the Corbino disk type with the electrodes on the center and on the periphery. For this shape there is less interference of the current-flow paths by the Hall effect than for any other shape. A fair agreement with Eq. (11.43) is found out to flux densities of 0.7 webers/meter^2. The other shapes require a solution of the current flow throughout the body produced by the presence of a force on the electrons due to the magnetic flux. Such a solution has been given by Wick,[10] but it is not discussed here. It is evident from Fig. 11.6 that a rectangular shape with the width larger than the length in the direction of current flow will give a larger magnetoresistance effect than one for which the current-flow direction is much larger than the width. It has been suggested by Welker and Weiss[11] that if a long, thin specimen has to be used, a greatly enhanced effect can be obtained by plating short-circuiting rings of silver around the slab so that in effect the length is broken up into a number of short sections.

11.3 Applications of the Hall and Magnetoresistance Effects

A number of uses have been proposed for both the Hall and the magnetoresistance effects, and some of them have been applied in practical devices.

[10] See Wick [8].
[11] See Welker and Weiss [7].

These include devices for measuring magnetic flux, nonreciprocal networks, networks having propagation in only one direction, displacement-measuring devices, microphones and pickup devices, modulators, amplifiers, and variable-resistance devices.

11.31 Magnetic Flux Measurements

One of the simplest uses for the Hall effect is the measurement of magnetic flux densities. By sending the current along the dimension marked i_1 in Fig. 11.7, through electrodes 1 and 2, and applying a flux density B perpendicular

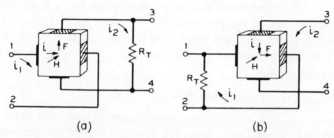

(a)   (b)

FIG. 11.7   (a) Relations of input and output current in a Hall-effect plate for a forward propagation of the current; (b) similar relations for the backward flow.

to the surface, a voltage is generated across electrodes 3 and 4. Because a crystal can be made very thin, such crystals can be put in narrow spaces, and they have been used[12] in measuring the flux density in relays as a function of time. The sensitivity of the germanium is sufficient to allow the unit to be connected directly to a conventional cathode-ray oscillograph using only the amplifiers that are present in such instruments. For example,[13] a crystal 6.0 mm long, 1.25 mm wide in the voltage-generation direction, and 0.15 mm thick will generate a voltage of 0.024 volts for a flux density of 0.1 weber/meter^2 when the resistivity is 1,000 ohm-meter and the input current 0.001 amp. This follows directly from Eq. (11.38) since

$$V_2 = R_0 w I_1 B_3 = \frac{R_0 w i B_3}{wt} = \frac{R_0 i B_3}{t} = \frac{3.58 \times 10^{-5} \rho i B_3}{t} = 0.024 \text{ volts}$$

(11.44)

where w is the width and t the thickness. Since $\rho = 1,000$ ohm-meter, $i = 0.001$ amp, $B_3 = 0.1$ webers/meter^2, and $t = 1.5 \times 10^{-4}$ meter, it is seen that the voltage is 0.024 volts. This value will give a deflection of about 0.8 in. on a standard type of oscillograph. By using the orientation of the

---

[12] See Mason *et al.* [4].
[13] *Ibid.*

first line of Table 27, the sensitivity remains constant within 2 per cent out to fluxes of 2 webers/meter^2.

If one wishes to measure a very small flux, the enhanced sensitivity of the high-mobility indium antimonide is an advantage. Since, from Table 26, the high mobility occurs only for n-type material, this is universally used. On account of the small effective mass of the electron of about $\frac{1}{77}m_0$, the degeneracy temperature is considerably raised for the same doping. Following the discussion of Chapter 10, a degeneracy temperature of 350° K is obtained for a doping of 3.5×10^{22} impurity atoms/meter^3. The two curves of Fig. 11.1 have doping values of 2 and 4.8×10^{22} so that the top curve is degenerate for all temperatures and as a consequence has a resistivity that is nearly independent of the temperature. By using such crystals of indium antimonide together with mu metal rods to concentrate the field, Ross and Saker[14] were able to measure the earth's magnetic field and hence produced an electrical compass.

11.32 Miscellaneous Hall-Effect Devices

A number of other devices have been suggested, although none of them appear to be in general use. Since the voltage output E_2 is proportional to the product of the magnetic field times the input current, then if one varies as the sine of a carrier frequency f_c and the other according to a signal frequency f_s the output voltage is proportional to the product

$$E_2 = A \sin (f_c t) \sin (f_s t) = \frac{A[\cos (f_c - f_s)t - \cos (f_c + f_s)t]}{2} \quad (11.45)$$

Hence the output consists of a sum frequency and a difference frequency— called the side band frequencies—without the addition of other frequencies. Such devices have been suggested[15] in connection with direct-current amplifiers.

By putting a Hall-effect plate in a magnetic field that varies with the displacement of the plate one can measure the displacement by measuring the change in the Hall voltage. Displacement measurements as small as 1 angstrom[16] have been observed. By connecting such a Hall plate to a diaphragm that moves in a large flux gradient, one can obtain a sensitive microphone.

Perhaps the most interesting application is the use of a Hall-effect plate as a nonreciprocal circuit element called a *gyrator*. The nonreciprocal action is evident from Eq. (11.37). When a current i_2 is drawn from the

[14] See Ross and Saker [9]; see also Moss [10] for a description of applications of Hall and magnetoresistance effects.

[15] See Mason *et al.* [4].

[16] See Ross and Saker [9] and Moss [10].

plate, the quantities E_1, I_1, E_2, and I_2 are functions of the position in the plate, and it requires a complicated solution of Maxwell's equations together with the boundary conditions to evaluate the input and output voltages and currents. Such a solution was given by Wick,[17] and it was shown that

$$V_1 = i_1 R_{11} + i_2 R_{12} ; \qquad V_2 = -i_1 R_{12} + i_2 R_{22} \qquad (11.46)$$

where the self-resistances R_{11} and R_{22} depend on the size of the magnetic flux, the position of the electrodes, and so on. The value of the coupling

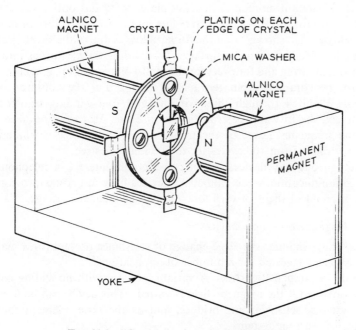

FIG. 11.8 "Gyrator" using a Hall-effect plate.

resistance R_{12} increases as the magnetic flux B and the mobility become larger. However, Wick has shown that for a rectangular plate with two sets of electrodes, R_{12}^2 cannot be larger than $R_{11}R_{22}$. Direct calculation shows that the transmission loss through such a gyrator cannot be less than 7.66 decibels. Transmission through this Hall-effect plate is nonreciprocal since, as shown by Fig. 11.7a, a current i_1 produces a current i_2 with the same direction as i_1, but when transmission occurs from the terminals 3 and 4, the current i_1 has the reverse direction to i_2. Such a gyrator arrangement is shown by Fig. 11.8.

By shunting a resistance R_s from terminal 1 to 3 and an equal one from 2 to 4 the resistance can be adjusted so that no current in the termination

[17] See Wick [8].

R_T , across the 1, 2 terminals, will flow for an input current i_2 ; but when the current i_1 is sent through the plate in the other direction, current will flow in the output resistance R_T . Hence the gyrator can be made a one-way transmission system. The resistance required for this adjustment is

$$R_s = \frac{R_{11}R_{22} + R_{12}^2}{2R_{12}} \tag{11.47}$$

Using indium antimonide as the Hall plate, Arlt[18] has obtained a forward loss of 9.2 db for a 0.92 weber/meter^2 flux and a backward loss of 60 db or more. He has also shown[19] that the forward loss is due to the short-circuiting effect of the electrodes on the surface. By dividing the electrodes into three sets for transmitting and four sets for the Hall-voltage pickup, and connecting them to the common input terminals 1 and 2 and the common output terminals 3 and 4 through three-winding and four-winding transformers, respectively, the forward loss can be cut to 2 db. However, on account of the finite frequency range of the transformers, this device is limited to a frequency range from 100 cycles/sec to 100 kilocycles/sec.

Although one-way devices are of considerable interest in telephone and radio communication, these Hall-effect devices have not come into common use on account of their forward loss and limited frequency range.

11.33 Magnetoresistance Devices

In a similar manner the large change of resistance present for a magnetic flux has been suggested for a number of applications.

11.331 VARIABLE RESISTOR. A variable resistor with no sliding contacts can be obtained if the magnetic field is varied. This device can be useful for servo systems, self-balancing bridges, and as the series element for high-voltage–low-current systems.

11.332 DISPLACEMENT METER AND MICROPHONE. The use of high-mobility indium antimonide crystals in a nonuniform magnetic field has been suggested for measuring displacements and for sensitive microphones.[20] In these uses the magnetoresistance element is employed as one arm of a Wheatstone bridge and the unbalanced voltage produces a signal proportional to the displacement. For the microphone, the magnetoresistance element is mounted on a diaphragm actuated by air waves. The magnetic flux varies with displacement and generates an unbalanced voltage whose power output can be made quite large. According to Moss,[21] the output can be made

---

[18] See Arlt [11].
[19] See Arlt [12].
[20] See Ross and Saker [9] and Moss [10].
[21] See Moss [10].

considerably larger than the input, and ideally an amplification is possible from sound energy to electrical energy.

RESUME CHAPTER 11

I. Introduction

When a magnetic field is applied in a direction normal to electric current flow, it is found that there is a voltage developed at right angles to the current flow and to the magnetic field. This effect is known as the Hall effect. It reverses sign when the electric current is carried by holes. At the same time it is found that the resistance of the sample has increased, an effect known as the magnetoresistance effect.

A simple derivation of the Hall effect can be obtained by equating the force on the electron caused by the vector product of the charge times velocity—that is, the electric current—and the magnetic flux density B to the electric field E_2 times the charge. This results in a ratio of the sidewise field E_2 to the product of the current density times the flux equal to

$$\frac{E_2}{I_1 B_3} = R = \frac{1}{Nq}$$

Hence a measurement of the Hall constant R determines the number and sign of the carriers of electricity. The current I is equal to the product of the number of charges times the value of the charge times the drift velocity V_1, or

$$I_1 = NqV_1 = Nq\mu E_1$$

where the drift velocity for a field of one volt per meter (μ) is known as the mobility. By calculating the mobility as determined by the acceleration of the electron for a period of time τ, determined as the average time between collisions, it is shown that

$$\mu = \frac{q\tau}{m^*} \quad \text{and} \quad \sigma = \frac{I_1}{E_1} = \frac{Nq^2\tau}{m^*}$$

where m^* is the effective mass and σ the conductivity. For electrons in metals the effective mass is approximately equal to the electron mass 9.1×10^{-31} kgm. Hence, for a metal such as copper with a resistivity of 1.68×10^{-8} ohm-meters, the average time between collisions is in the order of 3.4×10^{-14} seconds and the mobility is in the order of 4.4×10^{-3} meters^2/volt-sec. For semiconductors the mobility is much larger because τ may be in the order of 2.5×10^{-13} sec and the effective mass is smaller. Values are given in Table 26, and certain III-V semiconductors can have a mobility of 7.5 meters^2/volt-sec. For a given electrical conductivity, the number of carriers can be less and hence the Hall effect becomes larger. Semiconductors of this type are generally used for Hall-effect transducers.

The change in resistivity for an applied magnetic flux is given approximately by the equation

$$\frac{\Delta\rho}{\rho_0} = \frac{9\pi}{16}\left(\frac{4-\pi}{4}\right)(\mu B)^2$$

Hence, the change in resistance is proportional to the square of the mobility times the flux density.

2. Phenomenological Derivation of the Hall and Magnetoresistance Effects

When it is desired to investigate the effect of crystal symmetry and crystal orientation it is usually simpler to employ phenomenological treatments rather than physical calculations. A development has been carried up to fifth derivatives in order to evaluate measurements of the Hall effect and magnetoresistance effect made for germanium. It follows from Onsager's principle that the relations between the electric fields, electric current densities, and the magnetic flux can be written in the form

$$E_i = \rho_{ij}(B)I_j + \varepsilon_{ijk}I_jR_k(B)$$

where the resistivities ρ_{ij} are even functions of the magnetic flux while the Hall coefficients R_k are odd functions of the flux. By expanding them in derivatives to the fifth rank and replacing the partial derivatives by tensors of rank up to six, the electric fields can be written in the form

$$E_i = I_j[\rho_{ij_0} + \xi_{ijkl}B_kB_l + \zeta_{ijklmn}B_kB_lB_mB_n]$$
$$+ \varepsilon_{ijk}I_j[R_{km}B_m + \gamma_{kmno}B_mB_nB_o + \chi_{kmnopq}B_mB_nB_oB_pB_q]$$

For crystals of the cubic classes, $m3m$ or $\overline{4}3m$, tensors of rank two have the single components ρ_0 and R_0. The fourth-rank tensors have the components, in reduced notation, ξ_{11}, ξ_{12}, and ξ_{44} and γ_{11}, $\gamma_{12} = \gamma_{44}$. The sixth-rank tensor ζ has six independent coefficients given by (11.24), while the sixth-rank tensor χ has only three independent values. The result of combining all these terms is shown for the field along x_1 by Eq. (11.27). Rules are given for extending the results to the other two axes.

For a field applied along the x_1 axis and a magnetic flux along the x_3 axis, the equations reduce to the relatively simple form

$$E_1 = I_1[\rho_0 + \xi_{12}B_3^2 + \zeta_{122}B_3^4] + I_2B_3[R_0 + \gamma_{11}B_3^2 + \chi_{111}B_3^4]$$
$$E_2 = I_2[\rho_0 + \xi_{12}B_3^2 + \zeta_{122}B_3^4] - I_1B_3[R_0 + \gamma_{11}B_3^2 + \chi_{111}B_3^4]$$
$$E_3 = I_3[\rho_0 + \xi_{11}B_3^2 + \zeta_{111}B_3^4]$$

Hence if I_2 and I_3 are zero, the first equation determines the magnetoresistance effect while the second one determines the Hall effect. The third one determines the ξ_{11} and ζ_{111} magnetoresistance constants.

To determine the remaining fourth- and sixth-rank tensor terms requires the use of a set of axes rotated with respect to the crystallographic axes. In general, such orientations will have cross fields E_2' which involve the magnetoresistance effect as well as the Hall effect. It is shown that there are five general orientations, given by Fig. 11.3b to f, inclusive, for which the magnetoresistance does not produce any field E_2'. Using the orientation of Fig. 11.3c with $\theta = 0$ (E_1 along x_1 axis, E_2 measured along x_2 axis, and flux along x_3 axis) and $\theta = 45°$, which gives E_1' along $\langle 10\overline{1}\rangle$, E_2' along $\langle 010\rangle$ and B_3' along the $\langle 101\rangle$ direction, two equations are given in (11.39) for the Hall effect. These allow an evaluation of the two constants γ_{11} and γ_{12} as being

$$\frac{\gamma_{11}}{R_0} = 0; \qquad \frac{\gamma_{12}}{R_0} = -5.0 \times 10^{-2}$$

By using the orientations of Fig. 11.3d with $\varphi = -45°$, $\theta = 90°$ and ψ variable—that is, letting the axis of the cylinder lie along the $\langle 110\rangle$ axis, while the direction of the magnetic flux and the Hall voltage rotate around the cylinder—five conditions are obtained, whose equations are given by Table 27, and whose evaluated constants are given by Eq. (11.40). Two of these equations duplicate the results of Eq. (11.39)—although the orientations are different—but three new relations are obtained that allow a determination of all second- and third-order constants. Since the γ_{11} term turns out to be zero, the best orientations for measuring flux densities are the first and third of Table 27. For these orientations, the Hall voltage is proportional to the flux out to flux densities of 1.2 webers/meter^2. A slightly better orientation for very large fluxes is shown by the second curve of Fig. 11.4. Magnetoresistance values have been measured using the orientations of Table 28. These have been used to evaluate all three second-order moduli and four of the six third-order moduli for germanium.

For the higher-mobility III-V indium antimonide semiconductor, the energy surface is spherical. This is shown by the fact that there is no magnetoresistance effect when the magnetic flux is in the same direction as the current flow. Phenomenologically, this means that the constants ξ_{11} and ζ_{111} are zero. For such a material the increase in resistance is independent of the orientation if the flux is perpendicular to the major plane of the sample. The magnetoresistance does depend on the shape of the sample on account of the currents generated by the Hall effect. As shown by Fig. 11.6, the shape having the largest magneto-resistance effect is the "Corbino" disk for which the current flows from the center to the periphery.

3. Applications of the Hall Effect and the Magnetoresistance Effect

A number of applications have been suggested for both the Hall effect and the magneto-resistance effect. The simplest effect that has been fairly widely used is a direct measurement of magnetic flux. Germanium, with a crystal cut perpendicular to one crystallographic axis, currents flowing along another, and Hall voltages measured along the third axis, has a very linear response and a sensitivity large enough to measure small fluxes. By using the more sensitive indium antimonide, with μ metal concentrators, the earth's magnetic field can be measured and hence one can obtain an electrical compass.

Pure product modulators and displacement meters can be obtained by using the Hall effect. By using a four-electrode plate in a magnetic field a "gyrator" can be produced that allows transmission of electric currents in one direction but not in the other. With a single plate the best result obtained is a forward loss of 9 db but a backward loss of over 60 db. By using a series of transformer-connected electrodes the forward loss can be considerably reduced.

Magnetoresistance devices have been used as variable resistors with no sliding contacts, where the resistance is controlled by the value of the magnetic flux. They have also been used as displacement meters and as microphones.

PROBLEMS CHAPTER 11

1. Determine the equations for the magnetoresistance and Hall effects for the orientations shown by Fig. 11.3a, assuming that the applied field is in the x_1' direction, the magnetic flux B_3' along the x_3' axis, and the Hall voltage measured along x_2' for terms up to fourth-rank tensors.

2. Show that the orientations of Fig. 11.3b to -f inclusive are the only orientations for which the magnetoresistance does not contribute to the field E_2'.

3. If an orientation is found for which the magnetoresistance does contribute to E_2', how can one eliminate its effect and obtain the true Hall voltage?

4. Derive the magnetoresistance equation for the case where the magnetic flux is in the same direction as the current flow. Show that the following is true for a spherical energy surface for which there is no increase in magnetoresistance when the magnetic flux is in the same direction as the current flow:

$$\xi_{11} = \xi_{12} + 2\xi_{66} = \zeta_{111} = 6\zeta_{112} + \zeta_{122} + 8\zeta_{442} = \zeta_{123} + 2\zeta_{441} = 0$$

REFERENCES CHAPTER 11

1. "American Institute of Physics Handbook," page 9–45. McGraw-Hill, New York, 1962.
2. W. P. Mason *in* "Physical Acoustics" (Warren P. Mason, ed.), Vol. 1B, Chapter X. Academic Press, New York, 1964.
3. C. A. Wert and R. M. Thomson, "Physics of Solids," Chapter XII. McGraw-Hill, New York, 1964.
4. W. P. Mason, W. H. Hewitt, and R. F. Wick, *J. Appl. Phys.* **24,** 166 (1953).
5. H. B. G. Casimir, *Rev. Mod. Phys.* **17,** 343 (1945).
6. G. L. Pearson and H. Suhl, *Phys. Rev.* **83,** 768 (1951).
7. H. Welker and H. Weiss *in* "Solid State Physics" (F. Seitz and D. Tumbull, eds.), Vol. 3, Chapter 1. Academic Press, New York, 1952.
8. R. F. Wick, *J. Appl. Phys.* **25,** 741 (1954).
9. I. M. Ross and E. N. Saker, *J. Electron.* **1,** 223 (1955).
10. J. M. Moss, "Solid State Devices Using Indium Antimonide," *Proc. Symposium on the Role of Solid-State Phenomena in Electrical Circuits.* Wiley (Interscience), New York, 1957.
11. G. Arlt, *Solid-State Electron.* **1,** 75 (1960).
12. G. Arlt, Solid-State Circuit Conference, IRE. Philadelphia (Feb. 1961).

CHAPTER 12 • Thermoelectricity

12.1 Introduction

Thermoelectricity is an interaction process between the thermal variables, temperature gradient and heat flow, and the electrical variables, current densities and electric fields. The voltage produced between the hot and cold junctions of two dissimilar metals has long been used in the measurement of temperature. Recently the much higher thermoelectric efficiency obtained with certain semiconductors and semimetals has improved the possibilities of direct conversion of heat to electricity and of the refrigeration effect caused by a flow of current through a semiconductor junction between p and n semiconductors. The figure of merit z for both power conversion and refrigeration is the ratio of the square of the thermoelectric power, denoted by the letter S, to the product of the resistivity times the thermal conductivity. The thermoelectric power is defined as the difference in the electromotive forces between the hot and cold ends per degree centigrade. The thermo-electric power is important because it determines the thermal-electrical conversion. The lower the electrical resistivity, the less electrical power is turned into heat, while the lower the thermal conductivity, the greater the temperature separation that can be maintained in the sample. This figure of merit can be made high for such semiconductors as lead telluride and bismuth telluride. Wolfe[1] and Smith and Wolfe[2] have found that alloys of the semimetals bismuth and antimony can have a high figure of merit for low temperatures—that is, around 100° K—and at 200° K they are still higher than semiconductors. These semimetals have a very high mobility, which may be 10,000 times higher than that for a metal. As a result they have a very large magnetoresistance effect. Also, since about half the heat

[1] See Wolfe [1].
[2] See Smith and Wolfe [2].

conductivity is due to electrons and holes, a magnetic flux has an appreciable effect on the thermal conductivity. Wolfe and Smith[3] have also shown that the thermoelectric power increases in a magnetic flux and that the figure of merit can be increased nearly a factor of two when semimetal alloys are put in a magnetic flux.

A magnetic flux also produces some other effects that are potentially interesting. In the Ettinghausen effect a transverse magnetic field can produce a temperature gradient at right angles to the electric current and the magnetic field. In the Nernst effect an electric field is produced at right angles to the heat flow and the magnetic flux. In the Righi-Leduc effect a temperature gradient is produced at right angles to a heat current flow when a magnetic flux is exerted at right angles to the other two. This last effect is a direct analogue of the Hall effect, and in fact all of these effects result from the same force that produces the Hall effect—namely, the sidewise force due to the vector product of the flux and the electron or hole velocity.

The thermoelectric effect consists of two parts, the effect at the junctions and the effect in the medium. The junction parts are usually treated by considering two thermal reservoirs of slightly different temperatures with two sets of junctions immersed in them and assuming that the effects at the junctions are large compared to the effects in the medium in between. The treatment can be generalized by considering the effects in the medium primarily by introducing the chemical potential $\bar{\mu}$ and the absolute thermoelectric coefficients (Thomson) Σ for each medium. The effect at the junction represents a difference between these two sets of functions for the two media. The gradient of $\bar{\mu}$ and the complete tensor for $\sum_{ij}$ are then general enough to take account of any crystal effects. Differences in the normal components represent the junction effects. By considering $\bar{\mu}$ and $\sum_{ij}$ as functions of B and introducing terms similar to the Hall-effect term of Eq. (11.16) for both sets of variables, the magnetic-flux effects can be accounted for.

12.2 Derivation of Thermoelectric Effects in Isotropic Solids

For isotropic solids in the absence of magnetic fluxes, there are three thermoelectric effects.

1. *The Seebeck effect* results in a potential difference between the hot and cold junctions of two different metals a and b. If a condenser is inserted in conductor a, as shown by Fig. 12.1, it becomes charged. The Seebeck constant S_{ab} is the ratio $\Delta\varphi/\Delta\Theta$, where φ is the electrical potential.
2. *The Peltier heat.* When current is flowing through a junction between two metals it is found that heat must be continuously added or subtracted

---

[3] See Wolfe and Smith [3].

from the junction in order to keep the junction at a constant temperature. If the current is reversed, the sign of the heat production is reversed. The Peltier equation is

$$\dot{Q}_{ab} = \pi_{ab}i \tag{12.1}$$

where $\dot{Q}_{ab}$ is the rate at which heat is absorbed at the junction when the current i passes from metal a to metal b. π_{ab} is the Peltier coefficient, which depends on the two metals, a and b.

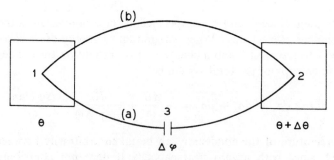

FIG. 12.1 Elements of the thermocouple.

3. *The Thomson heat* is a heating or cooling effect in a homogeneous conductor carrying a current in the direction of a thermal gradient. The heat that must be supplied to a piece of the wire in which the temperature rise $d\Theta$ is given by the equation

$$d\dot{Q} = \tau i \, d\Theta \tag{12.2}$$

τ is the Thomson coefficient.

The relation between the Seebeck constant S_{ab}, the Peltier constant π_{ab}, and the Thomson heat τ is usually proved[4] by considering the circuit of Fig. 12.1. The thermocouple is made of two metals, a and b, with junctions at temperatures Θ and $\Theta + \Delta\Theta$ in the two heat reservoirs 1 and 2. In the wire a a condenser 3 is inserted. Due to the action of the thermoelectric effect, a potential difference $\Delta\varphi$ is formed over the condenser. It is assumed that the condenser has no heat capacity and that the wires a and b are at every point in contact with a heat reservoir of large heat capacity with the same temperature as the wire at the point of contact.

The thermodynamic function that is of greatest interest in thermoelectric derivations is the Gibbs function G. This follows from the fact that G is constant for any phase change. (See, for example, the discussions in reference [1] of Chapter 4.) It is seen, then, that across the boundary between two different materials G does not change and hence $dG = 0$. The Gibbs function

[4] See DeGroot [4].

for each electron can be written in the form

$$G = U - \sigma\Theta - \varphi e \tag{12.3}$$

where U is the internal energy, Θ the absolute temperature, σ the entropy, φ the electrical potential, and e the charge of the electron. The differential of G can be written in the form

$$dG = dU - \Theta \, d\sigma - \varphi \, de = 0 \quad \text{or} \quad \Theta \, d\sigma = dU - \varphi e \, dn \tag{12.4}$$

for every point since Θ and φ are specified at these points. dn is the change in the number of electrons. When an amount of energy dU is carried from reservoir 1 to reservoir 2 and a charge $e \, dn$ is brought on the condenser, the change in entropy of the total system is

$$d\sigma = d\sigma_1 + d\sigma_2 + d\sigma_3 = -\frac{dU}{\Theta} + \frac{dU}{\Theta + \Delta\Theta} - \frac{\Delta\varphi \, e \, dn}{\Theta} \tag{12.5}$$

The temperature of the condenser can be taken arbitrarily between Θ and $\Theta + \Delta\Theta$, since with a zero heat capacity it does not contribute to the calculation. Equation (12.5) can be written in the form for an entropy that varies with time

$$\frac{d\sigma}{dt} = -\frac{dU}{dt} \times \frac{\Delta\Theta}{\Theta^2} - e\frac{dn}{dt}\frac{\Delta\varphi}{\Theta} \tag{12.6}$$

Hence for this case the forces are respectively $\Delta\Theta/\Theta^2$ and $\Delta\varphi/\Theta$. We can, however, multiply through by Θ since the result

$$\Theta \frac{d\sigma}{dt} = \dot{Q} \tag{12.7}$$

is proportional to the change of entropy and hence will satisfy the condition of Eq. (9.4). Then the forces and flows are respectively

$$\text{Forces:} \quad -\frac{\Delta\Theta}{\Theta} ; \quad -\Delta\varphi \quad \text{flows:} \quad \frac{dU}{dt} = j; \quad e\frac{dn}{dt} = i \tag{12.8}$$

where j is the total heat flow in the wire and i the total current flow.

Hence by Onsager's relations, Eqs. (9.2), we have

$$i = -L_{11}\Delta\varphi - L_{12}\frac{\Delta\Theta}{\Theta}$$

$$j = -L_{21}\Delta\varphi - L_{22}\frac{\Delta\Theta}{\Theta} \qquad \text{where} \quad L_{21} = L_{12} \tag{12.9}$$

To determine the three coefficients and their relationships, we first assume that i vanishes. Then from the first of (12.9) we have

$$\frac{\Delta \varphi}{\Delta \Theta} = - \frac{L_{12}}{L_{11}\Theta} = S_{ab} \qquad (12.10)$$

the Seebeck coefficient, which is also called the thermoelectric power.

Another special case results if we take a fixed value of potential difference $\Delta \varphi$ and assume that $\Delta \Theta = 0$. Then the ratio

$$\frac{j}{i} = \frac{L_{12}}{L_{11}} = \pi_{ab} \qquad (12.11)$$

This is the Peltier coefficient, which relates the evolution of heat at the junction to the flow of current i in the isothermal state. The sign of the Peltier coefficient is taken in such a way that π_{ab} is the heat absorbed by the junction when the unit current passes from metal a to metal b.

The relation between S_{ab} and π_{ab} can be obtained from (12.10) and (12.11) by solving for L_{12} in both equations and eliminating it. The result is

$$S_{ab} = - \frac{\pi_{ab}}{\Theta} \qquad (12.12)$$

which is known as Thomson's second relation. It follows directly from Onsager's reciprocal relation.

Thomson's first relation is a consequence of the first law of thermodynamics. If we join the two sides of the a wire and send a current around the closed loop, the thermal and electrical energies are conserved when

$$\pi_{ab} i - (\pi_{ab} + \Delta \pi_{ab})i + (\tau_b - \tau_a)i\,\Delta \Theta = i\,\Delta \varphi \qquad (12.13)$$

where $\Delta \pi_{ab}$ is the change in π_{ab} at the higher temperature $\Theta + \Delta \Theta$. Cancelling out common terms,

$$-\Delta \pi_{ab} + (\tau_b - \tau_a)\,\Delta \Theta = \Delta \varphi \qquad (12.14)$$

When $\Delta \varphi$ is eliminated by the use of Eq. (12.10) and S_{ab} by the use of Eq. (12.12), we have

$$\tau_b - \tau_a = \frac{d\pi_{ab}}{d\Theta} - \frac{\pi_{ab}}{\Theta} \qquad (12.15)$$

as the connection between the Thomson heats and the Peltier heat. Alternatively, we can eliminate the Peltier heat by using the relation

$$\Theta S_{ab} = \left(\Theta \frac{d\varphi}{d\Theta} \right) = -\pi_{ab} \quad \text{or} \quad \Delta \left(\Theta \frac{d\varphi}{d\Theta} \right) = -\Delta \pi_{ab} \qquad (12.16)$$

Dividing through by $\Delta \Theta$, we have

$$\tau_a - \tau_b = \Theta \frac{d^2 \varphi}{d\Theta^2} \qquad (12.17)$$

12.3 Thermoelectric Effects in Continuous Isotropic Media

In order to determine the thermoelectric effects for crystalline media it is necessary to derive the relations for continuous media rather than isolated wires and junctions. Such a theory was worked out by DeGroot[5] and Callen[6] and has been generalized in a manner to apply to crystals by Domenicali.[7] In this theory the chemical potential $\bar{\mu}$ is used in place of the electrical potential. The chemical potential can be written in the form

$$\bar{\mu} = \mu - e\varphi \qquad (12.18)$$

where μ is the part of U of the Gibbs function G of Eq. (12.3) that depends on the composition and temperature of the medium, and $-e\varphi$ is the electrical part, where e is the charge on the electron. This function is used, since from (12.4), at an interface where the temperature is the same on both sides, the difference of the entropies is equal to the difference of the chemical potentials of the two sides. The fluxes considered are the number of electrons per square meter carried across the surface. This flux is designated by J^e. The product of J^e times the charge $-e$ represents the current density I. The other flux is the heat-current density h. The two current densities can then be represented in the form

$$J_i^e = -\alpha \frac{\partial \bar{\mu}}{\partial x_i} - \beta \frac{\partial \Theta / \partial x_i}{\Theta}\,; \qquad h_i = -\beta \frac{\partial \bar{\mu}}{\partial x_i} - \gamma \frac{\partial \Theta / \partial x_i}{\Theta} \qquad (12.19)$$

On account of the Onsager relations, the constant β can be written for the two cross terms. When μ does not vary with position, then the gradient of $\bar{\mu}$ reduces to e times the gradient of φ and these equations are similar to the thermal and electrical relations discussed in Chapter 9. Since $J^e = -I/e$, we find that

$$\alpha e^2 = \sigma \qquad \text{or} \qquad \alpha^{-1} = e^2 \rho \qquad (12.20)$$

where ρ is the electrical resistivity. Similarly, γ is the thermal conductivity $k^{\bar{\mu}}$ divided by the absolute temperature Θ, in the absence of a gradient of chemical potential.

The rate of production of heat energy can be obtained by eliminating $\partial \bar{\mu} / \partial x_i$ from (12.19). We have from the first of (12.19)

$$-\frac{\partial \bar{\mu}}{\partial x_i} = \alpha^{-1} J_i^e + \alpha^{-1}\beta \frac{\partial \Theta / \partial x_i}{\Theta} \qquad (12.21)$$

[5] See DeGroot [4].
[6] See Callen [5].
[7] See Domenicali [6].

Substituting this value in the last of Eqs. (12.19) this becomes

$$h_i = \alpha^{-1}\beta J_i{}^e + (\alpha^{-1}\beta^2 - \gamma)\frac{\partial\Theta/\partial x_i}{\Theta} \tag{12.22}$$

A substitution is now made of

$$\Sigma = \frac{\alpha^{-1}\beta}{\Theta} \tag{12.23}$$

It will be shown later that $-\Sigma/e$ is the absolute thermoelectric power of the conductor at the particular point and temperature under consideration. Using this substitution, (12.21) and (12.22) can be written as

$$-\frac{\partial\bar{\mu}}{\partial x_i} = e^2\rho J_i{}^e + \Sigma\frac{\partial\Theta}{\partial x_i}$$
$$h_i = \Theta\sum J_i{}^e - k^J\frac{\partial\Theta}{\partial x_i} \tag{12.24}$$

where $k^J = (\gamma - \alpha^{-1}\beta^2)/\Theta$ is the thermal conductivity at zero current flow. The two thermal conductivities are related by the equation

$$k^J = k^{\bar{\mu}} - \frac{\Sigma^2\,\Theta}{\rho e^2} = k^{\mu}\left[1 - \frac{(\Sigma/e)^2\Theta}{\rho k^{\bar{\mu}}}\right] \tag{12.25}$$

As with the dielectric constants of (6.22) or the compliance constants of (6.26), the ratio of the thermal conductivity at constant $\bar{\mu}$ to the ratio at constant J^e is equal to unity minus a coupling factor squared. Hence the thermoelectric coupling factor squared for a single medium is

$$\frac{(\Sigma/e)^2\Theta}{\rho k^{\bar{\mu}}} \tag{12.26}$$

Since it will be shown that $-\Sigma/e$ is equal to an absolute thermoelectric power S, the square of the coupling is equal to the figure of merit z times Θ, or

$$\frac{S^2\Theta}{\rho k^{\bar{\mu}}} = z\Theta \tag{12.27}$$

When there is no electric current flow, the rate of energy flow across a surface, which we designate by the letter $J_i{}^U$, is equal to the thermal energy flow h_i. When current is flowing, chemical energy $(\mu - \varphi e) = \bar{\mu}$ is being transported per unit electron across the boundary, so that we have

$$J_i{}^U = h_i + \bar{\mu}J_i{}^e \tag{12.28}$$

since J_i^e determines the number of units of charge e being transported across the boundary. As a result, the last equation of (12.24) can be generalized to the form

$$J_i^U = (\Theta \sum + \bar{\mu})J_i^e - k^J \frac{\partial \Theta}{\partial x_i}$$

$$-\frac{\partial \bar{\mu}}{\partial x_i} = e^2 \rho J_i^e + \sum \frac{\partial \Theta}{\partial x_i} \tag{12.29}$$

The last equation is repeated from (12.24) in order to give the two equations of interest in energy convergence.

The rate at which heat is evolved per unit volume for steady-state conditions is the negative of the divergence of J_i^U. Hence

$$-\frac{\partial J_i^U}{\partial x_i} = -J_i^e \left[\frac{\partial}{\partial x_i}(\Theta \sum) + \frac{\partial \bar{\mu}}{\partial x_i} \right] - (\Theta \sum + \bar{\mu})\frac{\partial J_i^e}{\partial x_i} + \frac{\partial}{\partial x_i}\left(k^{J^e} \frac{\partial \Theta}{\partial x_i} \right) \tag{12.30}$$

Under steady-state conditions, the number of charged particles entering a volume is the same as that leaving the volume so that the divergence of J_i^e is zero and the second term on the right disappears.

Expanding out the first term, three types of heat accumulations appear. These are

$$-\frac{\partial J_i^U}{\partial x_i} = -J_i^e \left[\sum \frac{\partial \Theta}{\partial x_i} + \frac{\partial \bar{\mu}}{\partial x_i} \right] - \Theta J_i^e \left(\frac{\partial \sum}{\partial x_i} \right) + \frac{\partial}{\partial x_i}\left(k^{J^e} \frac{\partial \Theta}{\partial x_i} \right) \tag{12.31}$$

The gradient of $\bar{\mu}$ can be eliminated by using the last of Eqs. (12.29), and Eq. (12.31) can be divided into three types of heats: the Joule heat, the thermoelectric heat, and the conduction heat. These have the forms

$$-\frac{\partial J_i^U}{\partial x_i} = (J_i^e e)^2 \rho - \Theta J_i^e \left(\frac{\partial \sum}{\partial x_i} \right) + \frac{\partial}{\partial x_i}\left(k \frac{\partial \Theta}{\partial x_i} \right) \tag{12.32}$$

where k is the thermal conductivity at constant J^e. Since $-J_i^e e$ is the current density I_i, the first term becomes

$$(I_1^2 + I_2^2 + I_3^2)\rho = \text{Joule heat} \tag{12.33}$$

The last term represents the convergence of heat due to thermal conduction, while the middle term

$$\dot{Q} = -\Theta J_i^e \frac{\partial \sum}{\partial x_i} \tag{12.34}$$

is the thermoelectric heat. If the state of the body is steady, there can be no evolution of heat, and if a Joule heat exists it must be cancelled out by the other two terms.

All of these equations pertain to a single medium, but since they are set up in terms of functions that are continuous across boundaries, they can be used to derive the relations of Section 12.2. The first relation involving the Seebeck coefficient S_{ab} follows from the last of Eqs. (12.29) when we set $J_i^e = 0$. Then

$$\frac{\partial \bar{\mu}}{\partial x_i} = \frac{\partial}{\partial x_i}(\mu - \varphi e) = -\sum \frac{\partial \Theta}{\partial x_i} \tag{12.35}$$

If we integrate around the path of Fig. 12.1 from one side of the condenser to the other, then by Eq. (2.98), this results in the integration

$$\int d\mu - e \int d\varphi = -\int \sum d\Theta \tag{12.36}$$

If we start from one side of the condenser and integrate to the other side, μ has the same initial and final values since we have the same metal and same temperature, and hence the first integral vanishes. The second one becomes $-e \, \Delta\varphi$. For the integral on the right-hand side, the metal a and the metal b have constant values of Σ, and hence the integral from the condenser to B and from B back to the condenser results in the terms

$$-e \, \Delta\varphi = -(\Sigma^b - \Sigma^a) \, \Delta\Theta \tag{12.37}$$

Hence the Seebeck coefficient

$$S_{ab} = \frac{\Delta\varphi}{\Delta\Theta} = \frac{\Sigma^b - \Sigma^a}{e} \tag{12.38}$$

The terms $-\Sigma^b/e$ and $-\Sigma^a/e$ are absolute thermoelectric powers of the two materials. For semiconductors of the n and p types, these thermoelectric powers are of opposite sign so that they can be made to add.

The Peltier and Thomson heats can be derived directly from Eq. (12.34). Since the divergence of J_i^e is zero, then $\dot{Q}$ can be written in the form

$$\dot{Q} = -\Theta \frac{\partial}{\partial x_i}(J_i^e \Sigma) \tag{12.39}$$

At an interface (12.39) gives the evolution of heat at the junction as being due to the difference of the normal components of $J_i^e \Sigma$ times the temperature, or

$$\dot{Q} = -\Theta[\Sigma^a - \Sigma^b]J_3^e = \Theta\left[\frac{\Sigma^a - \Sigma^b}{e}\right]I_3 = \pi_{ab}I_3 \tag{12.40}$$

where J_3^e is the component along the axis of the medium, which is assumed to lie along the $z = x_3$ axis, and $I_3 = -J_3^e e$. The last relation follows from the definition of the Peltier effect given by Eq. (12.1). Since by (12.38), $S_{ab} = -[(\Sigma^a - \Sigma^b)/e]$, the relation (12.12) follows from (12.40) and (12.38).

The Thomson heat arises directly from Eq. (12.34) when we limit consideration to a single medium. If we take the axis of the wire along the $x_3 = z$ direction and consider the current flow to occur along the direction of increasing z, then J_i^e will be in the direction of decreasing z. Hence (12.34) gives

$$\dot{Q} = \Theta J_3^{\,e} \frac{\partial \Sigma}{\partial x_3} = \Theta(J_3^{\,e} e) \frac{\partial (\Sigma/e)}{\partial x_3} = -\Theta I_3 \frac{\partial (\Sigma/e)}{\partial x_3} \qquad (12.41)$$

The heat evolved per unit length is, for the area A,

$$\dot{Q} A \, dx_3 = -\Theta i_3 \, d(\Sigma/e) = \tau i_3 \, d\Theta \qquad (12.42)$$

Hence

$$\tau = -\Theta \left(\frac{\partial (\Sigma/e)}{\partial \Theta} \right) \qquad (12.43)$$

The difference between the Thomson heats for the two media is

$$(\tau_a - \tau_b) = -\Theta \left[\frac{\partial}{\partial \Theta} \left(\frac{\Sigma^a - \Sigma^b}{e} \right) \right] = \Theta \frac{\partial}{\partial \Theta} (S_{ab}) = \Theta \frac{\partial^2 \varphi}{\partial \Theta^2} \qquad (12.44)$$

and hence all the thermodynamic relations of Section 12.2 follow from the continuous-media development.

12.4 Practical Applications of the Thermoelectric Effects

As shown in Section (12.5) these results can be generalized to crystal media by introducing second-rank tensors in place of the constants α, Σ, and k of Eqs. (12.24). For isotropic materials and cubic crystals, however, there is only one component of a second-rank tensor, and hence the formulation of Section (12.3) is general enough for these cases. Since many applications use isotropic metals or cubic semiconductors, these formulas are sufficient. For the semimetals bismuth and antimony, however, the symmetry is trigonal, and it requires the more general formulas of Section (12.5) to provide all the necessary relations.

One of the largest uses for the thermoelectric effect is the measurement of temperature. For this purpose two different small wires, such as copper and constantan, are joined together at their two ends. One junction is usually immersed in an ice water bath, and the other is put on the device whose temperature is to be measured. One of the pairs of wires is broken in the manner of the condenser of Fig. 12.1 and connected to the terminals of a balancing potentiometer. An inverse voltage is put on by the potentiometer and is adjusted until no current flows through the galvanometer. Under these conditions the potentiometer voltage just equals the thermocouple voltage, and by knowing the calibration versus temperature difference, a measurement of the temperature is obtained.   Figure 12.2 shows a calibration voltage versus temperature for a copper-constantan thermocouple. It will be observed that at temperatures close to absolute zero, the curve is very flat, and hence thermoelectric means cannot be used to measure temperatures below 20° K with any certainty. However, certain gold-cobalt alloys[8] have been used down to the helium range.

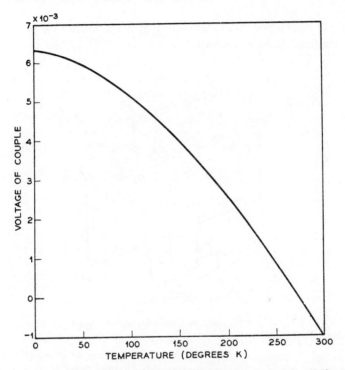

FIG. 12.2   Calibration voltage of a copper-constantan thermocouple.   Reference temperature is 273° K (0° C) maintained by ice-water bath.

Other uses for thermoelectric materials are in the direct production of electric power from heat sources and in the refrigeration of small chambers by means of electric current flowing through the junction. The application of thermoelectricity to power generation dates back over a hundred years to the work of Seebeck in 1822.[9] In fact the original work, which included some measurements on semiconductors, produced efficiencies of conversion of the same order as other contemporary methods of power generation.

[8] See White [7].
[9] See Seebeck [8].

However, this type of work was not investigated further. It was not until recent work on semiconductors was carried out in connection with device developments such as transistors that interest in thermoelectric power production was renewed. The efficiencies obtained have been on the order of 10 to 15 per cent, which makes them of interest for power supplies on satellites that are continuously exposed to sunlight. Radioisotopes and nuclear piles have also been used as heat sources. These do not require sunlight.

A simple calculation of the efficiency of a thermoelectric power source can be obtained from a consideration of Fig. 12.3. In this power source,

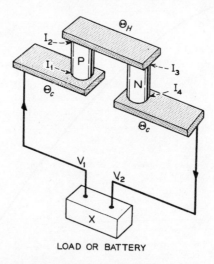

LOAD OR BATTERY

FIG. 12.3   Elements of a thermoelectric couple for converting heat into electrical energy or for refrigeration (after R. Wolfe).

which is made up of n- and p-type semiconductors, heat H_1 is absorbed at the hot end at temperature Θ_H and flows down the conductors and is lost to the sink at temperature Θ_c. Current is generated by the thermoelectric effect and delivers power to the load R_L. The conversion efficiency E_F is defined by the equation

$$E_F = \frac{\text{power supplied to the load}}{\text{heat absorbed in thermocouple per unit time}} \qquad (12.45)$$

The useful power is $i^2 R_L$, where i is the current flowing around the loop. If we neglect the Thomson heat[10]—which in any case is zero if we assume S_{ab} is independent of the temperature—the heat absorbed at the hot junction

[10] See Cohen and Abeles [9] for a discussion of the case of temperature-varying parameters.

at temperature Θ_H consists of the Peltier heat $\pi_{ab}i$ plus the ordinary heat absorbed by the material. Then the input heat $\dot{Q}_1$ is

$$\dot{Q}_1 = K\Theta_H - i\pi_{ab} = iS_{ab}\Theta_H + K\Theta_H$$

The output heat flow $\dot{Q}_2$ is equal to

$$\dot{Q}_2 = +K\Theta_c + \tfrac{1}{2}i^2R \tag{12.46}$$

where K is the thermal conductivity of the two branches in parallel and R the series resistance of both arms. It is necessary to subtract half the Joule heat in the conductor because this quantity is returned to the hot junction. Hence E_F is

$$E_F = \frac{i^2R_L}{iS_{ab}\Theta_H + K(\Theta_H - \Theta_c) - \tfrac{1}{2}i^2R} \tag{12.47}$$

For a small difference in temperature—which results in a small current i—the last term in the denominator can be neglected. Since the open-circuit voltage of the thermocouple is, by (12.10),

$$V = S_{ab}(\Theta_H - \Theta_c) \tag{12.48}$$

then the current i is given by

$$i = \frac{V}{R + R_L} = \frac{S_{ab}(\Theta_H - \Theta_c)}{(R + R_L)} \tag{12.49}$$

The output power is then

$$P_o = i^2R_L = \frac{S_{ab}^2(\Theta_H - \Theta_c)^2}{(R + R_L)^2} R_L \tag{12.50}$$

Direct differentiation shows that this power is a maximum when

$$R_L = R \tag{12.51}$$

Inserting the value of i in (12.47) and cancelling out common terms,

$$E_F = \frac{(\Theta_H - \Theta_c)}{\Theta_H} \left(\frac{S_{ab}^2\Theta_H}{4RK(1 + S_{ab}^2\Theta_H/2RK)} \right) \tag{12.52}$$

The thermal conductivity K is measured with the current i suppressed, and hence it should have the superscript J. It is seen by comparing the denominator of (12.52) with the two types of thermal coefficients of Eq. (12.25) that the term in the denominator of (12.52) is the thermal conductivity when the current is flowing through the termination. This we designate by K_0. The first term of (12.52) is simply the thermodynamic efficiency of a reversible Carnot cycle. The second factor represents the decrease in efficiency resulting from the irreversible heat conduction along the branches and the power dissipation due to Joule heat. We denote the expression of (12.53) as the figure of merit z

$$z = \frac{S_{ab}^2}{R K_0} \tag{12.53}$$

and hence the efficiency relative to a reversible Carnot cycle is

$$\frac{z \Theta_H}{4} \tag{12.54}$$

for small temperature differences. For large temperature differences, for which the last term in the denominator has to be taken account of, it has been shown[11] that the efficiency under optimum load conditions takes the form

$$E_F = \left(\frac{\Theta_H - \Theta_c}{\Theta_H}\right) \frac{\sqrt{1 + z\overline{\overline{\Theta}}} - 1}{\sqrt{1 + z\overline{\overline{\Theta}}} + \Theta_c/\Theta_H} \; ; \qquad \overline{\overline{\Theta}} = \left(\frac{\Theta_H + \Theta_c}{2}\right) \tag{12.55}$$

For small values of $z\Theta$, this expression reduces to (12.52).

If the resistances R_a and R_b of the two branches are matched to give z a maximum value, it is found that

$$\frac{R_a}{R_b} = \left(\frac{\rho_a k_a}{\rho_b k_b}\right)^{1/2} \tag{12.56}$$

where ρ_a and ρ_b are the resistivities and k_a and k_b the thermal conductivities measured with current flowing for the two samples. For this case the figure of merit reduces to

$$z = \frac{(S_a - S_b)^2}{(\sqrt{\rho_a k_a} + \sqrt{\rho_b k_b})^2} \tag{12.57}$$

where S_a and S_b are thermoelectric power with respect to some third medium.

For most semiconductors, n-type doping produces an opposite sign for the thermoelectric constant than does p-type doping. For dopings that give approximately the same resistivities, the thermoelectric coefficients are about equal and opposite. Even for high doping levels, the thermal conductivity is due mainly to heat waves (phonons) in the lattice rather than to transport of charges, and hence the conductivities of both samples are nearly the same. Hence

$$z = z_a = \frac{S_a^2}{\rho_a k_a} \tag{12.58}$$

is a good approximation for p and n semiconductors.

---

[11] See Altenkirch [10].

Many semiconductors have been investigated[12] as a function of doping level, temperature, and other variables. In the range up to 250° C solid-solution alloys of Bi_2Te_3, Bi_2Se_3, Sb_2Te_3, and Sb_2Se_3 provide the best p- and n-type materials. Dopings in the order of 10^{25} carriers/meter³ produce the best results. Figure 12.4 shows the figure of merit z for a number of

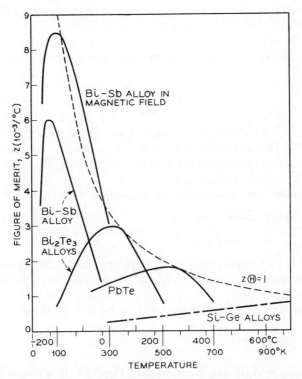

FIG. 12.4   The variation of z with temperature for various materials (after R. Wolfe).

alloys as a function of temperature. The figure also shows a similar curve for a bismuth-antimony alloy at low temperatures and the increased figure of merit obtainable by the use of a magnetic field. The dashed curve $z\Theta = 1$ is an effective figure of merit independent of the temperature.

By using bismuth telluride in the temperature range from 250° C to 25° C and the average value of $z = 3 \times 10^{-3}$ shown by Fig. 12.4, the efficiency relative to the Carnot efficiency is 25.2 per cent. Since the Carnot efficiency is itself equal to 43 per cent, the theoretical over-all efficiency from thermal to electrical is 10.8 per cent. This is ideal, since it does not take account of

[12] See Rose et al. [11] and Wolfe and Smith [12].

thermal losses through insulation, and so on. The efficiency can be increased by covering a wide temperature range and by cascading the thermocouples. For the same figure of merit this process produces a higher efficiency,[13] and it also allows the use of the best material for the temperature range of interest.

The other thermoelectric device of interest is the thermoelectric cooling (or heating) device. When an electric current is flowing through two thermoelectric junctions, heat is absorbed from one junction and delivered to the other by the Peltier effect. If we define the coefficient of performance of the device as the ratio of the heat transferred from one junction to the other to the electrical work expended, this ratio can be greater than unity and ideally is

$$\eta = \frac{\Theta_c}{\Theta_H - \Theta_c} \tag{12.59}$$

where Θ_H is the temperature of the sink in $°K$ and Θ_c is the temperature of the source. The quantities of interest in refrigeration are the coefficient of performance and the maximum temperature difference that can be maintained between the hot and cold junction.

This problem was first studied by Altenkirch,[14] who found the expression

$$\eta = \frac{\Theta_c \sqrt{1 + z\overline{\Theta}} - \Theta_H}{(\Theta_H - \Theta_c)[1 + \sqrt{1 + z\overline{\Theta}}]} \quad \text{where} \quad \overline{\Theta} = \left(\frac{\Theta_H + \Theta_c}{2}\right) \tag{12.60}$$

When $z\overline{\Theta}$ is very large, this expression reduces to (12.59). Hence again the same figure of merit times the average temperature controls the coefficient of performance.

The maximum temperature ratio Θ_H/Θ_c turns[15] out to be

$$\frac{\Theta_H}{\Theta_c} = \sqrt{1 + z\overline{\Theta}} \tag{12.61}$$

For example, with the semimetal data of Fig. 12.4, in a magnetic field, with a starting temperature of $200° K$ $(-73° C)$ a lowering of temperature of $62° K$ can be obtained. By using a cascade of such devices,[16] an even lower temperature can be realized.

12.5 Thermoelectric Effects in Continuous Anisotropic Media

It was first shown by Domenicali[17] that Eqs. (12.24) can be generalized in such a way that they apply to crystals. The only modification required

---

[13] See Rose *et al.* [11] and Wolfe and Smith [12].
[14] See Altenkirch [10].
[15] See Altenkirch [10].
[16] See Rose *et al.* [11] and Wolfe and Smith [12].
[17] See Domenicali [6].

is that the constants $\alpha^{-1} = e^2\rho$, Σ, and k^J be replaced by the second-rank tensors

$$\alpha_{ij}^{-1} = e^2\rho_{ij}, \quad \Sigma_{ij}, \quad \text{and} \quad k_{ij}^J \qquad (12.62)$$

Then Eqs. (12.24) can be written in the form

$$-\frac{\partial\bar{\mu}}{\partial x_i} = e^2\rho_{ij}J_j^e + \Sigma_{ij}\frac{\partial\Theta}{\partial x_j}$$

$$h_i = \Theta\sum_{ij}J_j^e - k_{ij}^J\frac{\partial\Theta}{\partial x_j} \qquad (12.63)$$

Using the extended definition of energy flow given by (12.28), the last equation becomes

$$J_i{}^U = (\Theta\sum_{ij} + \bar{\mu}\,\delta_{ij})J_j^e - k_{ij}^J\frac{\partial\Theta}{\partial x_j} \qquad (12.64)$$

It follows from Onsager's principle that the resistivity tensor ρ_{ij} and the thermal conductivity tensor k_{ij}^J are symmetrical. Since $\sum_{ij}$ relates two different flows, however, it is not symmetrical. Crystal symmetries determine the possible constants, as shown by Table B.2 of Appendix B.

The rate of convergence of heat is

$$-\frac{\partial J_i{}^U}{\partial x_i} = -\left(\sum_{ij}\frac{\partial\Theta}{\partial x_j} + \delta_{ij}\frac{\partial\bar{\mu}}{\partial x_j}\right)J_j^e - \Theta\frac{\partial}{\partial x_i}(J_j^e\sum_{ij}) + \frac{\partial}{\partial x_i}\left(k_{ij}^J\frac{\partial\Theta}{\partial x_j}\right)$$

$$(12.65)$$

The first of Eqs. (12.63) shows that the first term is $e^2\rho_{ij}J_i^eJ_j^e$, which is the Joule heat

$$\rho_{ij}I_iI_j \qquad (12.66)$$

The last term represents the convergence of energy due to heat flow, and the middle term, which can be written

$$\dot{Q} = -\Theta\frac{\partial}{\partial x_i}(J_j^e\sum_{ij}) \qquad (12.67)$$

represents the heat evolved by thermoelectric effects.

Since $\dot{Q}/\Theta$ represents the convergence of the entropy and the right-hand term represents the convergence of the vector $J_j^e\sum_{ij}$ it is seen that this vector is not in general parallel to the current density $(-eJ_j^e)$. At an interface (12.67) gives the evolution of heat at the junction as being due to the difference of the normal components of $J_j^e\sum_{ij}$ times the temperature Θ. Alternatively, we can use the current $I_j = -eJ_j^e$ in place of J_j^e, and we obtain

$$\Theta\left[I_j\left(\frac{\sum_{j3}^a - \sum_{j3}^b}{e}\right)\right] = I_j(\pi_{j3}^a - \pi_{j3}^b) \qquad (12.68)$$

if the junction is perpendicular to the x_3 direction. The Peltier heat is then

$$\pi_{ij} = \Theta\left(\frac{\sum_{ij}}{e}\right) = -\Theta S_{ij} \tag{12.69}$$

where S_{ij} is the form of the Seebeck coefficient for crystals. It is obvious that in the general case a flow of heat can occur perpendicular to the current-flow direction by virtue of the cross-coupling coefficients in $\sum_{ij}$.

If we carry through the differentiation of Eq. (12.67), remembering that as in the isotropic case $\sum_{ij}$ varies not only with distance but also with a nonuniformity of the temperature distribution, there are three resulting terms. These are

$$\dot{Q} = -J_i^e \Theta\left(\frac{\partial \sum_{ij}}{\partial x_j}\right)_\Theta - J_i^e \Theta\left(\frac{\partial \sum_{ij}}{\partial \Theta}\right)_x \frac{\partial \Theta}{\partial x_j} - \Theta \sum_{ij} \frac{\partial J_i^e}{\partial x_j} \tag{12.70}$$

The first term at constant temperature represents the change in $\sum_{ij}$ across a junction and is the Peltier heat. The fact that J_i^e and x_j have different subscripts in this term shows that changes in $\sum_{ij}$ can occur in directions normal to the current flow. This represents the transverse Peltier effect, whereas the normal effect is given by terms for which $i = j$.

The second term is the Thomson heat, which is proportional to the current and the temperature gradient. Since temperature gradients not in the direction of the current flow are relevant, we have transverse Thomson heats as well as normal Thomson heats. The complete expression for the Thomson heat evolved per unit volume is, by definition,

$$\dot{Q} = -I_i \tau_{ij} \frac{\partial \Theta}{\partial x_j} \tag{12.71}$$

Hence, comparing this term with the second term of (12.70), we find

$$\tau_{ij} = -\frac{\Theta}{e}\left(\frac{\partial \sum_{ij}}{\partial \Theta}\right)_x \tag{12.72}$$

The last term of (12.69) represents an effect that is not present in an isotropic solid or a cubic crystal. It was first discovered by Bridgman[18] and is known as the Bridgman heat. It is connected with the nonuniformity of the current distribution. As an example of the Bridgman heat, consider the L-shaped crystal with arms of equal cross section shown by Fig. 12.5. Let us assume that it is made up of one of the crystal classes that has two components $\sum_{11}$ and $\sum_{33}$. We take one arm along the x_1 direction and the other along the x_3 direction. The whole crystal is at the same temperature Θ and has the current i flowing through the two branches. With equal areas

[18] See Bridgman [14, 15, 16].

for the two arms, a uniform current density flows in the x_1 direction in the first arm and then changes direction in the second arm. Applying Eq. (12.66) to the shaded area, it is seen that $-\dot{Q}/\Theta$ is the divergence of $J_j^e \sum_{ij}$ integrated over the volume. By Gauss' theorem—Eq. (2.89)—this is equal to the flux of the outward normal components of $J_j^e \sum_{ij}$ crossing the sides. The outward normal components of the two end surfaces are $-J^e \sum_{11}$ and

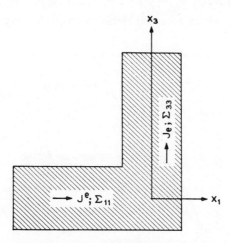

FIG. 12.5   Figure illustrating an example of Bridgman heat. Result is due to the change in direction of current in a crystal having the constants $\Sigma_{11} = \Sigma_{22}$; Σ_{33}.

$J^e \sum_{33}$. Near the surfaces of the crystal the electron current flows parallel to the surface, and outside the crystal it is zero. Hence there are no normal components along these surfaces, and

$$\dot{Q} = \Theta J^e (\textstyle\sum_{11} - \sum_{33}) \qquad (12.73)$$

This is an expression for the Bridgman heat, which in this case is due to a change in the current direction.

In general any nonuniform distribution of current density will give rise to nonvanishing terms

$$\frac{\partial J_i^e}{\partial x_j} \qquad (12.74)$$

in (12.70) and hence to a Bridgman heat.

12.6   Effect of Magnetic Flux on Thermoelectric Properties

When a magnetic flux is applied to a conductor carrying electric and thermal currents, new phenomena occur that have important uses. When the currents are entirely electrical, the new effects are the Hall effect and the magnetoresistance effect discussed in the preceding chapter. Since part of the heat flow is carried by electrons and holes, entirely analogous effects are the Righi-Leduc effect (analogue of the Hall effect) and the increase in thermal resistance corresponding to the magnetoresistance effect. The magnetothermal resistance effect accounts partly for the increased value of the figure of merit z of Fig. 12.4 for the semimetals. The remainder is due to an increased Seebeck constant resulting from the magnetic flux. On the

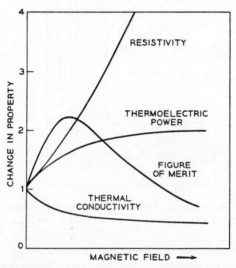

FIG. 12.6 Effect of magnetic flux on the three factors entering the expression for the figure of merit z. Each property is referred to its value at zero magnetic flux (after R. Wolfe).

other hand, the increased magnetoresistance constant tends to reduce z, and as shown by Fig. 12.6, there is an optimum value of the flux density. On account of the high electron mobility, the magnetic effects are particularly large for semimetals.

The effects of magnetic fluxes on the properties of isotropic materials have been discussed by Callen[19] and DeGroot.[20] The method consists in expanding the flows in terms of the forces by using the self- and cross-coupling coefficients L_{ij} of Eq. (9.1). Since most of the measured results, such as the Seebeck effect, Nernst effect, Ettinghausen effect, and Righi-Leduc effect, involve the forces rather than the flows, it is more convenient to

[19] See Callen [5].
[20] See DeGroot [4].

use the flows as the independent variables and the forces as the dependent variables. Then in these terms the magnetoresistance and Hall effects can be written for two directions as

$$-\frac{\partial \varphi}{\partial x_1} = E_1 = l_{11}I_1 + l_{12}I_2 \; ; \qquad -\frac{\partial \varphi}{\partial x_2} = E_2 = l_{21}I_1 + l_{22}I_2 \quad (12.75)$$

The constants l_{11} and l_{22}, since they are diagonal terms in the coefficient matrix, depend only on even power of the magnetic flux. The terms l_{12} and l_{21} have to satisfy the relation

$$l_{12}(B) = l_{21}(-B) \tag{12.76}$$

Since these terms depend on odd powers of the flux,

$$\frac{\partial l_{12}}{\partial B_3} = R_{23}B_3 \; ; \qquad \frac{\partial l_{21}}{\partial B_3} = -R_{23}B_3 \tag{12.77}$$

and hence $l_{21} = -l_{12}$. In terms of the variables of Eq. (11.16) Eq. (12.75) takes the form

$$E_1 = \rho_0(B)I_1 + R_0(B)I_2B_3 \; ; \qquad E_2 = -R_0(B)I_1B_3 + \rho_0(B)I_2 \quad (12.78)$$

where $\rho_0(B)$ is an even function of the flux and R_0 an odd function.

The complete equation for the magnetoresistive and Hall effects show that Eq. (12.75) is not general enough to agree with the effects in an anisotropic medium. If we take the current and field direction as x_1, the Hall-effect current and field direction as x_2, and the flux direction as x_3, the equations corresponding to (12.75) become

$$-\frac{\partial \varphi}{\partial x_1} = E_1 = \rho_{11}I_1 + (\rho_{12} + R_{33}B_3)I_2 \; ;$$

$$\tag{12.79}$$

$$-\frac{\partial \varphi}{\partial x_2} = E_2 = (\rho_{12} - R_{33}B_3)I_1 + \rho_{22}I_2$$

Hence the off-diagonal terms have a symmetric and an antisymmetric part. In practice it is better to write the two parts separately as in (11.16) rather than to try to incorporate them in a single matrix as in (12.75).

If we consider a specimen for which there is a temperature gradient between the two ends, but no electric current flow—that is, the two ends are not connected together electrically—it is evident that an equation similar to (11.16) can be written, in which heat flow takes the place of electric current flow. This equation takes the form

$$-\frac{\partial \Theta}{\partial x_i} = r_{ij}(B)h_j + \varepsilon_{ijk}h_j(L_{km}B_m) \tag{12.80}$$

where, for lack of other symbols, the constant L_{km} is designated as the Righi-Leduc coefficient measured for constant electric current. r_{ij} are the thermal resistivity functions, which vary as an even function of the applied magnetic flux. Variations may also occur in the constant L_{km}, but they are here neglected. As an example let us apply a temperature gradient in the x_1 direction, a magnetic flux in the x_3 direction, and consider the effect of B_3 on the temperature gradient and heat flow in the x_2 direction. Writing out the terms for the x_1 and x_2 directions we have

$$-\frac{\partial \Theta}{\partial x_1} = r_{11}h_1 + (r_{12} + L_{33}B_3)h_2 \; ; \qquad -\frac{\partial \Theta}{\partial x_2} = (r_{12} - L_{33}B_3)h_1 + r_{22}h_2$$

(12.81)

Hence there is a temperature gradient along the x_2 axis when a heat flow occurs along the x_1 axis. A temperature gradient can also occur due to the thermal resistance term r_{12}. It is always possible to transform the equations to the principal axes of the thermal resistance tensor, and for such axes the cross terms r_{ij}, $i \neq j$, disappear. For all crystal systems up to the monoclinic, these principal axes coincide with the crystallographic axes and the Righi-Leduc constants L_{ij} can be determined. As with the Hall coefficients R_{ij}, they are symmetrical. Sometimes the Righi-Leduc constants are expressed in terms of the applied temperature gradient. Then for $h_2 = 0$ we have

$$-\frac{\partial \Theta}{\partial x_2} = -\left(\frac{r_{12} - L_{33}B_3}{r_{11}}\right)\frac{\partial \Theta}{\partial x_1} = \left(\frac{L_{33}B_3}{r_{11}}\right)\frac{\partial \Theta}{\partial x_1}$$

(12.82)

if the cross thermal resistance term is eliminated. In general if sidewise heat flow h_2 occurs, Eqs. (12.81) refer only to an infinitesimal volume, and a partial differential equation with boundary conditions has to be solved before the heat flows and temperature gradients can be determined.

The remaining two thermoelectric effects occurring as a result of the magnetic field are the Nernst effect, which relates a cross voltage gradient to a heat flow, and the Ettinghausen effect, which relates a cross thermal gradient to a current flow. To evaluate these effects requires using both the electrical and thermal variables. Since the only interest at present is in the cubic semiconductors, the trigonal semimetals bismuth and antimony, and hexagonal crystals such as cadmium sulphide, zinc oxide, and so on, a simplified treatment that neglects any cross coefficients is possible if the equations refer to the crystallographic axes. This follows, since the symmetries are $m3m$, $\bar{4}3m$, $\bar{3}m$ and $6mm$ respectively, all of which have only diagonal terms for second-rank polar tensors.

In the Nernst effect only heat flow with no current flow occurs, whereas for the Ettinghausen effect only current flow with no heat flow occurs. Hence we do not need the extended definition of energy flow given by Eq. (12.28).

The two flow variables are the particle flow J_i^e and the heat flow h_i. Starting with the last of Eqs. (12.63) we can obtain an expression for the temperature derivatives in the form of the equation

$$-\frac{\partial \Theta}{\partial x_j} = r_{ij}^J h_i - \Theta X_{ij} J_i^e \tag{12.83}$$

where r_{ij}^J are the thermal resistivities determined by $r_{ij}^J k_{ij}^J = 1$, and X_{ij} is given by the equation

$$X_{ij} = r_{im}^J \sum_{jm} = r_{i1}^J \sum_{j1} + r_{i2}^J \sum_{j2} + r_{i3}^J \sum_{j3} \tag{12.84}$$

Since the current density $I_i = -eJ_i^e$, it is seen that the tensor (X_{ij}/e) relates the thermal gradients to the current flow.

Next, introducing these values of the temperature gradients into the first of Eqs. (12.63), we can write the chemical potential gradient in the form

$$-\frac{\partial \bar{\mu}}{\partial x_i} = e^2 \left(\rho_{ij} - \frac{\Theta \sum_{im} X_{jm}}{e^2} \right) J_j^e + X_{ij} h_j \tag{12.85}$$

The term in parentheses is the electrical resistivity at constant current flow h, whereas the original values were at constant temperature gradient. Hence the two fundamental equations can be written in the form

$$-\frac{\partial \Theta}{\partial x_i} = r_{ij}^J h_j - \Theta X_{ij} J_j^e$$

$$-\frac{\partial \bar{\mu}}{\partial x_i} = e^2 \rho_{ij}^h J_j^e + X_{ij} h_j \tag{12.86}$$

In this form it is seen that the two coupling terms are of opposite sign. The forces are $-(\partial \Theta/\partial x_i)/\Theta$ and $-(\partial \bar{\mu}/\partial x_i)$. Hence, the coupling terms are equal and opposite when expressed in terms of the heat currents and particle-flow currents. If we replace J_j^e by $-I_j/e$ and divide the second by $1/e$, symmetry is restored. The equations become

$$-\frac{\partial \Theta}{\partial x_i} = r_{ij}^J h_j + \Theta \left(\frac{X_{ij}}{e} \right) I_j ; \qquad -\frac{1}{e}\frac{\partial \bar{\mu}}{\partial x_i} = -\rho_{ij}^h I_j + \left(\frac{X_{ij}}{e} \right) h_j \tag{12.87}$$

(X_{ij}/e) is then the thermoelectric constant relating temperature gradients to current densities.

To take account of the effects of magnetic flux, we have to add two antisymmetric tensors that relate the products of the flows times the magnetic fluxes to the temperature gradients and the chemical potential gradients. At the same time the diagonal terms of the electrical and thermal resistivities will be even functions of the magnetic flux. Antisymmetric terms, such as in the X_{ij} thermoelectric tensor, may also have components that are odd powers of the flux. This is the origin of the *umkehreffekt*,[21] which allows the Seebeck coefficients to be different for different directions of the magnetic flux while still satisfying the relation $\pi_{ij}(+B) = -\Theta S_{ij}(-B)$.

If we add these terms, Eqs. (12.87) can be written in the form

$$-\frac{\partial\Theta}{\partial x_i} = r_{ij}^J h_j + \Theta\left(\frac{X_{ij}}{e}\right)I_j + \varepsilon_{ijk}h_j(L_{km}B_m) + \varepsilon_{ijk}I_j(U_{km}B_m)$$

$$-\frac{1}{e}\left(\frac{\partial\bar{\mu}}{\partial x_i}\right) = -\rho_{ij}^h I_j + \left(\frac{X_{ij}}{e}\right)h_j + \varepsilon_{ijk}h_j(N_{km}B_m) + \varepsilon_{ijk}I_j(R_{km}B_m)$$

$$(12.88)$$

The tensors U_{km} and N_{km} are related to the Ettinghausen and Nernst effects. The other two tensors, L_{km} and R_{km}, are related to the Righi-Leduc effect and the Hall effect.

For the first of these effects a current density I_i is maintained in the ith direction and a thermal gradient is generated at right angles to the current flow. Since no heat flow occurs, we can set h_j equal to zero. If the thermoelectric tensor (X_{ij}/e) has finite values when $i \neq j$, the ordinary thermoelectric effect can produce a sidewise temperature gradient as well as the Ettinghausen tensor. However for the class of crystals considered, there are only two components of both tensors, and these are

$$X_{11} = X_{22}; \qquad X_{33}; \qquad U_{11} = U_{22}; \qquad U_{33}$$

We consider two cases. The first, for a current along the x_1 direction with the magnetic flux along both the x_2 and x_3 directions, results in temperature gradients along x_3 and x_2 respectively. The other case is for the current along the x_3 axis with the magnetic flux along x_1 (or x_2). For the first case the equations are

$$-\frac{\partial\Theta}{\partial x_1} = \Theta\left(\frac{X_{11}}{e}\right)I_1; \qquad -\frac{\partial\Theta}{\partial x_2} = -I_1U_{33}B_3; \qquad -\frac{\partial\Theta}{\partial x_3} = I_1U_{11}B_2$$

$$(12.89)$$

For a current flowing along the x_3 direction, with fields along x_1 or x_2, the three equations are

$$-\frac{\partial\Theta}{\partial x_1} = -I_3U_{11}B_2; \qquad -\frac{\partial\Theta}{\partial x_2} = I_3U_{11}B_1; \qquad -\frac{\partial\Theta}{\partial x_3} = \Theta\left(\frac{X_{33}}{e}\right)I_3$$

$$(12.90)$$

On account of the fact that the temperature gradient is proportional both to the current flow and the magnetic flux, rather large thermal gradients

[21] See Wolfe and Smith [3].

can be produced, and the Ettinghausen effect has been considered[22] in thermoelectric cooling. By using an exponentially shaped crystal of bismuth to produce a cascade effect, a lowering of temperature[23] from $302°$ K to $201°$ K occurred in a field of 11 webers/meter^2 (110,000 Gauss) when a current of 55 amperes flowed through the crystal. The large flat edge was at the high temperature, while the low temperature occurred at the small edge of the exponential taper. The largest effect was produced when the current flowed along the trigonal (x_3) axis with the flux along an x_1 axis and the temperature difference measured along an x_2 axes. According to Eq. (12.89) this makes use of the U_{11} Ettinghausen coefficient.

The Nernst effect occurs when the electric current is eliminated but a temperature gradient is established between the two ends of the crystal. For this case I_i is zero and all the heat is carried by the terms h_i. Since no chemical potential is transported $\bar{\mu}$ can be replaced by $-\varphi e$, where φ is the electric potential and $-e$ the charge carried by the electrons. For the case under consideration the last of Eqs. (12.86) can be written

$$\frac{\partial \varphi}{\partial x_i} = -E_i = \left(\frac{X_{ij}}{e}\right)h_j + \varepsilon_{ijk}h_j\left(\frac{N_{km}}{e}\right)B_m \qquad (12.91)$$

The two cases considered above result in the equations

$$-E_1 = \left(\frac{X_{11}}{e}\right)h_1 ; \qquad -E_2 = -h_1\left(\frac{N_{33}}{e}\right)B_3 ; \qquad -E_3 = h_1\left(\frac{N_{11}}{e}\right)B_2$$

$$(h_i = h_1)$$
$$(12.92)$$

$$-E_1 = -h_3\left(\frac{N_{11}}{e}\right)B_2 ; \qquad -E_2 = h_3\left(\frac{N_{11}}{e}\right)B_1 ; \qquad -E_3 = \left(\frac{X_{33}}{e}\right)h_3$$

$$(h_i = h_3)$$

Hence the terms (N_{ii}/e) relate the heat flow to the voltage gradient.

A relation between the Nernst coefficient (N_{ii}/e) and the Ettinghausen coefficient (U_{ii}/Θ) follows at once from Onsager's relation. The forces equivalent to h_i and I_i are, from (12.88), equal to

$$-\frac{\partial\Theta/\partial x_i}{\Theta} \quad \text{and} \quad -\frac{1}{e}\frac{\partial\bar{\mu}}{\partial x_i} \qquad (12.93)$$

Hence by Onsager's principle the cross-product terms have to be equal. From Eqs. (12.89), (12.91), and (12.92) the terms are

$$-\frac{1}{I_1}\frac{\partial\Theta/\partial x_1}{\Theta} = -\frac{U_{33}}{\Theta}B_3 \quad \text{and} \quad -\frac{1}{eh_1}\frac{\partial\bar{\mu}}{\partial x_2} = -N_{33}B_3 \qquad (12.94)$$

[22] See Wolfe [1], Smith and Wolfe [2], Wolfe and Smith [3], and O'Brien et al. [13].
[23] See O'Brien et al. [13].

Hence

$$U_{33} = \Theta N_{33} \; ; \qquad \text{Similarly,} \qquad U_{11} = \Theta N_{11} \qquad (12.95)$$

Since the Nernst effect is sometimes defined in terms of the temperature gradient $(\partial \Theta / \partial x_i)$ rather than h_i, the relation in this form will be

$$k_i U_{ii} = \Theta N'_{ii} \qquad (12.96)$$

Where N'_{ii} is the Nernst constant defined in terms of the temperature gradient rather than the heat flow. This relation was first proved by Bridgman[24] for isotropic materials. For the general case the relation of (12.95) can be written in the form

$$U_{ij} = \Theta N_{ij} \qquad (12.97)$$

as can be seen by expanding the general Eq. (12.88).

RESUME CHAPTER 12

I. Introduction

Thermoelectricity is an interaction process between the thermal variables—temperature gradients $\partial \Theta / \partial x_i$ and heat flows h_i—and the electrical variables—fields E_i and current densities I_i. Differences in voltages produced by differences in temperatures have long been used for the measurement of temperature. Figure 12.2 shows a typical curve for the thermocouple constantan and copper. Such voltage differences determine the thermoelectric power S_{ab}, which is defined as the voltage difference between two junctions of the conductor a and b when they differ by $1°$ C.

With the advent of suitable semiconductors and semimetals, thermoelectric devices such as those shown by Fig. 12.3 have become more efficient and now have applications in power supplies in satellites and as refrigeration sources for stabilizing temperatures below room temperatures. The figure of merit for both of these applications is

$$z = \frac{S_{ab}^2}{\rho K} \qquad (12.53)$$

where the electrical resistivity ρ and the thermal conductivity K are assumed to be the same for both media. This is usually true since p- and n-type semiconductors, for which the thermoelectric powers add, are normally used with similar doping values and hence similar resistivities and conductivities. Figure 12.4 shows the figures of merit for a number of materials plotted against the temperature. As will be shown, the efficiency of conversion is determined by the product of z times the average temperature $\overline{\Theta}$, and the dashed line of Fig. 12.4 represents a constant-efficiency curve for these materials. The semimetals are more advantageous for low temperatures, and their figure of merit can be increased by putting them in a magnetic flux. Figure 12.6 shows that there is an optimum value of flux for making the figure of merit z a maximum.

Without magnetic flux, the three thermoelectric coefficients are the thermoelectric power S_{ab} defined above, the Peltier coefficient, which determines the amount of heat added or subtracted from a junction when a current is flowing through it, and the Thomson heat,

[24] See Bridgman [14, 15, 16].

which determines the heat supplied to the current-carrying wire to keep the temperature distribution steady. These three effects can be described in terms of junction thermodynamics. However, in order to generalize the results to crystals it is necessary to consider the effects in each medium by introducing the absolute thermoelectric Thomson coefficients Σ and the chemical potential $\bar{\mu}$. The effect at a junction represents a difference between the two sets of functions for the two media. The gradient of $\bar{\mu}$ and the complete tensor $\sum_{ij}$ are then general enough to take account of any crystal effects.

By considering $\bar{\mu}$ and $\sum_{ij}$ as functions of the magnetic flux B and introducing terms similar to the Hall-effect terms of Chapter 11, the magnetic-flux effects can be accounted for. The new effects are the Ettinghausen effect, which represents a transverse temperature gradient, the Nernst effect, which represents a transverse electric field, the Righi-Leduc effect, which represents a temperature gradient at right angles to the heat flow and the magnetic flux, and the thermomagnetoresistance effect. All of these effects result from the force on the electrical carriers produced by the vector product of the magnetic flux and the drift velocity.

2. Thermodynamic Derivations for Isotropic Materials

A. Case for Two Isotropic Curves

All thermodynamic derivations of the thermoelectric effect make use of the Gibbs function G, which is constant for any phase change. Since $dG = 0$ across a junction, it follows that

$$\Theta \, d\sigma = dU - \varphi e \, dn \tag{12.4}$$

where φ is the electrical potential and $e \, dn$ is the change in the charge density. The chemical potential used in Section 12.3 can be written in the form

$$\bar{\mu} = \mu - e\varphi \tag{12.18}$$

where μ is the part of the internal energy U that depends on the composition and temperature of the medium.

If we consider the couple of Fig. 12.1 with a condenser in one wire, it follows from Onsager's relation that

$$i = -L_{11} \Delta\varphi - L_{12} \frac{\Delta\Theta}{\Theta}$$
$$\hspace{4cm} L_{21} = L_{12} \tag{12.9}$$
$$j = -L_{21} \Delta\varphi - L_{22} \frac{\Delta\Theta}{\Theta}$$

Here i is the current flow, j the heat flow, $\Delta\varphi$ the change in potential, $\Delta\Theta/\Theta$ the change in temperature divided by the absolute temperature, and the L_{ij} values are self- and cross-coupling parameters between the forces $\Delta\varphi$ and $\Delta\Theta/\Theta$ and the flows i and j. By assuming first that the current flow vanishes as in Fig. 12.1, we find

$$\frac{\Delta\varphi}{\Delta\Theta} = -\frac{L_{12}}{L_{11}\Theta} = S_{ab} \quad \text{(thermoelectric power)} \tag{12.10}$$

Another special case occurs when $\Delta\Theta = 0$. For this case

$$\frac{j}{i} = \frac{L_{12}}{L_{11}} = \pi_{ab} \quad \text{(Peltier heat)} \tag{12.11}$$

From the above equation there is a relation

$$S_{ab} = -\frac{\pi_{ab}}{\Theta} \tag{12.12}$$

The third relation for the Thomson heat, defined as

$$\Delta\dot{Q} = \tau i \, \Delta\Theta \tag{12.2}$$

where $\Delta\dot{Q}$ is the rate of heat flow into the wire carrying a current i, is given by the first law of thermodynamics in the following form, which sums all the heat flows:

$$\pi_{ab}i - (\pi_{ab} + \Delta\pi_{ab})i + (\tau_b - \tau_a)i \, \Delta\Theta = i \, \Delta\varphi$$

where $\Delta\pi_{ab}$ is the change in π_{ab} at the higher temperature $\Theta + \Delta\Theta$. After using Eqs. (12.10), (12.12), and (12.17) this results in the two forms

$$\tau_b - \tau_a \equiv \frac{d\pi_{ab}}{d\Theta} - \frac{\pi_{ab}}{\Theta} \quad \text{or} \quad \tau_b - \tau_a = -\Theta \frac{d^2\varphi}{d\Theta^2} \tag{12.17}$$

B. Thermoelectric Effects in Continuous Isotropic Media

To derive the thermoelectric effects for a continuous isotropic media it is necessary to replace the electrical potential by the chemical potential $\bar{\mu}$ which is written in the form applicable to a single electron

$$\bar{\mu} = \mu - e\varphi \tag{12.18}$$

where μ is the part of the internal energy U that depends on the composition and temperature of the medium. At an interface for which the temperature is the same on either side, the differences in the entropies are the differences between the chemical potentials of the two sides. The forces used are $\partial\bar{\mu}/\partial x_i$ and $(\partial\Theta/\partial x_i)/\Theta$, while the flows are the number of electrons per square meter carried across the surface, J_i^e, and the heat current density h_i. It follows from Onsager's relations that

$$J_i^e = -\alpha \frac{\partial\bar{\mu}}{\partial x} - \beta \frac{\partial\Theta/\partial x_i}{\Theta} ; \qquad h_i = -\beta \frac{\partial\bar{\mu}}{\partial x_i} - \gamma \frac{\partial\Theta/\partial x_i}{\Theta} \tag{12.19}$$

If there is no temperature gradient; then because $J_i^e e = -I_i$, the current density, and $\partial\bar{\mu}/\partial x_i = -e \, \partial\varphi/\partial x_i$, since μ does not vary at constant temperature, we find

$$\alpha e^2 = \sigma \quad \text{or} \quad \alpha^{-1} = e^2\rho \tag{12.20}$$

where ρ is the electrical resistivity. Similarly, γ is the thermal conductivity at constant $\bar{\mu}$ divided by the absolute temperature Θ. We can introduce the absolute thermoelectric power (Σ/e) for the material through the equations

$$\left(-\frac{\Sigma}{e}\right) = -\left(\frac{\alpha^{-1}\beta}{\Theta}\right)$$

Inserting this definition in (12.19) and changing the form of (12.19) we can write

$$-\frac{\partial\bar{\mu}}{\partial x_i} = e^2\rho J_i^e + \Sigma \frac{\partial\Theta}{\partial x_i} ; \qquad h_i = \Theta \sum J_i^e - k^J \frac{\partial\Theta}{\partial x_i} \tag{12.24}$$

where the thermal conductivity for constant current flow is

$$k^J = k^{\bar{\mu}}\left[1 - \frac{(\Sigma/e)^2\Theta}{\rho k^{\bar{\mu}}}\right] = k^{\bar{\mu}}\left[1 - \frac{S^2\Theta}{\rho k^{\bar{\mu}}}\right] \tag{12.25}$$

The term in brackets is the figure of merit z times the absolute temperature Θ. $z\Theta$ represents the square of a coupling factor between thermal and electrical systems.

When both electrical and thermal energies flow across the boundary, the energy flow is generalized to

$$J_i^U = h_i + \bar{\mu} J_i^e \tag{12.28}$$

By using this generalized energy flow it is shown that the convergence of this energy flow represents a Joule heat, a convergence of the thermal energy, and a thermoelectric heat gives by

$$\dot{Q} = -\Theta J_i^e \frac{\partial \sum}{\partial x_i} \tag{12.34}$$

The relations of Section 12.2 can be derived by taking the differences for these functions across boundaries between two different mediums. The Seebeck coefficient can be obtained from the first of Eqs. (12.24), with $J_i^e = 0$, by integrating the chemical potential around a closed path from one side of the condenser of Fig. 12.1 to the other. The result is

$$S_{ab} = \frac{\Delta\varphi}{\Delta\Theta} = \frac{\sum^b - \sum^a}{e} \tag{12.38}$$

The terms $-\sum^b/e$ and $-\sum^a/e$ are absolute thermoelectric powers of the two materials. The Peltier and Thomson heats can be derived directly from (12.34). Since the divergence of J_i^e is zero, (12.34) can be written

$$\dot{Q} = -\Theta \frac{\partial}{\partial x_i} (J_i^e \sum) \tag{12.39}$$

At the interface the evolution of heat is determined by the difference between normal components, and hence

$$\dot{Q} = -\Theta[\sum^a - \sum^b]J_3^e = +\Theta\left(\frac{\sum^a - \sum^b}{e}\right)I_3 = -\Theta S_{ab}I_3 = \pi_{ab}I_3$$

The Thomson heat arises directly from (12.34) when consideration is limited to a single medium. By considering the current flow to occur in the direction of increasing x_3, it is shown that

$$\dot{Q} A \, dx_3 = -\Theta i_3 \, d(\sum/e) = i_3 \tau \, d\Theta \tag{12.42}$$

where τ is the Thomson coefficient. Hence

$$\tau = -\Theta \frac{\partial(\sum/e)}{\partial\Theta} \tag{12.43}$$

The difference between the Thomson heats for the two media is

$$(\tau_a - \tau_b) = -\Theta \frac{\partial}{\partial\Theta}\left(\frac{\sum^a - \sum^b}{e}\right) = \Theta \frac{\partial}{\partial\Theta}(S_{ab}) = \Theta \frac{\partial^2\varphi}{\partial\Theta^2} \tag{12.44}$$

3. Efficiency Equations for Thermoelectric Devices

The principal uses of thermoelectric materials are as devices for measuring temperatures, as transducers for converting thermal energy into electrical energy, and as refrigerating devices for pumping heat from one junction to the other. Temperature-measuring devices employ two junctions between dissimilar metals, one of which is held at a fixed temperature such as the ice-water temperature 0° C, while the other junction is placed at the position whose temperature is to be determined. This device makes use of the Seebeck effect, and the voltage difference represents an integration of the Seebeck coefficient from Θ_1 to Θ_2—that is,

$$V_2 - V_1 = \int_{\Theta_1}^{\Theta_2} S_{ab}\, d\Theta$$

Near absolute zero the voltage difference per degree becomes very small, and thermocouples can usually not be used below 20° K.

Thermoelectric energy converters absorb energy at a hot junction at temperature Θ_H and deliver thermal and electric energy to the cold junction at temperature Θ_c. For small temperature differences, the efficiency of conversion is given by

$$E_F = \frac{\text{power supplied to load}}{\text{heat absorbed in thermocouple}} = \frac{\Theta_H - \Theta_c}{\Theta_H}\left(\frac{S_{ab}^2}{4RK_c}\Theta_H\right) = \frac{\Theta_H - \Theta_c}{\Theta_H}\left(\frac{z\Theta_H}{4}\right)$$

where the first term is the reversible Carnot efficiency and the second term involves the figure of merit z times the temperature Θ_H. For a large temperature difference the efficiency is given by

$$E_F = \left(\frac{\Theta_H - \Theta_c}{\Theta_H}\right)\left(\frac{\sqrt{1 + z\overline{\Theta}} - 1}{\sqrt{1 + z\overline{\Theta}} + \Theta_c/\Theta_H}\right) \quad \text{where} \quad \overline{\Theta} = \left(\frac{\Theta_H + \Theta_c}{2}\right) \quad (12.55)$$

Such devices are usually made from p- and n-type semiconductors, since their Seebeck coefficients are of opposite sign and hence add. Efficiencies of conversion in the order of 10 per cent are readily realized. These devices have applications as power supplies in satellites and for other special purposes.

When electrical energy is supplied to the thermoelectric device of Fig. 12.3, heat can be pumped from a cold junction at a temperature Θ_H to a hot junction at Θ_c, and hence a refrigerating or alternatively a heating device can be obtained. The coefficient of performance η, which is defined as

$$\eta = \frac{\text{heat transferred from one junction to the other}}{\text{electrical energy supplied}}$$

$$= \left(\frac{\Theta_c}{\Theta_H - \Theta_c}\right)\left(\frac{\sqrt{1 + z\overline{\Theta}} - \Theta_H}{1 + \sqrt{1 + z\overline{\Theta}}}\right) \quad (12.60)$$

The first term represents the ideal performance with $z\overline{\Theta} = \infty$, while the second term gives the change due to the finite values of $z\overline{\Theta}$. The maximum temperature difference that can be obtained is

$$\frac{\Theta_H}{\Theta_c} = \sqrt{1 + z\overline{\Theta}} \quad (12.61)$$

Such devices are used as thermostatic devices good for temperatures under room temperatures.

4. Changes in Thermoelectric Equations for Anisotropic Media

The only modification required for the equations of (12.24) of Chapter 12, to take account of anisotropic effects is to introduce the tensor terms $\alpha_{ij}^{-1} = e^2\rho_{ij}$, $\sum_{ij}$, and k_{ij}^J in place of α^{-1}, Σ, and k^J. The two fundamental equations, equivalent to (12.24), become

$$-\frac{\partial \bar{\mu}}{\partial x_i} = e^2\rho_{ij}J_j^e + \sum_{ij}\frac{\partial \Theta}{\partial x_j}; \quad h_i = \Theta\sum_{ij}J_j^e - k_{ij}^J\frac{\partial \Theta}{\partial x_j} \quad (12.63)$$

Using the extended definition of energy flow (12.28) the last equation becomes

$$J_i^U = (\Theta \sum_{ij} + \bar{\mu}\, \delta_{ij}) J_j^e - k_{ij}^J \frac{\partial \Theta}{\partial x_j} \tag{12.64}$$

ρ_{ij} and k_{ij}^J are symmetric but $\sum_{ij}$ is not. By forming the convergence of the energy J_i^U it is found that there are three terms, the Joule heat, the convergence of the heat flow, and the term

$$\dot{Q} = -\Theta \frac{\partial}{\partial x_i}(J_j^e \sum_{ij}) \tag{12.67}$$

which represents the heat evolved by the thermoelectric effect.

At an interface the evolution of heat is the difference between the normal components of (12.67). Using the current density $I_j = -e^J i^e$, we have

$$-\Theta[I_j(\sum_{j3}^a - \sum_{j3}^b)] = I_j(\pi_{j3}^a - \pi_{j3}^b) \tag{12.68}$$

If $j = 3$ this represents the Peltier coefficient along the junction. There are, however, Peltier heats at right angles to the junction normal. The generalized Seebeck coefficient S_{ij} is related to the Peltier heat by

$$\pi_{ij} = \Theta\left(\frac{\sum_{ij}}{e}\right) = -\Theta S_{ij} \tag{12.69}$$

The Thomson heat is given by

$$\tau_{ij} = -\frac{\Theta}{e}\left(\frac{\partial \sum_{ij}}{\partial \Theta}\right)_x$$

where the subscript x denotes that $\sum_{ij}$ varies with distance. A third term not present for isotropic materials is the Bridgman heat caused by a nonuniform current distribution. Whenever there are nonvanishing terms

$$\frac{\partial J_i^e}{\partial x_j}$$

there are corresponding Bridgman heats.

5. Effect of a Magnetic Flux on Thermoelectric Properties

For isotropic materials and cubic crystals, the effect of a magnetic flux B can be taken account of by letting the phenomenological constants of Eq. (12.9) be functions of B. The self constants are even functions of the magnetic flux while the interaction constants are odd functions of the flux with

$$l_{ij}(B) = l_{ji}(-B)$$

To take account of complete anisotropic media it is necessary to introduce two anti-symmetric tensors similar to the Hall-effect tensor of Eq. (11.16). For the most general case, the equations become

$$-\frac{\partial \Theta}{\partial x_i} = r_{ij}h_i + \Theta\left(\frac{X_{ij}}{e}\right)I_j + \varepsilon_{ijk}h_j(L_{km}B_m) + \varepsilon_{ijk}I_j(U_{km}B_m)$$

$$\tag{12.88}$$

$$-\frac{1}{e}\left(\frac{\partial \bar{\mu}}{\partial x_i}\right) = -\rho_{ij}^h I_j + \left(\frac{X_{ij}}{e}\right)h_j + \varepsilon_{ijk}h_j(N_{km}B_m) + \varepsilon_{ijk}I_j(R_{km}B_m)$$

The tensors U_{km} and N_{km} are related to the Ettinghausen and Nernst effects. The other two tensors, L_{km} and R_{km}, are related to the Righi-Leduc effect and the Hall effect. These tensors are odd functions of the magnetic flux B, while the thermal resistivity r_{ij}, the electrical resistivity ρ_{ij}, and the thermoelectric functions X_{ij}, which relate the current flow to the force $(\partial \Theta / \partial x_i)/\Theta$, are all even functions of the flux.

Examples are given for the Ettinhausen effect—which has been considered for thermoelectric cooling—and for the Nernst effect. A relation between the Nernst coefficients and the Ettinghausen coefficients is

$$U_{ij} = \Theta N_{ij}$$

PROBLEMS CHAPTER 12

1. The heat absorbed from a cold junction of a thermoelectric refrigerator is $-\pi_{pn}i = S_{pn}(\overline{\Theta} - \Delta\Theta/2)i$. Half the Joule heat—that is, $(\frac{1}{2})i^2R$—flows to this junction, as does the conduction heat from the hot junction. Hence

$$\dot{Q}_c = S_{pn}(\overline{\Theta} - \Delta\Theta/2)i - \tfrac{1}{2}i^2R - K\,\Delta\Theta$$

where K is the thermal conductivity for both branches. The electrical power supplied to the thermocouple is $S_{ab}\,\Delta\Theta\,i + i^2R$. Hence the coefficient of performance η, after multiplying numerator and denominator by R, is

$$\eta = \frac{S_{pn}(\overline{\Theta} - \Delta\Theta/2)iR - \tfrac{1}{2}(iR)^2 - KR\,\Delta\Theta}{S_{pn}\,\Delta\Theta\,(iR) + (iR)^2}$$

Show that the optimum value of the current in the thermoelectric refrigerator is given by

$$(iR) = \frac{S_{pn}\,\Delta\Theta}{(\sqrt{1 + z\overline{\Theta}} - 1)}$$

where

$$z = \frac{S_{pn}^2}{KR}$$

2. The dimensions of the p and n branches of the thermoelectric refrigerator are A_p and l_p and A_n and l_n where A is the area and l the length of each branch. If ρ_p and ρ_n are the resistivities and k_p and k_n the thermal conductivities, show that the optimum value of z occurs when

$$\frac{l_pA_n}{l_nA_p} = \left(\frac{k_p}{\rho_p} \times \frac{\rho_n}{k_n}\right)^{1/2}$$

and that

$$KR = \{(k_p\rho_p)^{1/2} + (k_n\rho_n)^{1/2}\}^2$$

3. With these values of optimum current and resistivities, show that the optimum performance coefficient η is given by Eq. (12.60).

4. Since the value of $\dot{Q}$, the heat absorbed from the cold junction, must drop to zero when the maximum temperature difference is reached, show that this maximum difference is given by Eq. (12.61).

REFERENCES CHAPTER 12

1. R. Wolfe, *Sci. Am.* **210**, 70 (June 1964).
2. G. E. Smith and R. Wolfe, *J. Appl. Phys.* **33**, 841 (1962).
3. R. Wolfe and G. E. Smith, *Phys. Rev.* **129**, 1086 (1963).

4. S. R. DeGroot, "Thermodynamics of Irreversible Processes." North-Holland Publ., Amsterdam, 1951.
5. H. B. Callen, *Phys. Rev.* **73,** 1349 (1948).
6. Charles A. Domenicali, *Phys. Rev.* **92,** 877 (1953).
7. G. K. White, "Experimental Techniques in Low Temperature Physics," p. 136. Oxford Univ. Press, London and New York, 1959.
8. T. J. Seebeck, *Rept. Prussian Acad. Sci.* (1822–1823).
9. R. W. Cohen and B. Abeles, *J. Appl. Phys.* **34,** 131 (1963).
10. E. Altenkirch, *Phys. Z.* **12,** 920 (1911).
11. F. D. Rose, E. F. Hockings, and N. E. Lindenblad, *RCA Rev.* **22,** 82 (1961).
12. R. Wolfe and G. E. Smith, *Semicond. Prod.* **6,** 23–33 (1963).
13. B. J. O'Brien, C. S. Wallace, and K. Landecker, *J. Appl. Phys.* **27,** 820 (1956).
14. "The Thermodynamics of Electrical Phenomena in Metals." P. W. Bridgman, McGraw-Hill, New York, 1934.
15. ——, *Phys. Rev.* **55,** 845 (1940).
16. ——, *Rev. Mod. Phys.* **22,** 56 (1950).
17. T. C. Harman, J. M. Honig, S. Fischler, A. E. Paladino, and M. Jane Button, *Appl. Phys. Letters* **4,** 77 (1964).

APPENDIX A • Electromagnetic Wave Propagation in a Crystalline Medium

A.1 Solution of Maxwell's Equations

The derivation of Maxwell's equations was discussed in Chapter 3, Section 3.3. All the applications for wave transmission in crystals are for insulating materials. Hence the charge density ρ can be set equal to zero. With this simplification the equations in tensor form become

$$\varepsilon_{ijk}\frac{\partial H_k}{\partial x_j} = \frac{\partial D_i}{\partial t}\ ; \qquad \varepsilon_{ijk}\frac{\partial E_k}{\partial x_j} = -\mu_0\frac{\partial H_i}{\partial t}\ ; \qquad \frac{\partial D_i}{\partial x_i} = 0; \qquad \frac{\partial B_j}{\partial x_j} = 0 \qquad (3.40)$$

We assume the electric vector E_k to be representable by a plane wave whose planes of equal phase are taken normal to the unit vector n_j. Then

$$E_k = E_{0_k} \exp\left[j\omega(t - x_j l_j/v)\right] \qquad (A.1)$$

where E_{0_k} are constants representing the maximum values of the field along the three rectangular axes, $j = \sqrt{-1}$, and v is the velocity of the wave in the crystal in the l_j direction. Substituting (A.1) in (3.40), noting that E_{0_k} are not functions of the space coordinates, we have

$$\frac{\partial H_i}{\partial t} = \frac{j\omega}{\mu_0 v}\left[\varepsilon_{ijk}E_{0_k}l_j\right] \exp\left[j\omega(t - x_j l_j/v)\right] \qquad (A.2)$$

Integrating with respect to the time

$$H_i = \frac{1}{\mu_0 v}\left[\varepsilon_{ijk}E_{0_k}l_j\right] \exp\left[j\omega(t - x_j l_j/v)\right] = H_{0_i} \exp\left[j\omega(t - x_j l_j/v)\right] \qquad (A.3)$$

and hence

$$H_{0_i} = \frac{1}{\mu_0 v}\left[\varepsilon_{ijk}E_{0_k}l_j\right] \qquad (A.4)$$

301

From the discussion of the vector product of two vectors given in Section 2.5 it is seen that the magnetic vector is normal to the plane determined by the electric field E_{0_k} and the direction of wave propagation l_j. Figure A.1 shows the relation for H_i, l_j, and E_{0_k} when H is directed along the x_1 axis.

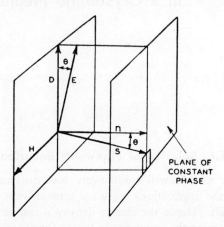

FIG. A.1 Position of electric, magnetic, and normal vectors for an electromagnetic wave in a crystal.

Next, using the first equations of (3.40) and the solution for H, we find that

$$\frac{\partial D_i}{\partial t} = \varepsilon_{ijk}\frac{\partial H_k}{\partial x_j} = -\frac{j\omega}{v}\left[\varepsilon_{ijk}H_{0_k}l_j\right]\exp\left[j\omega(t - x_jl_j/v)\right] \qquad (A.5)$$

Integrating this equation with respect to the time t,

$$D_i = -\frac{1}{v}\left[\varepsilon_{ijk}H_{0_k}l_j\right]\exp\left[j\omega(t - x_jl_j/v)\right] \qquad (A.6)$$

Inserting the value of H_{0_k} from (A.4), this equation takes the form

$$D_i = -\frac{1}{\mu_0v^2}\left[\varepsilon_{ijk}(\varepsilon_{ijk}E_{0_k}l_j)l_j\right]\exp\left[j\omega(t - x_jl_j/v)\right] \qquad (A.7)$$

and in general

$$D_i = -\frac{1}{\mu_0v^2}\left[\varepsilon_{ijk}(\varepsilon_{ijk}E_kl_j)l_j\right] \qquad (A.8)$$

Expanding the inner parenthesis, we have the components

$$(E_3l_2 - E_2l_3); \qquad (E_1l_3 - E_3l_1); \qquad (E_2l_1 - E_1l_2) \qquad (A.9)$$

Then

$$D_1 = - \frac{1}{\mu_0 v^2} \left[\varepsilon_{1jk}[E_3 l_2 - E_2 l_3, E_1 l_3 - E_3 l_1, E_2 l_1 - E_1 l_2] l_j \right] \quad \text{(A.10)}$$

gives

$$D_1 = - \frac{1}{\mu_0 v^2} \left[l_2(E_2 l_1 - E_1 l_2) - l_3[E_1 l_3 - E_3 l_1] \right]$$

$$\text{(A.11)}$$

$$= - \frac{1}{\mu_0 v^2} \left[l_1[E_1 l_1 + E_2 l_2 + E_3 l_3] - E_1(l_1^2 + l_2^2 + l_3^2) \right]$$

after adding and subtracting $E_1 l_1^2$. Since $l_1^2 + l_2^2 + l_3^2 = 1$, because it is a unit vector, the electric displacement D_i can be written in the general form

$$D_i = \frac{1}{\mu_0 v^2} [E_i - (E_j l_j) l_i] \quad \text{or} \quad \mu_0 v^2 D_i - E_i + (E_j l_j) l_i = 0 \quad \text{(A.12)}$$

This equation shows that D_i, E_i, and l_i are all in the same plane. As shown previously H_i is in a plane normal to this plane. Figure A.1 shows the relative positions of E_i, D_i, l_i, and H_i. The angle θ between the field and the electric displacement depends on the dielectric properties of the medium. The direction of energy flow is determined by the Poynting vector S_i defined by the equation

$$S_i = \varepsilon_{ijk} E_i H_k \quad \text{(A.13)}$$

Since this is in the form of a vector product it is seen that S_i is in a direction perpendicular to the electric and magnetic fields and hence it will be the same angle from l_i that E_i is from D_i—that is, the angle θ as shown by Fig. A.1.

Equation (A.12) is general for any dielectric medium and to proceed further it is necessary to specify the relation between D and E. These relations can be written in the form

$$D_i = \varepsilon_{ij} E_j \quad \text{or inversely} \quad E_j = \beta_{ji} D_i \quad \text{(A.14)}$$

where ε_{ij} are the dielectric constants measured at optical frequencies and β_{ji} the impermeability constants determined from the relations

$$\beta_{ji} = (-1)^{i+j} \frac{\Delta^{ji}}{\Delta^\varepsilon} \quad \text{(A.15)}$$

where

$$\Delta^\varepsilon = \begin{vmatrix} \varepsilon_{11} & \varepsilon_{12} & \varepsilon_{13} \\ \varepsilon_{12} & \varepsilon_{22} & \varepsilon_{23} \\ \varepsilon_{13} & \varepsilon_{23} & \varepsilon_{33} \end{vmatrix}$$

and Δ^{ji} is the minor obtained by suppressing the jth row and ith column.

The impermeability constants are more convenient to use since they allow the fields to be eliminated from (A.12). The relative impermeability constants $\beta_{ij}/\beta_0 = \beta_{ij}\varepsilon_0$ are more useful than the values given by (A.15). These will be designated by the terms B_{ij} to parallel the use of K_{ij} for relative dielectric constants. Furthermore the product

$$\varepsilon_0\mu_0 = \frac{1}{V^2} \tag{A.16}$$

where V is the velocity of light waves in a vacuum—that is, $V = 3 \times 10^8$ meters/sec. With these substitutions the three Eqs. (A.12) can be put into the form

$$\frac{v^2}{V^2}D_1 = B_{11}D_1 + B_{12}D_2 + B_{13}D_3 - \varepsilon_0(E_jl_j)l_1$$

$$\frac{v^2}{V^2}D_2 = B_{12}D_1 + B_{22}D_2 + B_{23}D_3 - \varepsilon_0(E_jl_j)l_2 \tag{A.17}$$

$$\frac{v^2}{V^2}D_3 = B_{13}D_1 + B_{23}D_2 + B_{33}D_3 - \varepsilon_0(E_jl_j)l_3$$

From these equations one can solve for D_1, D_2, and D_3, obtaining

$$D_1 = \left[\left(B_{22} - \frac{v^2}{V^2}\right)\left(B_{33} - \frac{v^2}{V^2}\right) - B_{23}^2\right]\varepsilon_0(E_jl_j)l_1$$

$$D_2 = \left[\left(B_{11} - \frac{v^2}{V^2}\right)\left(B_{33} - \frac{v^2}{V^2}\right) - B_{13}^2\right]\varepsilon_0(E_jl_j)l_2 \tag{A.18}$$

$$D_3 = \left[\left(B_{11} - \frac{v^2}{V^2}\right)\left(B_{22} - \frac{v^2}{V^2}\right) - B_{12}^2\right]\varepsilon_0[E_jl_j]l_3$$

Now D and l are at right angles as shown by Fig. A.1, and hence

$$D_1l_1 + D_2l_2 + D_3l_3 = 0 \tag{A.19}$$

Introducing (A.18) in these equations and dividing out the common factor $\varepsilon_0(E_jl_j)$ this equation reduces to the equation of a quadric surface

$$0 = \left[\left(B_{22} - \frac{v^2}{V^2}\right)\left(B_{33} - \frac{v^2}{V^2}\right) - B_{23}^2\right]l_1^2$$

$$+ \left[\left(B_{11} - \frac{v^2}{V^2}\right)\left(B_{33} - \frac{v^2}{V^2}\right) - B_{13}^2\right]l_2^2 \tag{A.20}$$

$$+ \left[\left(B_{11} - \frac{v^2}{V^2}\right)\left(B_{22} - \frac{v^2}{V^2}\right) - B_{12}^2\right]l_3^2$$

As discussed in Section 2.6 a quadric surface can be referred to three principal axes at right angles to each other for which $B_{12} = B_{13} = B_{23} = 0$ and with $B_{11} = B_1$, $B_{22} = B_2$, $B_{33} = B_3$. From the table of symmetrical second-rank tensors given in Appendix B, the principal axes coincide with the crystallographic axes except for monoclinic and triclinic crystals. For monoclinic crystals the b crystallographic axis is a principal axis, and the position of the other two can be found from a Mohr circle diagram. Hence we shall assume that the principal axis directions are known and shall call them x_1, x_2, and x_3. With $B_{12} = B_{13} = B_{23} = 0$, Eq. (A.20) reduces to

$$0 = \frac{l_1^2}{B_1 - v^2/V^2} + \frac{l_2^2}{B_2 - v^2/V^2} + \frac{l_3^2}{B_3 - v^2/V^2} \qquad \text{(A.21)}$$

For transmission along the x_1 axis $l_1 = 1$, $l_2 = l_3 = 0$, and from Eq. (A.20) the two velocities are given by

$$v^2 = B_2 V^2 = b^2 \; ; \qquad v^2 = B_3 V^2 = c^2 \qquad \text{(A.22)}$$

Similarly the third velocity $v^2 = B_3 V^2 = a^2$ can be used and Eqs. (A.21) reduces to

$$\frac{l_1^2}{a^2 - v^2} + \frac{l_2^2}{b^2 - v^2} + \frac{l_3^2}{c^2 - v^2} = 0 \qquad \text{(A.23a)}$$

This is a quadric equation for the velocities v in terms of the principal velocities a, b, and c, which are usually taken so that $a > b > c$. The velocities in any direction z', which determine the wave normal, have been shown[1] to be given by the equation

$$2v^2 = a^2(\sin^2 \varphi \sin^2 \theta + \cos^2 \theta) + b^2(\cos^2 \varphi \sin^2 \theta + \cos^2 \theta) + c^2 \sin^2 \theta$$

$$\pm [(a^2 - b^2)^2(\cos^2 \theta \cos^2 \varphi + \sin^2 \varphi)^2 + 2(a^2 - b^2)(c^2 - b^2) \qquad \text{(A.23b)}$$

$$\times \sin^2 \theta(\cos^2 \theta \cos^2 \varphi - \sin^2 \varphi) + (c^2 - b^2)^2 \sin^4 \theta]^{1/2}$$

This equation is independent of the angle ψ (shown by Fig. A.2), which determines the position of the x', y' axes with respect to the plane shown.

A.2   Derivation of the Indicatrix and Fresnel's Ellipsoids from Maxwell's Equations

An elegant construction for the wave normal and dielectric displacements is the indicatrix ellipsoid. Another ellipsoid that specifies the ray direction— the direction of energy transmission—and the electric fields is the Fresnel index ellipsoid. Both of these ellipsoids can be derived from Maxwell's

[1] See Mason [1].

equations. The indicatrix ellipsoid is defined in the equivalent ways

$$\frac{x_1^2}{n_1^2} + \frac{x_2^2}{n_2^2} + \frac{x_3^2}{n_3^2} = 1 = B_1 x_1^2 + B_2 x_2^2 + B_3 x_3^2 \qquad (A.24)$$

where n_1 is the index of refraction $V/a = 1/\sqrt{B_1}$. As shown by Fig. 7.1, if one draws a straight line OP in an arbitrary direction, the surface perpendicular to it—which is an ellipse—contains the directions of electric displacement possible for the two waves propagating along OP as major and minor axes of the ellipse.

To prove this assertion we can transform the impermeability ellipsoid of Eq. (A.24) to the set of coordinates x', y', z' shown in Fig. A.2. z' is considered as the direction of wave propagation—that is, the direction OP. The direction cosines of the new axes with respect to those of the impermeability ellipsoid are

$$\begin{array}{c|ccc} & x & y & z \\ \hline x' & \alpha_{11} & \alpha_{12} & \alpha_{13} \\ y' & \alpha_{21} & \alpha_{22} & \alpha_{23} \\ z' & \alpha_{31} & \alpha_{32} & \alpha_{33} \end{array} \qquad (A.25)$$

where

$$\alpha_{11} = \cos\theta\cos\varphi\cos\psi - \sin\varphi\sin\psi$$

$$\alpha_{12} = \cos\theta\sin\varphi\cos\psi + \cos\varphi\sin\psi$$

$$\alpha_{13} = -\sin\theta\cos\psi$$

$$\alpha_{21} = -\cos\theta\cos\varphi\sin\psi - \sin\varphi\cos\psi$$

$$\alpha_{22} = \cos\varphi\cos\psi - \sin\varphi\sin\psi\cos\theta \qquad (A.26)$$

$$\alpha_{23} = \sin\theta\sin\psi$$

$$\alpha_{31} = \cos\varphi\sin\theta$$

$$\alpha_{32} = \sin\varphi\sin\theta$$

$$\alpha_{33} = \cos\theta.$$

The impermeability surface at right angles to the direction z' can be obtained by evaluating the constant B'_{11} as a function of the angle ψ. From Eq. (2.34) the value of B'_{11} is obtained by evaluating

$$B'_{11} = \alpha_{1i}\alpha_{1j}B_{ij} \qquad \text{where} \qquad B_{12} = B_{13} = B_{23} = 0 \qquad (A.27)$$

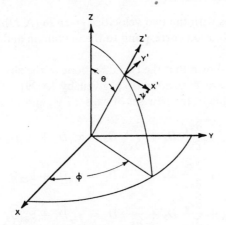

FIG. A.2 Rotated axes and angles relating them to the unrotated axes.

The result is

$$B'_{11} = \alpha_{11}^2 B_1 + \alpha_{12}^2 B_2 + \alpha_{13}^2 B_3$$

$$= B_1\left[\cos^2\theta\cos^2\varphi\cos^2\psi - \frac{\sin 2\varphi\sin 2\psi\cos\theta}{2} + \sin^2\varphi\sin^2\psi\right]$$

$$+ B_2\left[\cos^2\theta\sin^2\varphi\cos^2\psi + \frac{\sin 2\varphi\sin 2\psi\cos\theta}{2} + \cos^2\varphi\sin^2\psi\right] \quad \text{(A.28)}$$

$$+ B_3\sin^2\theta\cos^2\psi$$

To determine the maximum and minimum values of B'_{11}, we differentiate with respect to ψ and set the resultant equal to zero. Solving for the values of ψ that satisfy this condition,

$$\tan 2\psi = \frac{(B_2 - B_1)\sin 2\varphi\cos\theta}{(B_1 - B_2)(\cos^2\theta\cos^2\varphi - \sin^2\varphi) + (B_3 - B_2)\sin^2\theta} \quad \text{(A.29)}$$

For a solution of this equation having a positive or negative value on the right-hand side of the equation, there is another solution 90° larger having the same solution and hence there are two directions of vibration 90° apart that satisfy the condition of minimum or maximum values of B'_{11}. Inserting these values of B'_{11} into (A.28) we find

$$2B'_{11} = B_1(\sin^2\varphi\sin^2\theta + \cos^2\theta) + B_2(\cos^2\varphi\sin^2\theta + \cos^2\theta) + B_3\sin^2\theta$$

$$\pm \left[(B_1 - B_2)^2(\cos^2\theta\cos^2\varphi + \sin^2\varphi)^2\right.$$

$$+ 2(B_1 - B_2)(B_3 - B_2)\sin^2\theta(\cos^2\theta\cos^2\varphi - \sin^2\varphi) \quad \text{(A.30)}$$

$$\left.+ (B_3 - B_2)^2\sin^4\theta\right]^{1/2}$$

Since

$$B_1 = a^2/v^2 \;; \qquad B_2 = b^2/v^2 \;; \qquad B_3 = c^2/v^2 \;; \qquad B'_{11} = v^2/V^2$$

this equation agrees with the two velocities given in (A.23b) and shows that the directions of vibration correspond to the maximum and minimum values of B'_{11}.

It remains to be shown that the two directions of the electric displacement coincide with the two directions of ψ given by (A.29). Transforming the electric displacements to the new set of axes x', y', and z', we have

$$D_1' = \frac{\partial x_1'}{\partial x_1} D_1 + \frac{\partial x_1'}{\partial x_2} D_2 + \frac{\partial x_1'}{\partial x_3} D_3 = \alpha_{11} D_1 + \alpha_{12} D_2 + \alpha_{13} D_3$$

$$D_2' = \frac{\partial x_2'}{\partial x_1} D_1 + \frac{\partial x_2'}{\partial x_2} D_2 + \frac{\partial x_2'}{\partial x_3} D_3 = \alpha_{21} D_1 + \alpha_{22} D_2 + \alpha_{23} D_3 \quad \text{(A.31)}$$

$$D_3' = \frac{\partial x_3'}{\partial x_1} D_1 + \frac{\partial x_3'}{\partial x_2} D_2 + \frac{\partial x_3'}{\partial x_3} D_3 = \alpha_{31} D_1 + \alpha_{32} D_2 + \alpha_{33} D_3$$

Inserting the values of D_1, D_2, and D_3 from (A.18)—with $B_{12} = B_{13} = B_{23} = 0$—and noting that $l_1 = \alpha_{31}$, $l_2 = \alpha_{32}$, $l_3 = \alpha_{33}$ since these are the direction cosines of the z' axis, we have

$$\begin{aligned} D_1' &= \alpha_{31}\alpha_{11}[(B_2 - B'_{11})(B_3 - B'_{11})] + \alpha_{32}\alpha_{12}[(B_1 - B'_{11})(B_3 - B'_{11})] \\ &\quad + \alpha_{33}\alpha_{13}[(B_1 - B'_{11})(B_2 - B'_{11})] \\ D_2' &= \alpha_{31}\alpha_{21}[(B_2 - B'_{11})(B_3 - B'_{11})] + \alpha_{32}\alpha_{22}[(B_1 - B'_{11})(B_3 - B'_{11})] \\ &\quad + \alpha_{33}\alpha_{23}[(B_1 - B'_{11})(B_2 - B'_{11})] \\ D_3' &= \alpha_{31}^2[(B_2 - B'_{11})(B_3 - B'_{11})] + \alpha_{32}^2[(B_1 - B'_{11})(B_3 - B'_{11})] \\ &\quad + \alpha_{33}^2[(B_1 - B'_{11})(B_2 - B'_{11})] \end{aligned} \quad \text{(A.32)}$$

The last equation is identical with (A.20) with $B_{12} = B_{13} = B_{23} = 0$ and hence D_3' vanishes. Hence the two values of electric displacement lie in a plane perpendicular to z'. By inserting the value of B'_{11} from (A.30), with the positive value of the square-root sign, it is found after some calculation that $D_2 = 0$ and hence the electric displacement lies along the direction of the largest value of B'_{11}. Similarly, taking the negative square root value, $D_1 = 0$, and the second wave is perpendicular to the first and in the direction of the smallest value of B'_{11}. Since D is perpendicular to the direction of wave propagation, the impermeability ellipsoid determines the direction of equal-phase propagation and the corresponding dielectric displacement directions.

The Fresnel ellipsoid is determined from the direction of energy transmission S_i—usually called the ray direction—and the corresponding electric fields. The Fresnel ellipsoid has the equation

$$\frac{x^2}{a^2/v^2} + \frac{y^2}{b^2/v^2} + \frac{z^2}{c^2/v^2} = 1 = K_1 x^2 + K_2 y^2 + K_3 z^2 \quad \text{(A.33)}$$

where $K_1 = 1/B_1$, $K_2 = 1/B_2$, $K_3 = 1/B_3$ are the relative dielectric constants along the principal axes. By going through a similar set of equations, it is readily shown that the plane perpendicular to the ray direction is an ellipse containing the two electric field vectors for rays propagated along z'. The maximum and minimum values of K'_{11} determine the two field directions for which single waves are transmitted along z'. For any other angle two independent waves are transmitted, and since they go with different velocities, they interfere with each other at certain thickness values. This phenomenon is known as birefringence.

A.3   Effect of Optical Activity on the Transmission of Light Waves in Crystals

A.31   Equations for Elliptically Polarized Electromagnetic Waves

As discussed in the last section, a light wave transmitted along a given direction in a crystal breaks up into two waves with the electric displacements at right angles to each other, which are transmitted with two different velocities v_1 and v_2. These velocities can be defined in terms of the index of refraction n which is given by

$$n_1 = \frac{V}{v_1} \; ; \qquad n_2 = \frac{V}{v_2} \tag{A.34}$$

where V is the velocity of light in a vacuum and v_1 and v_2 are respectively the velocities of the ordinary and extraordinary rays. The positions of the polarization vectors for these two rays are discussed in Chapter 7, Section 7.1. In both cases only plane waves are transmitted through the crystals.

In certain crystals such as quartz there is a rotation in the plane of polarization when a linearly polarized light is transmitted along the optic axis (see Section 7.1). This rotation of the plane is proportional to the thickness of the crystal. The cause of this effect, as discussed for example by Condon,[2] is that the effective electric field acting on individual molecules is the sum of the applied field plus an average field due to neighboring molecules. Phenomenologically the relation between the electric displacement and the electric and magnetic fields—as well as the magnetic flux B and the fields—can be written in the tensor form

$$E_j = \beta_{ij} D_i + \frac{g_{ij}}{\varepsilon_0 \mu_0} \frac{\partial B_i}{\partial t} \; ; \qquad H_i = \frac{B_i}{\mu_0} - \frac{g_{ij}}{\varepsilon_0 \mu_0} \frac{\partial D_j}{\partial t} \tag{A.35}$$

It is assumed that the magnetic permeability is essentially the same as that for a vacuum. g_{ij} is a new gyration tensor, which determines the rotation of the plane of polarization. When these equations are introduced in

[2] See Condon [2].

Maxwell's equations they show that in the most general case two elliptically polarized waves are transmitted through the crystal. Along an optic axis these waves become circularly polarized, and since they travel with slightly different velocities their superposition—which is equivalent to a plane polarized wave—causes the plane of polarization to rotate as the waves go through the crystal.

An elliptically polarized wave is one for which there is a relation between the two components of one of the variables—say, the electric displacement—that propagate along the z' axis. The propagation equation can be written in the form

$$D_1' = A_1 \exp\left[j\omega(t - nz'/V) + j\delta_1\right]; \quad D_2' = A_2 \exp\left[j\omega(t - nz'/V) + j\delta_2\right]$$

(A.36)

Taking the real parts of these equations we find

$$\frac{D_1'}{A_1} = \cos\Gamma\cos\delta_1 - \sin\Gamma\sin\delta_1 ; \qquad \frac{D_2'}{A_2} = \cos\Gamma\cos\delta_2 - \sin\Gamma\sin\delta_2$$

(A.37)

where $\Gamma = \omega[t - nz'/V]$. Taking the two differences

$$\frac{D_1'}{A_1}\sin\delta_2 - \frac{D_2'}{A_2}\sin\delta_1 = \cos\Gamma\sin(\delta_2 - \delta_1)$$

$$\frac{D_1'}{A_1}\cos\delta_2 - \frac{D_2'}{A_2}\cos\delta_1 = \sin\Gamma\sin(\delta_2 - \delta_1)$$

(A.38)

Squaring and adding

$$\left(\frac{D_1'}{A_1}\right)^2 + \left(\frac{D_2'}{A_2}\right)^2 - 2\left(\frac{D_1'}{A_1}\right)\left(\frac{D_2'}{A_2}\right)\cos\delta = \sin^2\delta \qquad \text{(A.39)}$$

where $\delta = \delta_2 - \delta_1$. This is the equation of an ellipse inscribed in a rectangle whose sides are parallel to the coordinate axes and whose lengths are $2a_1$ and $2a_2$ as shown by Fig. A.3. Hence Eqs. (A.36) describe the propagation of an elliptic wave. All the solutions obtained in crystal optics are of the form for which $\delta = \delta_2 - \delta_1 = \pi/2$. Hence

$$\left(\frac{D_1'}{A_1}\right)^2 + \left(\frac{D_2'}{A_2}\right)^2 = 1 \qquad \text{(A.40)}$$

In this case one major axis lies along the x_1' axis while the other lies along the x_2' axis. The ellipticity ξ, which is defined as the ratio of the minor axis to the major axis, is given by

$$\xi = \frac{A_2}{A_1} \qquad \text{if} \qquad A_2 < A_1 \qquad \text{(A.41)}$$

In terms of the complex solution of Eq. (A.36) the ratio of D_2' to D_1' is

$$\frac{D_2'}{D_1'} = \frac{A_2 \exp [j\delta_2]}{A_1 \exp [j\delta_2]} = \frac{A_2}{A_1} \exp [j(\delta_2 - \delta_1)] = \frac{A_2}{A_1} [\cos (\delta_2 - \delta_1) + j \sin (\delta_2 - \delta_1)]$$

(A.42)

Hence if $(\delta_2 - \delta_1) = \pi/2$ the ratio

$$\frac{D_2'}{D_1'} = j \frac{A_2}{A_1} = j\xi$$

(A.43)

Two special cases of the elliptic wave are of interest, one when $A_1 = A_2$ with $(\delta_2 - \delta_1) = \pi/2$ and the other where $(\delta_2 - \delta_1) = 0$ and $A_2/A_1 = 0$.

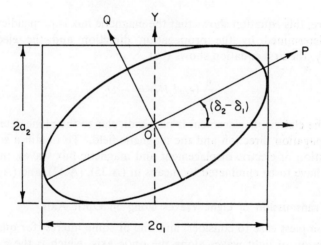

FIG. A.3   Elliptically polarized wave. The vibrational ellipse for the electric vector.

The first case is a circularly polarized wave, while the second is a plane polarized wave.

If we express Eqs. (A.36) in terms of the coordinates determining the principal axes of the indicatrix, Eqs. (A.36) can be written in the form

$$D_k = A_k \exp [j\delta_k] \times \exp [j\omega(t - nl_j x_j/V)]$$

(A.44)

If we apply the third equation of (3.40) to this equation, we find

$$0 = \frac{\partial D_k}{\partial x_k} = A_k \exp [j\delta_k] \times \left(-\frac{j\omega n l_j}{V} \frac{\partial x_j}{\partial x_k}\right) \exp [j\omega(t - nl_j x_j/V)]$$

$$= -\frac{j\omega n}{V} \left(D_k l_j \frac{\partial x_j}{\partial x_k}\right)$$

(A.45)

Since $\partial x_j/\partial x_k = 0$ unless $j = k$, it follows that $D_j l_j = 0$ and hence the dielectric displacement is perpendicular to the direction of propagation. The last of Eqs. (3.40) shows that the magnetic-flux vector B is also perpendicular to the propagation path as in Fig. A.1.

The second equation of (3.40) results in the relation

$$\frac{n}{V} E_{k_0} \exp\left[-j\delta_k\right](-j\omega l_j) \exp\left[-j\omega(t - nl_j x_j/V)\right]$$
$$= -j\omega B_i \exp\left[-\delta_i\right] \exp\left[-j\omega(t - nl_j x_j/V)\right] \tag{A.46}$$

or

$$\frac{n}{V} \varepsilon_{ijk} l_j E_k = B_i \tag{A.47}$$

As before, this equation shows that the magnetic flux is perpendicular to the plane determined by the propagation direction and the electric field. Similarly, the first equation shows that

$$\frac{n}{V} (\varepsilon_{ijk} l_j H_k) = -D_i \tag{A.48}$$

Hence the electric displacement is perpendicular to the plane determined by the propagation direction and the magnetic field. To obtain a solution for propagation of electric displacement and magnetic flux waves, the fields E_k and H_k have to be eliminated by means of (A.35), (A.47), and (A.48).

A.32 Transmission of Light Waves along an Optic Axis

The simplest case to consider, and one of some interest for quartz, is the transmission of light waves along the optic axis, which is the z axis. The indicatrix surface perpendicular to the z axis is circular, so that the relative impermeability constants are

$$B_{11} = \frac{\beta_{11}}{\beta_0} = B_{22} = \frac{\beta_{22}}{\beta_0} ; \qquad B_{33} = \frac{\beta_{33}}{\beta_0} ; \qquad B_{12} = B_{13} = B_{23} = 0 \tag{A.49}$$

where $\beta_0 = 1/\varepsilon_0$ is the impermeability constant of space. Hence along the principal axes

$$E_1 = \frac{B_1 D_1}{\varepsilon_0} + \frac{j\omega g_{11} \bar{B}_1}{\varepsilon_0 \mu_0} ; \qquad E_2 = \frac{B_1 D_2}{\varepsilon_0} + \frac{j\omega g_{22} \bar{B}_2}{\varepsilon_0 \mu_0} ; \qquad E_3 = 0$$

$$H_1 = \frac{\bar{B}_1}{\mu_0} - \frac{j\omega g_{11} D_1}{\mu_0 \varepsilon_0} ; \qquad H_2 = \frac{\bar{B}_2}{\mu_0} - \frac{j\omega g_{22} D_2}{\mu_0 \varepsilon_0} ; \qquad H_3 = 0 \tag{A.50}$$

In this equation bars are put over the magnetic-flux variables to distinguish them from the relative impermeability constants. As shown in Appendix B, Table B.6, $g_{22} = g_{11}$ for the symmetry of quartz. Also $g_{12} = g_{13} = g_{23} = 0$, so that no fluxes or displacements occur except the ones shown.

From the curl Eqs. (A.47) and (A.48) we have the following, since $l_3 = 1$, $l_1 = l_2 = 0$:

$$\frac{n}{V} E_1 = \bar{B}_2; \quad -\frac{n}{V} E_2 = \bar{B}_1; \quad \frac{n}{V} H_1 = -D_2; \quad \frac{n}{V} H_2 = D_1 \qquad (A.51)$$

Substituting in the values of E_1, E_2, H_1, and H_2 from (A.50), we have four equations for $\bar{B}_1$, $\bar{B}_2$, D_1, and D_2. These are

$$-\frac{n}{V}\left[\frac{B_1 D_2}{\varepsilon_0} + \frac{j\omega g_{11}\bar{B}_2}{\varepsilon_0\mu_0}\right] = \bar{B}_1; \quad \frac{n}{V}\left[\frac{B_1 D_1}{\varepsilon_0} + \frac{j\omega g_{11}\bar{B}_1}{\varepsilon_0\mu_0}\right] = \bar{B}_2$$

$$\frac{n}{V}\left[\frac{\bar{B}_2}{\mu_0} - \frac{j\omega g_{11} D_2}{\varepsilon_0\mu_0}\right] = D_1; \quad \frac{n}{V}\left[\frac{\bar{B}_1}{\mu_0} - \frac{j\omega g_{11} D_1}{\varepsilon_0\mu_0}\right] = -D_2 \qquad (A.52)$$

As a first step we can eliminate D_1 and $\bar{B}_1$, collect terms, and—after using $V = 1/\sqrt{\varepsilon_0\mu_0}$—have two relations between D_2 and $\bar{B}_2$ of the form

$$\bar{B}_2\left[\frac{1}{n^2} - \left(B_1 + \frac{\omega^2 g_{11}^2}{\varepsilon_0\mu_0}\right)\right] = -\frac{2j\omega g_{11} D_2 B_1}{\varepsilon_0}$$

$$D_2\left[\frac{1}{n^2} - \left(B_1 + \frac{\omega^2 g_{11}^2}{\varepsilon_0\mu_0}\right)\right] = \frac{2j\omega g_{11}\bar{B}_2}{\mu_0} \qquad (A.53)$$

Eliminating $\bar{B}_2$ from the two equations, the multiplier of D_2—which must be zero—is

$$\left[\frac{1}{n^2} - \left(B_1 + \frac{\omega^2 g_{11}^2}{\varepsilon_0\mu_0}\right)\right]^2 - \frac{4\omega^2 g_{11}^2 B_1}{\varepsilon_0\mu_0} = 0 \qquad (A.54)$$

Solving for the index of refraction n we find

$$\frac{1}{n} = \left[B_1^{1/2} \pm \frac{\omega g_{11}}{\sqrt{\varepsilon_0\mu_0}}\right] \qquad (A.55)$$

Making use of the fact that $\omega g_{11}/\sqrt{\varepsilon_0\mu_0}$ is small compared to $B_1^{1/2}$, we have

$$n = \frac{1}{B_1^{1/2} \pm (\omega g_{11})/\sqrt{\varepsilon_0\mu_0}} = \frac{1 \mp (\omega g_{11} B_1^{-1\,2})/\sqrt{\varepsilon_0\mu_0}}{B_1^{1/2}}$$

$$= K_1^{1/2} \mp \frac{\omega g_{11} K_1}{\sqrt{\varepsilon_0\mu_0}} \qquad (A.56)$$

The negative sign corresponds to a right circularly polarized wave and the positive sign to a left circularly polarized wave. These have respectively the values

$$n_r = n_0 - \frac{\omega g_{11} n_0^2}{\sqrt{\varepsilon_0 \mu_0}} \; ; \qquad n_l = n_0 + \frac{\omega g_{11} n_0^2}{\sqrt{\varepsilon_0 \mu_0}} \tag{A.57}$$

where n_0 is the index of refraction for polarized light, equal to $n_0 = K_1^{1/2}$. Inserting these values in the relation between D_1 and D_2 given in the third of (A.52) when the second relation of (A.53) is substituted for $\bar{B}_2$ we find

$$D_1 = \mp j D_2 \tag{A.58}$$

with the top sign for the right rotation and the bottom sign for the left rotation. Hence the ellipticity constant $\xi = 1$, and we are dealing with two circularly polarized waves that have slightly different velocities.

Summing the two components of the D_1 polarization for light transmission along the $z = x_3$ axis, we have

$$
\begin{aligned}
D_1 &= \frac{D_{10}}{2} \exp \left\{ j\omega \left[t - \left(n_0 - \frac{\omega g_{11} n_0^2}{\sqrt{\varepsilon_0 \mu_0}} \right) \frac{x_3}{V} \right] \right\} \\
&\quad + \frac{D_{10}}{2} \exp \left\{ j\omega \left[t - \left(n_0 + \frac{\omega g_{11} n_0^2}{\sqrt{\varepsilon_0 \mu_0}} \right) \frac{x_3}{V} \right] \right\} \\
&= D_{10} \cos \omega^2 g_{11} n_0^2 x_3 \exp \left\{ j\omega \left[t - \frac{n_0 x_3}{V} \right] \right\}
\end{aligned}
\tag{A.59}
$$

since $1/V = \sqrt{\varepsilon_0 \mu_0}$. Similarly, the D_2 component of the two waves becomes

$$
\begin{aligned}
D_2 &= -\frac{j D_{10}}{2} \exp \left\{ j\omega \left[t - \left(n_0 - \frac{\omega g_{11} n_0^2}{\sqrt{\varepsilon_0 \mu_0}} \right) \frac{x_3}{V} \right] \right\} \\
&\quad + \frac{j D_{10}}{2} \exp \left\{ j\omega \left[t - \left(n_0 + \frac{\omega g_{11} n_0^2}{\sqrt{\varepsilon_0 \mu_0}} \right) \frac{x_3}{V} \right] \right\} \\
&= D_{10} \left[\frac{\exp \left[j\omega^2 g_{11} n_0^2 x_3 \right] - \exp \left[-j\omega \, g_{11} n_0^2 x_3 \right]}{2j} \right] \exp \left\{ j\omega \left(t - \frac{n_0 x_3}{V} \right) \right\} \\
&= D_{10} \sin \left[\omega^2 g_{11} n_0^2 x_3 \right] \exp \left\{ j\omega \left[t - \frac{n_0 x_3}{V} \right] \right\}
\end{aligned}
\tag{A.60}
$$

Hence the plane of polarization rotates around the x_3 axis in the direction of a right-handed screw when viewed along the x_3 axis in the direction of the origin. The rotation per unit length is given by

$$\rho = \omega^2 g_{11} n_0^2 \text{ radians per meter} \tag{A.61}$$

Since $\omega\lambda = 2\pi V$ where V is the velocity of light in a vacuum and λ is the wavelength in a vacuum, the rotation per unit length is given by

$$\rho = \left(\frac{2\pi n_0 V}{\lambda}\right)^2 g_{11} = \left(\frac{\pi}{\lambda}\right)(n_l - n_r) \tag{A.62}$$

This formula implies that the rotation varies proportionally to the square of the frequency, which is in good agreement with experiment over a wide frequency range for quartz.[3] However, near the cutoff frequency for light transmission in quartz, the rotation varies much more rapidly than proportional to the square of the frequency. Chandrasekhar[4] suggests a formula of the type

$$\rho = \frac{kf^2}{(f_i^2 - f^2)^2} = \frac{k\lambda_i^4\lambda^2}{V^2[\lambda^2 - \lambda_i^2]^2} \tag{A.63}$$

where $\lambda_i = 9.26 \times 10^{-6}$ cm $= 926$ A. This formula will result if g_{11} is replaced by

$$g_{11}' = \frac{g_{11}f_i^4}{(f_i^2 - f^2)^2} \tag{A.64}$$

Such a formula is in good agreement with experiment from $\lambda = 6000$ A to 1525 A. For wavelengths larger than this, another term with frequency f_i in the infrared has to be used.

The order of magnitude of the rotation is also of interest. For quartz for sodium light (near $\lambda = 5893$ A), the rotation is $217°$ or 3.79 radians/cm. Hence the difference between the index of refraction for left and right quartz is 7.11×10^{-5} and the rotation term of (A.57) is 2.3×10^{-5} times as large as the index-of-refraction term.

A.33 Transmission of Light Waves along a General Axis

When the direction of propagation is taken along the z' axis of Fig. A.2, the light wave will in general be broken up into two elliptically polarized rays that travel with different velocities. It was shown by Eq. (A.28) that the relative dielectric impermeability in a plane perpendicular to z' is an ellipse with two major axes B_{11}' and B_{22}'. The relation of these directions to the major axes of the indicatrix ellipsoid was discussed in Eq. (A.29). The tensor g_{ij} is a second-rank tensor and has different values for different directions. From Eq. (2.34) it is seen that the value in any direction is

$$g_{11}' = g_{11}\alpha_{11}^2 + g_{22}\alpha_{12}^2 + g_{33}\alpha_{13}^2 \tag{A.65}$$

---

[3] See Chandrasekhar [3].
[4] *Ibid.*

since the cross terms are zero. Using the direction cosines of (A.25) and (A.26) and noting that for quartz—the only crystal for which both birefringence and optical rotation have been measured—$g_{11} = g_{22}$, we find from (A.29) that $\psi = 0$ or $\pi/2$. With these values

$$\psi = 0, \qquad g'_{11} = g_{11}\cos^2\theta + g_{33}\sin^2\theta; \qquad g'_{22} = g_{11}, \qquad (\psi = \pi/2) \quad \text{(A.66)}$$

All the cross terms g'_{12}, g'_{13}, and g'_{23} vanish for these values. For these two directions

$$B'_{11} = B_1\cos^2\theta + B_3\sin^2\theta, \qquad (\psi = 0); \qquad B'_{22} = B_1, \qquad (\psi = \pi/2) \tag{A.67}$$

With these equations, the relations of (A.35) can be written

$$\frac{n}{V}\left[\frac{B'_{11}D_1'}{\varepsilon_0} + \frac{j\omega g'_{11}\bar{B}_1'}{\varepsilon_0\mu_0}\right] = \bar{B}_2'; \qquad -\frac{n}{V}\left[\frac{B'_{22}D_2'}{\varepsilon_0} + \frac{j\omega g'_{22}B_2'}{\varepsilon_0\mu_0}\right] = \bar{B}_1'$$

$$\frac{n}{V}\left[\frac{\bar{B}'_{11}}{\mu_0} - \frac{j\omega g'_{11}D_1'}{\varepsilon_0\mu_0}\right] = -D_2'; \qquad \frac{n}{V}\left[\frac{\bar{B}'_{22}}{\mu_0} - \frac{j\omega g'_{22}D_2'}{\varepsilon_0\mu_0}\right] = D_1' \tag{A.68}$$

Eliminating the variables $\bar{B}_1'$, $\bar{B}_2'$, D_1', and D_2', one obtains an equation for the index of refraction n of the form

$$n^2 = \frac{\left[\dfrac{B'_{11} + B'_{22}}{2} + \dfrac{\omega^2 g'_{11}g'_{22}}{\varepsilon_0\mu_0}\right] \mp \sqrt{\left(\dfrac{B'_{11} - B'_{22}}{2}\right)^2 + \dfrac{\omega^2}{\varepsilon_0\mu_0}[B'_{11}g'_{22} + B'_{22}g'_{11}][g'_{11} + g'_{22}]}}{\left(B'_{11} - \dfrac{\omega^2 g'^2_{11}}{\varepsilon_0\mu_0}\right)\left(B'_{22} - \dfrac{\omega^2 g'^2_{22}}{\varepsilon_0\mu_0}\right)} \tag{A.69}$$

Since $\omega^2 g_{11}^2/\varepsilon_0\mu_0$ is much smaller than the relative impermeability constants B_1 or B_2 it can be neglected and the equation can be written

$$n = \sqrt{\frac{B'_{11} + B'_{22}}{2B'_{11}B'_{22}}}$$

$$\times \left[1 \mp \frac{1}{2}\sqrt{\left(\frac{B'_{11} - B'_{22}}{B'_{11} + B'_{22}}\right)^2 + \frac{4\omega^2}{\varepsilon_0\mu_0}\left[\frac{B'_{11}g'_{22} + B'_{22}g'_{11}}{(B'_{11} + B'_{22})^2}\right](g'_{11} + g'_{22})}\right] \tag{A.70}$$

As discussed in Chapter 8, the propagation constant can be regarded as the superposition of birefringence, determined by the first term in the square root, and the optical rotation, determined by the second term.

From Eq. (A.68) we find

$$\frac{D_2}{D_1} = -\frac{\dfrac{-j\omega}{\sqrt{\varepsilon_0\mu_0}}\sqrt{(g'_{11} + g'_{22})(B'_{11}g'_{22} + B'_{22}g'_{11})}}{\left(\dfrac{B'_{11} - B'_{22}}{2}\right) \pm \sqrt{\left(\dfrac{B'_{11} - B'_{22}}{2}\right)^2 + \dfrac{\omega^2}{\varepsilon_0\mu_0}(g'_{11} + g'_{22})(B'_{11}g'_{22} + B'_{22}g'_{11})}}$$

(A.71)

For the faster wave, the ellipticity constant ξ is

$$\xi_1 = \left| \frac{\dfrac{\omega}{\sqrt{\varepsilon_0\mu_0}}\sqrt{(g'_{11} + g'_{22})(B'_{11}g'_{22} + B'_{22}g'_{11})}}{\left(\dfrac{B'_{11} - B'_{22}}{2}\right) + \sqrt{\left(\dfrac{B'_{11} - B'_{22}}{2}\right)^2 + \dfrac{\omega^2}{\varepsilon_0\mu_0}(g'_{11} + g'_{22})(B'_{11}g'_{22} + B'_{22}g'_{11})}} \right|$$

(A.72)

For the slower wave, the ellipticity constant ξ is given by

$$\xi_2 = \frac{D_1}{D_2}$$

$$= \left| \frac{\dfrac{B'_{11} - B'_{22}}{2} - \sqrt{\left(\dfrac{B'_{11} - B'_{22}}{2}\right)^2 + \dfrac{\omega^2}{\varepsilon_0\mu_0}(g'_{11} + g'_{22})(B'_{11}g'_{22} + B'_{22}g'_{11})}}{\dfrac{\omega}{\sqrt{\varepsilon_0\mu_0}}\sqrt{(g'_{11} + g'_{22})(B'_{11}g'_{22} + B'_{22}g'_{11})}} \right|$$

(A.73)

is the same as that for the fast wave, as can be shown by direct substitution. The effect of both waves is to rotate the plane of the elliptic wave—without change of shape—according to the equation

$$\rho = \frac{\pi}{\lambda}(n_l - n_r) = \frac{\pi}{\lambda}\sqrt{\left(\frac{B'_{11} + B'_{22}}{2B'_{11}B'_{22}}\right)}$$

$$\times \sqrt{\left(\frac{B'_{11} - B'_{22}}{B'_{11} + B'_{22}}\right)^2 + \frac{4\omega^2}{\varepsilon_0\mu_0}\frac{(B'_{11}g'_{22} + B'_{22}g'_{11})}{(B'_{11} + B'_{22})^2}(g'_{11} + g'_{22})} \quad \text{(A.74)}$$

The rate of rotation of the plane is increased by the birefringence of the crystal.

Two limiting cases are of interest. When $B_1 = B_2$ the birefringent term disappears and the ellipticity becomes unity for both waves. The rate of rotation of the plane of polarization reduces to

$$\rho = \left[\frac{2\pi}{\lambda}n_0V\right]^2 g_{11}$$

in agreement with Eq. (A.62). The other case is when the rotation term vanishes. In quartz this happens since $g_{33} = -2.22 \times g_{11}$.[5] Hence the term $g'_{11} + g'_{22}$ can be expressed as

$$g'_{11} + g'_{22} = g_{11} \cos^2 \theta + g_{33} \sin^2 \theta + g_{11} = 0 \qquad \text{or}$$

$$\tan \theta = \sqrt{\frac{2}{2.22 - 1}} \qquad \text{or} \qquad \theta = 52° 5' \quad \text{(A.75)}$$

The rotation also vanishes when

$$B'_{11} g'_{22} + B'_{22} g'_{11} = 0 \tag{A.76}$$

From Table 1,

$$B_1 = \frac{1}{4.5} = 0.222; \qquad B_3 = \frac{1}{4.6} = 0.217 \tag{A.77}$$

(A.64) vanishes when

$$\tan \theta = \sqrt{\frac{2}{2.22 - B_3/B_1}} \qquad \text{or} \qquad \theta = 51° 50' \quad \text{(A.78)}$$

When the rotation term vanishes $\xi_1 = 0$; $\xi_2 = \infty$. The first term represents a plane wave with polarization D_1 along the x_1' axis, while the other is a plane wave with polarization D_2 along the x_2' axis. As discussed in Chapter 7, these waves interfere with each other and produce patterns that are determined by the birefringence of the crystal.

A.4 Rotation of Light Waves Due to the Faraday Effect

Another source of rotation of the plane of polarization of a light wave is the Faraday effect. In this effect the plane of polarization of a plane light wave is rotated when a magnetic field is applied in the direction of the light wave. The amount of rotation is proportional to the applied magnetic field, and since the magnetic field is an axial vector, the rotation doubles when light is reflected back over the same path. As discussed in Chapter 8, this fact has important consequences for microwave frequencies.

For microwave ferrites the nonreciprocal Faraday effect results from the interaction of the electron spins with a magnetic field. The relation is shown by Fig. A.4. When the ferrite is saturated by an applied magnetic field, the spin axes of these electrons are aligned in the direction of the field and they produce the saturation magnetization of the crystal. For fields less than saturation, these spinning particles will react to a perturbing force at right angles to their spin by precessing about the spin axis, as shown by the dashed line of Fig. A.4. This perturbing force is supplied by a radio-frequency wave of a suitable frequency. The closer the applied frequency is to the precessional frequency, the more marked is the interaction.

[5] See Szivessy and Münster [4].

When the radio wave is transmitted in the direction of the magnetic field, it breaks up into two circularly polarized waves that propagate with slightly different velocities, and this corresponds to the rotation of the plane of polarization of the radio wave. The nearer the frequency of the r-f wave is to the precessing frequency, the larger is the rotation of the plane of polarization. The effect was first investigated theoretically by D. Polder,[6] who

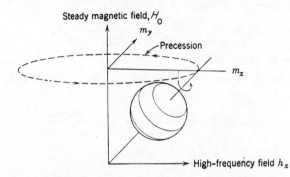

Fig. A.4 Electron precession under the influence of a steady and high-frequency magnetic field.

considered only the undamped motion. The equation of motion of the undamped spinning electron of angular momentum J_i and magnetic moment M_j in a magnetic field H_k is

$$\frac{dJ_i}{dt} = T_i \qquad (A.79)$$

where T_i is the torque exerted by the magnetic field. This is given by the equation

$$T_i = \varepsilon_{ijk} M_j H_k \qquad \text{and} \qquad J_j = \frac{M_j}{\gamma} \qquad (A.80)$$

where γ is the magnetomechanical ratio, and M_j is the magnetization per unit volume. The first part of (A.80) follows directly from the force equation (3.45) since the torque $T_i = F_i r$ and M_j is the product of charge q times the velocity V_j times the radius r of the spinning electron, all multiplied by μ_0, the permeability of space. Hence

$$\frac{dM_b{}'}{dt} = \gamma \varepsilon_{ijk} M_j H_k \qquad (A.81)$$

To solve this equation we divide M_j and H_k into constant parts M_0 and H_0, which are assumed to lie along the direction of wave propagation z, and the variable parts m_j and h_k. For variable parts small compared to the constant

[6] See Polder [5].

parts, and with the direction of propagation being taken as $z = x_3$, Eq. (A.81) becomes

$$j\omega m_1 = \gamma(m_2 H_0 - M_0 h_2); \qquad j\omega m_2 = \gamma(M_0 h_1 - m_1 H_0); \qquad m_3 = 0 \quad \text{(A.82)}$$

The solution of these equations for m_1 and m_2 becomes

$$m_1 = \frac{\gamma^2 M_0 H_0}{\gamma^2 H_0{}^2 - \omega^2} h_1 - \frac{j\omega\gamma M_0}{\gamma^2 H_0{}^2 - \omega^2} h_2$$

$$m_2 = \frac{\gamma^2 M_0 H_0}{\gamma^2 H_0{}^2 - \omega^2} h_2 + \frac{j\omega\gamma M_0}{\gamma^2 H_0{}^2 - \omega^2} h_1$$

(A.83)

But since $b_i = h_i + m_i$, we have

$$b_1 = \mu h_i - j\kappa h_2; \qquad b_2 = j\kappa h_1 + \mu h_2; \qquad b_3 = \mu_0 h_3$$

where

$$\mu = \mu_0 + \frac{\gamma^2 M_0 H_0}{\gamma^2 H_0{}^2 - \omega^2} = \mu_0 - \frac{M_0 \gamma \omega_0}{\omega_0{}^2 - \omega^2}$$

$$\kappa = \frac{\omega\gamma M_0}{\gamma^2 H_0{}^2 - \omega^2} = \frac{\omega\gamma M_0}{\omega_0{}^2 - \omega^2}$$

(A.84)

where $\omega_0 = +\gamma H_0$; In MKS units $\gamma = -0.204 \times 10^6$ radians per second/ amperes per meter. In these equations μ_0 is the permeability of free space and M_0 is the magnetism induced by the field H_0. In these equations H_0 is the internal field, which is determined by the sum of the applied field, the demagnetizing field, and the anisotropic field. The latter is usually neglected, and it can be shown[7] that the internal field is given by

$$H_0 = \{[H_z + (N_y - N_z)M_z][H_z + (N_x - N_z)M_z]\}^{1/2} \quad \text{(A.85)}$$

where H_z is the applied magnetic field and M_z is the resulting magnetism along the z axis. N_x, N_y, and N_z are the demagnetization factors for the x, y, and z directions. The resulting demagnetizing factors and resonant frequencies for a number of different shapes are given in (A.86). Wave transmission is along the z axis.

Form	Demagnetizing factors	ω_0
Slab, H_0 along z, y small	$N_x = N_z = 0$; $N_y = 1$	$-\gamma\sqrt{B_z H_z}$
Sphere, H_0 along z	$N_x = N_y = N_z = \frac{1}{3}$	$-\gamma H_z$
Cylinder, length and H_0 along x	$N_x = N_y = \frac{1}{2}$; $N_z = 0$	$-\gamma(H_z + M_z/2)$
Disk, thickness and H_0 along z	$N_x = N_y = 0$; $N_z = 1$	$-\gamma[H_z - M_z]$

(A.86)

---

[7] See Polder [5].

For the most general wave transmitted along the z axis, the permeability is a tensor and can be written in the form

$$b_i = \mu_{ij} h_j$$

where

$$\mu_{ij} = \begin{vmatrix} \mu & -j\kappa & 0 \\ j\kappa & \mu & 0 \\ 0 & 0 & 1 \end{vmatrix} \tag{A.87}$$

The question arises if there is any type of wave for which the tensor permeability can be reduced to the more familiar case of scalar permeability. This simplification is achieved for a circularly polarized wave. For a circularly polarized wave the y component of the field is at right angles to the x component. Hence

$$h_2 = \mp j h_1 \tag{A.88}$$

Substituting these values in (A.84), we have

$$b_1 = h_1[\mu \mp \kappa]; \qquad b_2 = \mp j h_1[\mu \mp \kappa] = \mp j b_1 \tag{A.89}$$

The negative sign refers to the right-handed (clockwise) rotation, while the positive sign refers to the left-handed rotation. For both cases the flux density b is also circularly polarized and is related to the magnetic field through the scalar permeabilities $\mu - \kappa$ and $\mu + \kappa$ for right- and left-handed waves respectively. The values of $\mu_r = \mu - \kappa$ and $\mu_l = \mu + \kappa$ are shown plotted by Fig. A.5. One component shows a resonant effect while the other does not. Actually, dissipation will cause a rounding off of the resonant component, but this is neglected in the present section.

If we insert the relation between h_i and b_i for a given wave in Maxwell's equations and assume that the dielectric constant is a scalar equal to ε—ε may have a relative value $K = \varepsilon/\varepsilon_0$ from 9 to 20 for ferrites—it is readily shown that the velocity of propagation v is given by

$$\frac{1}{v} = \sqrt{\varepsilon(\mu \mp \kappa)} \tag{A.90}$$

The angle of rotation of the plane of polarization of the plane wave epuivalent to the two circularly polarized waves is

$$\rho = \omega(n_r - n_l)l = \omega \sqrt{\frac{\varepsilon}{\varepsilon_0 \mu_0}} \, [\sqrt{\mu - \kappa} - \sqrt{\mu + \kappa}]l \tag{A.91}$$

where ε_0 and μ_0 are the dielectric constant and permeability of a vacuum. For frequencies low or high compared to ω_0, κ is small compared to μ and the rotation angles are respectively

$$\rho = -\omega\sqrt{\frac{\varepsilon\mu}{\varepsilon_0\mu_0}}\left[\frac{\kappa}{\mu}\right]l \doteq -\frac{\omega^2}{\omega_0{}^2}\sqrt{\frac{\varepsilon\mu}{\varepsilon_0\mu_0}}\left[\frac{\gamma M_0}{\mu_0 - \gamma M_0/\omega_0}\right]l \qquad \omega \ll \omega_0$$

$$\doteq \sqrt{\frac{\varepsilon\mu}{\varepsilon_0\mu_0}}\left[\frac{\gamma M_0}{\mu_0}\right]l \qquad\qquad \omega \gg \omega_0$$

$$(A.92)$$

In between these two limits an algebraic solution of (A.91) is required.

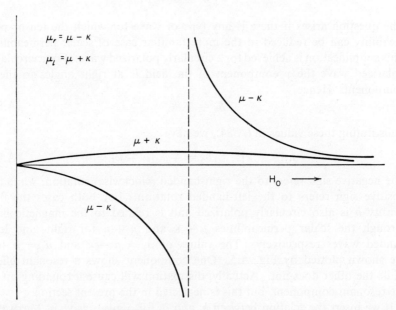

FIG. A.5   Variation of the permeabilities for right-handed (μ_r) and left-handed (μ_l) circularly polarized waves as a function of H_0.

The ordinary optical Faraday effect is similar in principle to the microwave Faraday effect except that it is due to the Zeeman effect—that is, the splitting of the resonant frequency, causing the dispersion, into two frequencies whose separation is proportional to the magnetic field. The dispersion equation can be written in the form

$$n^2 - 1 = \sum_i \frac{A_i N_i S_i}{f_i^2 - f^2} \qquad (A.93)$$

where A_i is a constant, N_i is the number, S_i the strength, and f_i the frequency of the dispersion electrons. For quartz a single frequency of 3.24×10^{15} cycles is sufficient to represent the dispersion curve in the visible light range.[8]

[8] See Chandrasekhar [3].

A magnetic field can affect N_i, S_i, and f_i, but the principal effect is to split f_i into two components by the Zeeman effect.[9] Hence the index of refraction can be expressed as

$$n^2 - 1 = \frac{A_1 N_1 S_1}{(f_1 \pm FH)^2 - f^2} \tag{A.94}$$

where, to agree with the derivation of the microwave Faraday effect, F has the form $(f/f_1)F_0$. If we expand this out in a series involving H, we have

$$n^2 = \left[1 + \frac{A_1 N_1 S_1}{(f_1{}^2 - f^2)}\right]\left[1 \mp \frac{2fF_0 A_1 N_1 S_1 H}{(f_1{}^2 - f^2)^2 \left/ \left[1 + \frac{A_1 N_1 S_1}{(f_1{}^2 - f^2)}\right]\right.}\right] \tag{A.95}$$

Since the first term in brackets is $n_0{}^2$, the square of the index of refraction when no magnetic field is present, this equation can be solved for n with the result

$$n = n_0\left[1 \mp \frac{fF_0 A_1 N_1 S_1 H}{(f_1{}^2 - f^2)^2 n_0{}^2}\right] = n_0 \mp \frac{fF_0 A_1 N_1 S_1 H}{(f_1{}^2 - f^2)^2 n_0} \tag{A.96}$$

The difference between the two solutions times (π/λ) is, from (A.62), the rotation per unit length. Hence

$$\rho = \frac{\pi}{\lambda}(n_l - n_r)l = \frac{Bf^2 H l}{(f_1{}^2 - f^2)^2} \tag{A.97}$$

where B combines all the other terms into a single constant. The usual way for describing the Faraday rotation is in terms of the Verdet constant $\bar{V}$

$$\rho = \bar{V} H l \cos\theta \qquad \text{where} \qquad \bar{V} = \frac{Bf^2}{(f_1{}^2 - f^2)^2} \tag{A.98}$$

This is the same form as the rotation along an optic axis (A.63), and in the case of quartz it is found that the extra rotation due to the magnetic field is proportional to the natural rotation caused by the structure of the crystal.

The form of Eq. (A.98) is in agreement with Becquerel's[10] classical formula for the Verdet constant, which in cgs units takes the form

$$\bar{V} = -\frac{e}{2mV^2}\lambda\frac{dn}{d\lambda} \tag{A.99}$$

where n is the refractive index in the absence of a magnetic field and λ is the vacuum wavelength of the light—that is, $\lambda = V/f$—while e and m are the charge and mass of the electron.

[9] See Joos [6].
[10] See Condon and Odishaw [7].

REFERENCES APPENDIX A

1. W. P. Mason, *Bell System Tech. J.* **29**, 161–188 (1950).
2. E. U. Condon, *Rev. Mod. Phys.* **9**, 432–457 (1937).
3. S. Chandrasekhar, *Proc. Indian Acad. Sci.* **35**, 103 (1952).
4. G. Szivessy and C. Münster, *Ann. Phys.* **20**, 703–736 (1934).
5. D. Polder, *Phil. Mag.* **40**, 99 (1949).
6. G. Joos, "Theoretical Physics." G. E. Stechert, New York, 1934.
7. E. U. Condon and H. Odishaw (eds.), "Handbook of Physics," pp. 6–116. McGraw-Hill, New York, 1958.

APPENDIX B • Tables of Tensor Terms for Polar and Axial Tensors

The tensor equations given in the preceding chapters are general for any crystal symmetries. To apply them to a specific crystal requires a knowledge of what terms remain for the symmetry applying to this crystal. The 32 point group have been discussed in Chapter 5, and direct inspection methods were given for calculating the resulting terms for a number of symmetries. Since most of the properties are expressed by tensors of rank four or less, the tables given here are limited to these ranks except for one sixth-rank tensor connected with the third-order moduli—that is, the deviations from linearity of the ordinary or second-order elastic constants.

In determining the number of tensor terms it is necessary to know whether the tensor is polar or axial. In general if the tensor is a ratio of a polar vector to an axial vector, the tensor is an axial second-rank tensor. An example is the gyration tensor g_{ij}, which is the ratio of the electric displacement to the time derivative of the magnetic field (see Eq. [A.35]). A third-rank axial tensor is the tensor d_{ijk}, occurring in the crystalline piezomagnetic effect, which is defined by the equation

$$S_{ij} = d_{mij}H_m \qquad (B.1)$$

When the tensor is a ratio of polar vectors or tensors, the tensor is polar. A tensor is also polar if it is determined by a ratio of a polar vector or tensor and the product of two axial vectors. An example is the magnetostrictive tensor

$$S_{ij} = M_{ijkl}B_kB_l \qquad (B.2)$$

of Chapter 4.

An interesting case is the Hall-effect tensor of Chapter 11, which can be written in the form

$$E_i = \varepsilon_{ijk}I_j(R_{km}H_m) \qquad (B.3)$$

325

The rotation tensor ε_{ijk} acting on two polar vectors produces an axial vector, while if it acts on the product of a polar and an axial vector it produces a polar vector. Hence it follows that the vector

$$J_k = R_{km}H_m \qquad (B.4)$$

has to be axial. Hence the tensor R_{km} is a polar tensor.

B.1 Tables of Polar Tensors

B.11 Zero-Rank Tensors (Scalars)

There are a number of effects that are independent of the handedness of the axes and of the crystal systems. Some of these effects are

Thermal Capacity:

$$\delta Q = \rho C_p \, \delta\Theta: \qquad\qquad\qquad \delta\Theta = \frac{\delta Q}{\rho C_v}$$

Heat of Deformation:

$$\delta Q = \Theta\alpha_{kl}T_{kl}: \qquad\qquad\qquad \delta\Theta = \Theta\lambda_{ij}S_{ij}$$

Electrocaloric Effect: $\qquad\qquad\qquad\qquad\qquad\qquad\qquad$ (B.5)

$$\delta Q = \Theta p_m{}^T E_m: \qquad\qquad\qquad \delta\Theta = -\frac{\Theta q_m{}^T D_n}{\rho C^{S,D}}$$

Compressibility:

$$\Delta = [s_{11} + s_{22} + s_{23} + 2(s_{12} + s_{13} + s_{23})]p = \kappa p; \quad B = \frac{1}{\kappa}$$

The magnetocaloric effect is also represented by a scalar. Multiplying equation (4.59), last line, through by the absolute temperature Θ the heat of deformation is the same as the second line of (B.5), while the magnetocaloric effect is

$$\delta Q = \Theta i_m{}^T H_m \qquad (B.6)$$

The second form of the magnetocaloric effect, similar to the second part of the third line of (B.5), is

$$\delta\Theta = -\frac{\Theta j_n{}^T B_n}{\rho C^{H,T}} \qquad (B.7)$$

where $j_n = i_n{}^T/\mu_{mn}^{T,\Theta}$. Since $i_m{}^T$ is the ratio of the axial vector to a scalar, it is an axial tensor. However, in (B.6) it is multiplied by another axial vector so that the expression for δQ is a scalar.

Another scalar is the Faraday rotation ρ, which is the same in either a right-handed or left-handed system of axes.

B.12   First-Rank Tensors

There are certain first-rank tensors, such as the pyroelectric effect and the vector ratio between the piezoelectric polarization and a hydrostatic pressure, that have in general the three components of a polar vector. Table B.1 shows these components for the various crystallographic systems.

Table B.1

Components of a Polar Rank-One Tensor for Various
Crystallographic Systems

1; $\lvert q_1, q_2, q_3 \rvert$; 2; $\lvert 0, q_2, 0 \rvert$; $m = \bar{2}$; $\lvert q_1, 0, q_3 \rvert$	
$mm2, 4, 4mm, 3, 3m, 6, 6mm$; $\lvert 0, 0, q_3 \rvert$	

All other classes give a zero result. All classes giving a positive result have a polar axis. Examples are

1. Pyroelectric effect:

$$D_n = p_n\, \delta\Theta; \qquad E_m = q_m\, \delta\Theta$$

2. Piezoelectric charge generated by a hydrostatic pressure:

$$D_n = -(d_{n1} + d_{n2} + d_{n3})p$$

B.13   Second-Rank Tensors

Second-rank tensors can arise as a relation between two vectors or as a relation between a scalar and a quantity expressed by a second-rank tensor. Examples of the first case are electric permittivity, dielectric impermeability, magnetic permeability and impermeability, electric conductivity and resistivity, thermal conductivity and resistivity, and Thomson thermoelectricity. Equations for these effects are given by (B.8), with a glossary of terms given in (B.14).

$$
\begin{aligned}
D_i &= \varepsilon_{ij}E_j & E_j &= \beta_{ij}D_i & B_i &= \mu_{ij}H_j \\
H_j &= \beta_{ij}B_i & I_i &= \sigma_{ij}E_j & E_j &= \rho_{ij}I_i \\
h_i &= -k_{ij}\left(\frac{\partial\Theta}{\partial x_j}\right) & \frac{\partial\Theta}{\partial x_j} &= -r_{ij}h_i & \frac{\partial\bar{\mu}}{\partial x_i} &= -\sum_{ik}\frac{\partial\Theta}{\partial x_k}
\end{aligned}
\tag{B.8}
$$

Examples of the second case, for which the tensors arise as a relation between a scalar and quantities expressed as second-rank tensors, are thermal expansions, stresses due to temperature changes, strain for a hydrostatic stress, and Peltier thermoelectric coefficients. These relations are given by Eq. (B.9):

$$S_{ij} = \alpha_{ij}\delta\Theta; \qquad T_{ij} = -\lambda_{ij}\delta\Theta; \qquad S_{ij} = s_{ijkl}T_{kl}; \qquad \pi_{ik} = \frac{\Theta}{e}\sum_{ik} \tag{B.9}$$

Table B.2

SECOND-RANK POLAR TENSORS

Crystal Systems	Tensor Terms			Reduction for Symmetric Form
Triclinic				
	π_{11}	π_{12}	π_{13}	If symmetric
1, $\bar{1}$	π_{21}	π_{22}	π_{23}	$\alpha_{12} = \alpha_{21}$; $\alpha_{13} = \alpha_{31}$
9 constants	π_{31}	π_{32}	π_{33}	$\alpha_{23} = \alpha_{32}$; 6 constants
Monoclinic				
	π_{11}	0	π_{13}	If symmetric
2, m, $2/m$	0	π_{22}	0	$\alpha_{13} = \alpha_{31}$
5 constants	π_{31}	0	π_{33}	4 constants
Trigonal, tetragonal, hexagonal				
	π_{11}	π_{12}	0	If symmetric
3, $\bar{3}$, 4, $\bar{4}$, $4/m$	$-\pi_{12}$	π_{11}	0	$\alpha_{12} = 0$
6, $\bar{6}$, $6/m$	0	0	π_{33}	2 constants
3 constants				
Trigonal, tetragonal, hexagonal				
	π_{11}	0	0	If symmetric the same number of constants
32, $3m$, $\bar{3}m$	0	π_{11}	0	
422, $4mm$, $\bar{4}2m$, $4/mmm$	0	0	π_{33}	
622, $6mm$, $\bar{6}m2$, $6/mmm$				
2 constants				
Cubic or isotropic				
	π_{11}	0	0	Same for symmetric tensor
23, $m3$, $\bar{4}3m$	0	π_{11}	0	
432, $m3m$	0	0	π_{11}	
1 constant				

All the second-rank tensors are symmetric except the thermoelectric tensors. Table B.2 shows the terms for the various crystal symmetries and the changes caused by the relations $\alpha_{ij} = \alpha_{ji}$.

B.14 Third-Rank Tensors

Third-rank tensors have been employed in expressing the direct and inverse piezoelectric effect with four different forms depending on the sets of variables used. They have also been employed in defining the electro-optical effect and the Hall effect. These relations are given by Eq. (B.10):

$$D_n = d_{nij}T_{ij} \; ; \qquad S_{ij} = d_{mij}E_m \; ; \qquad T_{kl} = -e_{mkl}E_m \; ; \qquad D_n = e_{nij}S_{ij}$$

$$T_{kl} = -h_{nkl}D_n \; ; \qquad E_m = -h_{mij}S_{ij} \; ; \qquad S_{ij} = g_{nij}D_n \; ; \qquad E_m = -g_{mij}T_{ij}$$

$$E_m = D_n(\beta_{mn}^S + r_{mno}^S D_0); \qquad E_i = \varepsilon_{ijk}I_j(R_{km}H_m) \qquad \text{(B.10)}$$

In all third-rank tensors, two pairs of indices can be interchanged—for example, ij in d_{nij}—since T_{ij} is a symmetric tensor with $T_{ij} = T_{ji}$. Hence it is usual to replace the two indices by a single one according to the convention

$$11 = 1; \quad 22 = 2; \quad 33 = 3; \quad 23 = 32 = 4; \quad 13 = 31 = 5; \quad 12 = 21 = 6$$

$$\text{(B.11)}$$

Table B.3 gives the resulting third-rank tensors for the various crystal symmetries.

B.15   Fourth-Rank Tensors

All the fourth-rank tensors in general use express relations between two second-rank tensors such as stress and strain or between a second-rank tensor and the product of two vectors. Examples are elasticity equations, piezo-optic relations, magnetostrictive and electrostrictive relations, magneto-resistance effects, piezoresistance effects, the quadratic electro-optic effect, and the Cotton-Mouton effect, expressed by the equations

$$S_{ij} = s_{ijkl}T_{kl} \; ; \qquad T_{kl} = c_{ijkl}S_{ij} \; ; \qquad E_m = D_n(\beta_{mn}^S + \beta_0 m_{ijmn}S_{ij})$$

$$S_{ij} = M_{ijkl}B_kB_l \; ; \qquad S_{ij} = q_{ijkl}D_kD_l \; ; \qquad E_i = \alpha_{ijkl}I_jH_kH_i \qquad \text{(B.12)}$$

$$E_i = (\rho_{ij} + \rho_0\pi_{ijkl}T_{kl})I_i \; ; \qquad (\beta_{mn}^S(D) - \beta_{mn}^S(0)) = \beta_0 \frac{f_{mnop}}{2} D_o{}^0D_p{}^0 \; ;$$

$$(\beta_{mn}^S(B) - \beta_{mn}^S(0)) = \beta_0 l_{mnop}B_oB_p$$

Except in the case of ferroelectric or ferromagnetic crystals (Chapter 3), it is generally believed that $T_{ij} = T_{ji}$, so that the compliance tensor s_{ijkl} and the elastic stiffness tensor c_{ijkl} would indicate 36 independent constants. On account of Maxwell-type relations, one can interchange the ij with the kl moduli, and this reduces the number to 21. When it is not permissible to interchange ij with kl, as in the magnetostrictive equations

$$S_{ij} = M_{ijkl}B_kB_l \qquad \text{(B.13)}$$

there are 36 possible constants. Table B.4 for fourth-rank tensors shows how the crystal symmetries affect the number and relations among the independent constants. Type c relations indicated are for the case that ij can be interchanged with kl. A third type of symmetry for fourth-rank tensors occurs

Table B.3
THIRD-RANK POLAR TENSORS

1

$$\begin{pmatrix} e_{11} & e_{12} & e_{13} & e_{14} & e_{15} & e_{16} \\ e_{21} & e_{22} & e_{23} & e_{24} & e_{25} & e_{26} \\ e_{31} & e_{32} & e_{33} & e_{34} & e_{35} & e_{36} \end{pmatrix}$$

$\bar{1}=0$;   **2**

$$\begin{pmatrix} 0 & 0 & 0 & e_{14} & 0 & e_{16} \\ e_{21} & e_{22} & e_{23} & 0 & e_{25} & 0 \\ 0 & 0 & 0 & e_{34} & 0 & e_{36} \end{pmatrix}$$

m

$$\begin{pmatrix} e_{11} & e_{12} & e_{13} & 0 & e_{15} & 0 \\ 0 & 0 & 0 & e_{24} & 0 & e_{26} \\ e_{31} & e_{32} & e_{33} & 0 & e_{35} & 0 \end{pmatrix}$$

$2/m=0$;   **222**

$$\begin{pmatrix} 0 & 0 & 0 & e_{14} & 0 & 0 \\ 0 & 0 & 0 & 0 & e_{25} & 0 \\ 0 & 0 & 0 & 0 & 0 & e_{36} \end{pmatrix}$$

mm2

$$\begin{pmatrix} 0 & 0 & 0 & 0 & e_{15} & 0 \\ 0 & 0 & 0 & e_{24} & 0 & 0 \\ e_{31} & e_{32} & e_{33} & 0 & 0 & 0 \end{pmatrix}$$

$mmm=0$;   **$\bar{4}$**

$$\begin{pmatrix} 0 & 0 & 0 & e_{14} & e_{15} & 0 \\ 0 & 0 & 0 & -e_{15} & e_{14} & 0 \\ e_{31} & -e_{31} & 0 & 0 & 0 & e_{36} \end{pmatrix}$$

4, 6

$$\begin{pmatrix} 0 & 0 & 0 & e_{14} & e_{15} & 0 \\ 0 & 0 & 0 & e_{15} & -e_{14} & 0 \\ e_{31} & e_{31} & e_{33} & 0 & 0 & 0 \end{pmatrix}$$

$4/m=0$;   **$\bar{4}2m$**

$$\begin{pmatrix} 0 & 0 & 0 & e_{14} & 0 & 0 \\ 0 & 0 & 0 & 0 & e_{14} & 0 \\ 0 & 0 & 0 & 0 & 0 & e_{36} \end{pmatrix}$$

Transverse isotropy

4mm, 6mm :
$$\begin{pmatrix} 0 & 0 & 0 & 0 & e_{15} & 0 \\ 0 & 0 & 0 & e_{15} & 0 & 0 \\ e_{31} & e_{31} & e_{33} & 0 & 0 & 0 \end{pmatrix} \quad ; \quad 4/mmm = 0$$

32 :
$$\begin{pmatrix} e_{11} & -e_{11} & 0 & e_{14} & 0 & 0 \\ 0 & 0 & 0 & 0 & -e_{14} & -e_{11} \\ 0 & 0 & 0 & 0 & 0 & 0 \end{pmatrix} \quad ; \quad \bar{3} = 0$$

3m :
$$\begin{pmatrix} 0 & 0 & 0 & 0 & e_{15} & -e_{22} \\ -e_{22} & e_{22} & 0 & e_{15} & 0 & 0 \\ e_{31} & e_{31} & e_{33} & 0 & 0 & 0 \end{pmatrix} \quad ; \quad \bar{3}m = 0$$

23, $\bar{4}$3m :
$$\begin{pmatrix} 0 & 0 & 0 & e_{14} & 0 & 0 \\ 0 & 0 & 0 & 0 & e_{14} & 0 \\ 0 & 0 & 0 & 0 & 0 & e_{14} \end{pmatrix} \quad ; \quad \left.\begin{matrix} 6/mmm,\ 6/m \\ m3,\ m3m,\ 432 \end{matrix}\right\} = 0$$

422, 622 :
$$\begin{pmatrix} 0 & 0 & 0 & e_{14} & 0 & 0 \\ 0 & 0 & 0 & 0 & -e_{14} & 0 \\ 0 & 0 & 0 & 0 & 0 & 0 \end{pmatrix} \quad ;$$

3 :
$$\begin{pmatrix} e_{11} & -e_{11} & 0 & e_{14} & e_{15} & -e_{22} \\ -e_{22} & e_{22} & 0 & e_{15} & -e_{14} & -e_{11} \\ e_{31} & e_{31} & e_{33} & 0 & 0 & 0 \end{pmatrix} \quad ;$$

2 :
$$\begin{pmatrix} e_{11} & -e_{11} & 0 & 0 & 0 & -e_{22} \\ -e_{22} & e_{22} & 0 & 0 & 0 & -e_{11} \\ 0 & 0 & 0 & 0 & 0 & 0 \end{pmatrix} \quad ;$$

$\bar{6}$m2 :
$$\begin{pmatrix} e_{11} & -e_{11} & 0 & 0 & 0 & 0 \\ 0 & 0 & 0 & 0 & 0 & -e_{11} \\ 0 & 0 & 0 & 0 & 0 & 0 \end{pmatrix} \quad ;$$

Table B.4

Fourth-Rank Polar Tensors

(Type M_{ijkl}, $i \to j$, $k \to l$; Type c_{ijkl}, $i \to j$, $k \to l$, $ij \to kl$; Type K_{ijkl}, $i \to j \to k \to l$)

Group I Triclinic 1, $\bar{1}$ 36 constants

M_{11}	M_{12}	M_{13}	M_{14}	M_{15}	M_{16}
M_{21}	M_{22}	M_{23}	M_{24}	M_{25}	M_{26}
M_{31}	M_{32}	M_{33}	M_{34}	M_{35}	M_{36}
M_{41}	M_{42}	M_{43}	M_{44}	M_{45}	M_{46}
M_{51}	M_{52}	M_{53}	M_{54}	M_{55}	M_{56}
M_{61}	M_{62}	M_{63}	M_{64}	M_{65}	M_{66}

c constants the same except $c_{ab} = c_{ba}$, resulting in 21 constants.
K constants the same as c except $K_{44} = K_{23}$, $K_{55} = K_{13}$, $K_{66} = K_{12}$, $K_{46} = K_{25}$, $K_{56} = K_{14}$, $K_{45} = K_{36}$, resulting in 15 constants.

Group II Monoclinic 2, m, 2/m 20 constants y = unique axis

M_{11}	M_{12}	M_{13}	0	M_{15}	0
M_{21}	M_{22}	M_{23}	0	M_{25}	0
M_{31}	M_{32}	M_{33}	0	M_{35}	0
0	0	0	M_{44}	0	M_{46}
M_{51}	M_{52}	M_{53}	0	M_{55}	0
0	0	0	M_{64}	0	M_{66}

c constants the same except $c_{ab} = c_{ba}$, resulting in 13 constants.
K constants the same as c, except $K_{44} = K_{23}$, $K_{55} = K_{13}$, $K_{66} = K_{12}$, $K_{46} = K_{25}$, resulting in 9 constants.

Group III Orthorhombic $mm2$, 222, mmm 12 constants

M_{11}	M_{12}	M_{13}	0	0	0
M_{21}	M_{22}	M_{23}	0	0	0
M_{31}	M_{32}	M_{33}	0	0	0
0	0	0	M_{44}	0	0
0	0	0	0	M_{55}	0
0	0	0	0	0	M_{66}

c constants the same except $c_{ab} = c_{ba}$, resulting in 9 constants.
K constants the same as c except $K_{44} = K_{23}$, $K_{55} = K_{13}$, $K_{66} = K_{12}$, resulting in 6 constants.

Group IV Trigonal 3, $\bar{3}$ 12 constants

M_{11}	M_{12}	M_{13}	M_{14}	$-M_{25}$	$2M_{62}$
M_{12}	M_{11}	M_{13}	$-M_{14}$	M_{25}	$-2M_{62}$
M_{31}	M_{31}	M_{33}	0	0	0
M_{41}	$-M_{41}$	0	M_{44}	M_{45}	$2M_{52}$
$-M_{52}$	M_{52}	0	$-M_{45}$	M_{44}	$2M_{41}$
$-M_{62}$	M_{62}	0	M_{25}	M_{14}	$M_{11} - M_{12}$

c constants the same, except $c_{ab} = c_{ba}$ and $c_{62} = c_{45} = 0$.
$c_{46} = c_{25}$; $c_{56} = c_{14}$, resulting in 7 constants.
K constants the same as c constants except $K_{44} = K_{23}$, resulting in 6 constants.

Group V Trigonal 3m, 32, $\bar{3}m$ 8 constants

M_{11}	M_{12}	M_{13}	M_{14}	0	0
M_{12}	M_{11}	M_{13}	$-M_{14}$	0	0
M_{31}	M_{31}	M_{33}	0	0	0
M_{41}	$-M_{41}$	0	M_{44}	0	0
0	0	0	0	M_{44}	M_{14}
0	0	0	0	$2M_{41}$	$M_{11} - M_{12}$

c constants the same except $c_{ab} = c_{ba}$ except c_{56}. 6 constants.
K constants the same as the c constants except $K_{44} = K_{23}$. 5 constants.

Group VI Tetragonal 4, $\bar{4}$, 4/m 10 constants

M_{11}	M_{12}	M_{13}	0	0	M_{16}
M_{12}	M_{11}	M_{13}	0	0	$-M_{16}$
M_{31}	M_{31}	M_{33}	0	0	0

c constants the same except $c_{13} = c_{31}$, $c_{16} = c_{61}$, $c_{45} = 0$. 7 constants.
K constants the same except $K_{44} = K_{23}$, $K_{66} = K_{12}$. 5 constants.

Group VII
Tetragonal
$4mm$, $\bar{4}2m$, 422, $4/mmm$
7 constants

$$\begin{bmatrix} M_{11} & M_{12} & M_{13} & 0 & 0 & 0 \\ M_{12} & M_{11} & M_{13} & 0 & 0 & 0 \\ M_{31} & M_{31} & M_{33} & 0 & 0 & 0 \\ 0 & 0 & 0 & M_{44} & 0 & 0 \\ 0 & 0 & 0 & 0 & M_{44} & 0 \\ 0 & 0 & 0 & 0 & 0 & M_{66} \end{bmatrix}$$

c constants the same except $c_{13} = c_{31}$.
6 constants.
K constants the same except $K_{44} = K_{23}$, $K_{66} = K_{12}$.
4 constants.

Group VIII
Hexagonal
$\bar{6}$, 6, $6/m$
8 constants

$$\begin{bmatrix} M_{11} & M_{12} & M_{13} & 0 & 0 & 2M_{61} \\ M_{12} & M_{11} & M_{13} & 0 & 0 & -2M_{61} \\ M_{31} & M_{31} & M_{33} & 0 & 0 & 0 \\ 0 & 0 & 0 & M_{44} & M_{45} & 0 \\ 0 & 0 & 0 & -M_{45} & M_{44} & 0 \\ M_{61} & -M_{61} & 0 & 0 & 0 & M_{11}-M_{12} \end{bmatrix}$$

c constants the same except $c_{13} = c_{31}$, $c_{61} = 0$, $c_{45} = 0$. 5 constants.
K constants the same as c constants except $K_{44} = K_{23}$.
4 constants.

Group IX
Hexagonal
$\bar{6}m2$, 622, $6mm$, $6/mmm$
6 constants

$$\begin{bmatrix} M_{11} & M_{12} & M_{13} & 0 & 0 & 0 \\ M_{12} & M_{11} & M_{13} & 0 & 0 & 0 \\ M_{31} & M_{31} & M_{33} & 0 & 0 & 0 \\ 0 & 0 & 0 & M_{44} & 0 & 0 \\ 0 & 0 & 0 & 0 & M_{44} & 0 \\ 0 & 0 & 0 & 0 & 0 & M_{11}-M_{12} \end{bmatrix}$$

c constants the same except $c_{13} = c_{31}$.
5 constants.
K constants the same as c constants, except $K_{44} = K_{23}$.
4 constants.

Group X
Cubic
23, $m3$
4 constants

$$\begin{bmatrix} M_{11} & M_{12} & M_{12} & 0 & 0 & 0 \\ M_{13} & M_{11} & M_{12} & 0 & 0 & 0 \\ M_{12} & M_{13} & M_{11} & 0 & 0 & 0 \\ 0 & 0 & 0 & M_{44} & 0 & 0 \\ 0 & 0 & 0 & 0 & M_{44} & 0 \\ 0 & 0 & 0 & 0 & 0 & M_{44} \end{bmatrix}$$

c constants the same except $c_{12} = c_{13}$.
3 constants.
K constants the same as c constants except $K_{44} = K_{12}$.
2 constants.

Group XI
Cubic
$\bar{4}3m$, 432, $m3m$
3 constants

$$\begin{bmatrix} M_{11} & M_{12} & M_{12} & 0 & 0 & 0 \\ M_{12} & M_{11} & M_{12} & 0 & 0 & 0 \\ M_{12} & M_{12} & M_{11} & 0 & 0 & 0 \\ 0 & 0 & 0 & M_{44} & 0 & 0 \\ 0 & 0 & 0 & 0 & M_{44} & 0 \\ 0 & 0 & 0 & 0 & 0 & M_{44} \end{bmatrix}$$

c constants the same. 3 constants.
K constants the same as c constants except $K_{44} = K_{12}$.
2 constants.

Group XII
Isotropic
2 constants

$$\begin{bmatrix} M_{11} & M_{12} & M_{12} & 0 & 0 & 0 \\ M_{12} & M_{11} & M_{12} & 0 & 0 & 0 \\ M_{12} & M_{12} & M_{11} & 0 & 0 & 0 \\ 0 & 0 & 0 & M_{11}-M_{12} & 0 & 0 \\ 0 & 0 & 0 & 0 & M_{11}-M_{12} & 0 \\ 0 & 0 & 0 & 0 & 0 & M_{11}-M_{12} \end{bmatrix}$$

c and K constants the same. 2 constants.

when all the indices i, j, k, and l are interchangeable. Such a case occurs when the elastic moduli satisfy the Cauchy relationship. This is denoted by type K symmetry in Table B.4.

(B.14), which follows, gives a glossary of the terms appearing in the polar tensors from rank 0 to 4.

Symbol	Meaning	Symbol	Meaning
δQ	Increment of heat	r_{mno}	Electro-optic constants
$\delta \Theta$	Increment of temperature	R_{kl}	Hall-effect constants
$\delta \sigma$	Increment of entropy	s_{ijkl}	Compliance constants
B_i	Magnetic flux density	S_{ij}	Strain components
c_{ijkl}	Elastic stiffness constants	Θ	Absolute temperature
$C^{S,D}$	Specific heat	T_{kl}	Stress components
D_n	Electric displacements	x_i	Length variable
D_o	Electric displacement at optical frequencies	z_{mno}	Electro-optic constants
d_{nij}	Piezoelectric constants	α_{ij}	Temperature-expansion coefficients
e	Electronic charge	α_{ijkl}	Magnetoresistive constants
e_{mkl}	Piezoelectric constants	β_{ij}	Dielectric or magnetic impermeabilities
E_m	Electric fields		
f_{mnop}	Quadratic electro-optic constants	ε_{ij}	Dielectric constants
g_{nij}	Piezoelectric constants	ε_{ijk}	Rotation tensor (Chapter 2)
h_i	Flow of heat per unit area	λ_{ij}	Temperature coefficients of stress at constant volume
h_{nkl}	Piezoelectric constants		
H_i	Magnetic fields	$\bar{\mu}$	Electrochemical potential
I_i	Electric current densities	μ_{ij}	Magnetic permeability constants
k_{ij}	Thermal conductivities		
l_{mnop}	Cotton-Mouton constants	π_{ijkl}	Piezoresistive constants
m_{ijmn}	Piezo-optic constants	Π_{ik}	Peltier thermoelectric coefficients
M_{ijkl}	Magnetostrictive constants		
p_n , i_n	Pyroelectric or pyromagnetic constants	ρ	Optical or Faraday rotation
		ρ_{ij}	Electrical resistivity constants
P_i	Polarization	σ_{ij}	Electrical conductivity constants
q_n , j_n	Pyroelectric or pyromagnetic constants		
q_{ijkl}	Electrostrictive constants	$\sum_{ik}$	Thermoelectric coefficients (Thomson)
r_{ij}	Thermal resistive constants		

$$(B.14)$$

B.2 Tables of Axial Tensors

B.21 Pseudo Scalars

Since the rotation ρ for an optically active crystal changes sign when the direction of light propagation is reversed, then ρ is a pseudo scalar having a value of $+1$ in a right-handed system of axes and a value -1 in a left-handed system.

B.22   First-Rank Axial Tensors

A first-rank axial tensor is the pyromagnetic tensor of Eq. (4.59). It is the ratio of an axial vector to a scalar quantity. The components of such a tensor are given by Table B.5. Another example is the magnetic flux due to a hydrostatic pressure p. These two equations are

$$B_n = i_n{}^T \delta\Theta; \qquad B_n = d^{\Theta}_{nkl}T_{kl} = (d_{n1} + d_{n2} + d_{n3})p \qquad (B.15)$$

All the rest of the crystal systems give a zero result.

Table B.5

COMPONENTS OF AN AXIAL RANK-ONE TENSOR
FOR VARIOUS CRYSTAL SYSTEMS

$1, \bar{1},$	$|i_1, i_2, i_3|;$ $2, \bar{2} = m, 2/m,$	$|0, i_2, 0|$
	$3, \bar{3}, \bar{4}, 4, 4/m, 6, \bar{6}, 6/m;$	$|0, 0, i_3|$

B.23   Second-Rank Axial Tensors

The only example for a second-rank axial tensor found is the gyration tensor, g_{ij}, which, as shown by Chapter 8, determines the optical activity of a crystal. As shown by Nye[1] this tensor has the components of Table B.6. This tensor is symmetric. All systems with a center of symmetry produce a zero result, so that for an isotropic material this tensor is zero.

B.24   Third-Rank Axial Tensors

The components of the piezomagnetic effect were completely worked out by Voigt.[2] Although no piezomagnetic constant has been definitely shown to be greater than zero, the tensor is of theoretical interest, and the components are given in Table B.7.

In these tables the three-index tensor terms of Eq. (4.59) have been replaced by the two-index symbols. However, account has not been taken of the difference between the engineering shear strains and the tensor strains. If this is done any longitudinal component that appears in a position to generate a shear has to be doubled. For symmetries 32, 3m, and $\bar{3}m$, for example, the constant d_{26} has to be written $-2d_{11}$. Similarly, for the symmetry 3, $\bar{3}$ the constants d_{16} and d_{26} have to be written as $-2d_{22}$ and $-2d_{11}$ respectively. All other symmetries remain unchanged.

[1] See Nye [1].
[2] See Voigt [2].

Table B.6

COMPONENTS OF SECOND-RANK AXIAL TENSOR FOR VARIOUS CRYSTAL SYSTEMS

Triclinic

$$1; \quad \begin{vmatrix} g_{11} & g_{12} & g_{13} \\ g_{12} & g_{22} & g_{23} \\ g_{13} & g_{23} & g_{33} \end{vmatrix}; \quad \bar{1} = 0$$

Monoclinic

$$2 \parallel x_2 \quad \begin{vmatrix} g_{11} & 0 & g_{13} \\ 0 & g_{22} & 0 \\ g_{13} & 0 & g_{33} \end{vmatrix}; \quad m \perp x_2; \quad \begin{vmatrix} 0 & g_{12} & 0 \\ g_{12} & 0 & g_{23} \\ 0 & g_{23} & 0 \end{vmatrix}; \quad 2/m = 0$$

Orthorhombic

$$222; \quad \begin{vmatrix} g_{11} & 0 & 0 \\ 0 & g_{22} & 0 \\ 0 & 0 & g_{33} \end{vmatrix}; \quad mm2; \quad \begin{vmatrix} 0 & g_{12} & 0 \\ g_{12} & 0 & 0 \\ 0 & 0 & 0 \end{vmatrix}; \quad mmm = 0$$

Tetragonal

$$4, 422; \quad \begin{vmatrix} g_{11} & 0 & 0 \\ 0 & g_{11} & 0 \\ 0 & 0 & g_{33} \end{vmatrix}; \quad \bar{4}; \quad \begin{vmatrix} g_{11} & g_{12} & 0 \\ g_{12} & -g_{11} & 0 \\ 0 & 0 & 0 \end{vmatrix}; \quad \bar{4}2m, 2 \parallel x_1; \quad \begin{vmatrix} 0 & g_{12} & 0 \\ g_{12} & 0 & 0 \\ 0 & 0 & 0 \end{vmatrix}$$

$$4/m; \quad 4mm; \quad 4/mmm = 0$$

Trigonal

$$3, 32; \quad \begin{vmatrix} g_{11} & 0 & 0 \\ 0 & g_{11} & 0 \\ 0 & 0 & g_{33} \end{vmatrix}; \quad 3m, \bar{3}, \bar{3}m = 0$$

Hexagonal

$$6, 622; \quad \begin{vmatrix} g_{11} & 0 & 0 \\ 0 & g_{11} & 0 \\ 0 & 0 & g_{33} \end{vmatrix}; \quad 6mm, \bar{6}, \bar{6}m2, 6/m, 6/mmm = 0$$

Cubic

$$432, 23 \quad \begin{vmatrix} g_{11} & 0 & 0 \\ 0 & g_{11} & 0 \\ 0 & 0 & g_{11} \end{vmatrix}; \quad m3, \bar{4}3m, m3m = 0$$

Table B.7

Triclinic

$$1, \bar{1}; \quad \begin{vmatrix} d_{11} & d_{12} & d_{13} & d_{14} & d_{15} & d_{16} \\ d_{21} & d_{22} & d_{23} & d_{24} & d_{25} & d_{26} \\ d_{31} & d_{32} & d_{33} & d_{34} & d_{35} & d_{36} \end{vmatrix}$$

Monoclinic

2, $\bar{2}$, 2/m; twofold axis parallel to x_2

$$\begin{vmatrix} 0 & 0 & 0 & d_{14} & 0 & d_{16} \\ d_{21} & d_{22} & d_{23} & 0 & d_{25} & 0 \\ 0 & 0 & 0 & d_{34} & 0 & d_{36} \end{vmatrix}$$

Orthorhombic

$$222,\ mm2,\ mmm; \quad \begin{vmatrix} 0 & 0 & 0 & d_{14} & 0 & 0 \\ 0 & 0 & 0 & 0 & d_{25} & 0 \\ 0 & 0 & 0 & 0 & 0 & d_{36} \end{vmatrix}$$

Tetragonal

$\bar{4}2m$, 422, 4mm, 4/mmm;

$$\begin{vmatrix} 0 & 0 & 0 & d_{14} & 0 & 0 \\ 0 & 0 & 0 & 0 & -d_{14} & 0 \\ 0 & 0 & 0 & 0 & 0 & 0 \end{vmatrix} \ ; \quad \bar{4},\ 4,\ 4/m; \quad \begin{vmatrix} 0 & 0 & 0 & d_{14} & d_{15} & 0 \\ 0 & 0 & 0 & d_{15} & -d_{14} & 0 \\ d_{31} & d_{31} & d_{33} & 0 & 0 & 0 \end{vmatrix}$$

Trigonal

32, 3m, $\bar{3}m$;

$$\begin{vmatrix} d_{11} & -d_{11} & 0 & d_{14} & 0 & 0 \\ 0 & 0 & 0 & 0 & -d_{14} & -d_{11} \\ 0 & 0 & 0 & 0 & 0 & 0 \end{vmatrix} \ ; \quad \begin{matrix} 3, \\ \bar{3}; \end{matrix} \quad \begin{vmatrix} d_{11} & -d_{11} & 0 & d_{14} & -d_{15} & -d_{22} \\ -d_{22} & d_{22} & 0 & d_{15} & -d_{14} & -d_{11} \\ d_{31} & d_{31} & d_{33} & 0 & 0 & 0 \end{vmatrix}$$

Hexagonal

622, $\bar{6}m2$, 6/mmm, 6mm;

$$\begin{vmatrix} 0 & 0 & 0 & d_{14} & 0 & 0 \\ 0 & 0 & 0 & 0 & -d_{14} & 0 \\ 0 & 0 & 0 & 0 & 0 & 0 \end{vmatrix} \ ; \quad 6/m,\ 6,\ \bar{6}; \quad \begin{vmatrix} 0 & 0 & 0 & d_{14} & d_{15} & 0 \\ 0 & 0 & 0 & d_{15} & -d_{14} & 0 \\ d_{31} & d_{31} & d_{33} & 0 & 0 & 0 \end{vmatrix}$$

Cubic

$$\bar{4}3m,\ 23; \quad \begin{vmatrix} 0 & 0 & 0 & d_{14} & 0 & 0 \\ 0 & 0 & 0 & 0 & d_{14} & 0 \\ 0 & 0 & 0 & 0 & 0 & d_{14} \end{vmatrix} \ ; \quad 432,\ m3,\ m3m = 0$$

Table B.8

THIRD-ORDER ELASTIC COEFFICIENTS OF CRYSTALS

1 1̄	2 2/m m	222 mm2 mmm	3 3̄	3m 3̄m 32	4 4/m 4̄	4mm 4̄2m 422 4/mmm	23 m3	4̄3m 432 m3m	6 6̄ 6/m	622 6/mmm 6̄m2 6mm
111	111	111	111	111	111	111	111	111	111	111
112	112	112	112	112	112	112	112	112	112	112
113	113	123	113	112	113	113	113	112	113	113
114	0	0	114	114	0	0	0	0	0	0
115	115	0	115	115	116	0	0	0	0	0
116	0	0	116	115	0	0	0	0	116	0
122	122	122	113	112	112	112	113	112	$111 - 222 + 112$	$111 - 222 + 112$
123	123	123	123	123	123	123	123	123	123	123
124	0	0	124	124	0	0	0	0	0	0
125	125	0	125	124	0	0	0	0	0	0
126	0	0	126	126	0	0	0	0	-116	0
133	133	133	112	112	133	133	112	112	133	133
134	0	0	125	124	0	0	0	0	0	0
135	135	0	126	126	0	0	0	0	0	0
136	0	0	124	124	136	0	0	0	0	0
144	144	144	144	144	144	144	144	144	144	144
145	0	0	145	145	145	0	0	0	145	0
146	146	0	146	145	0	0	0	0	0	0
155	155	155	155	155	155	155	155	155	155	155
156	0	0	156	156	0	0	0	0	0	0
166	166	166	166	155	166	166	166	155	$(\tfrac{3}{4}\cdot 222 - \tfrac{1}{2}\cdot 111 - \tfrac{1}{4}\cdot 112)$	$(\tfrac{3}{4}\cdot 222 - \tfrac{1}{2}\cdot 111 - \tfrac{1}{4}\cdot 112)$
222	222	222	111	111	111	111	111	111	222	222
223	223	223	112	112	113	113	112	112	113	113
224	0	0	116	115	0	0	0	0	0	0

225	225	225	0	114	114	0	0	0	0
226	0	0	0	115	115	−116	0	116	0
233	233	233	233	113	112	133	133	133	133
234	0	0	0	126	126	0	0	0	0
235	235	235	0	124	124	−136	0	0	0
236	0	244	244	125	124	155	155	155	155
244	244	244	0	166	155	155	166	−145	0
245	0	0	0	146	145	−145	0	0	0
246	246	246	255	156	156	0	0	0	0
255	255	255	0	144	144	144	144	$(\tfrac{1}{2}\cdot 111 - \tfrac{1}{4}\cdot 222 - \tfrac{1}{4}\cdot 112)$	$(\tfrac{1}{2}\cdot 111 - \tfrac{1}{4}\cdot 222 - \tfrac{1}{4}\cdot 112)$
256	0	0	266	145	145	166	155	155	144
266	266	266	333	111	111	333	333	333	333
333	333	333	0	115	115	0	0	0	0
334	0	0	0	116	115	0	0	0	0
335	335	335	335	114	115	0	0	0	0
336	0	0	0	114	115	0	0	0	0
344	344	344	344	155	155	344	155	344	344
345	0	0	0	156	155	0	0	0	0
346	346	346	355	145	145	344	155	344	344
355	355	355	0	155	166	344	155	344	344
356	0	0	366	145	145	0	0	$\tfrac{1}{2}(113 - 123)$	$\tfrac{1}{2}(113 - 123)$
366	366	366	0	144	144	366	144	0	344
444	0	0	0	444	444	0	0	0	0
445	445	445	445	444	444	445	445	145	0
446	0	0	0	445	446	0	0	0	0
455	455	455	355	445	446	0	0	0	0
456	456	456	456	456	456	456	456	$\tfrac{1}{2}(155 - 144)$	$\tfrac{1}{2}(155 - 144)$
466	0	0	0	445	445	0	0	0	0
555	555	555	0	444	444	−446	0	0	0
556	0	0	0	445	444	0	0	−145	0
566	566	566	0	446	445	0	0	0	0
666	0	0	0	444	444	0	0	−116	0

B.3 Third-Order Elastic Moduli

On account of the interest in determining the nonlinearities of the ordinary elastic moduli, one sixth-rank polar tensor is included. This expresses the third-order elastic moduli that account for the anelastic effects in wave propagation. The energy storage due to these moduli can be expressed in the form

$$C_{ijklmn}\eta_{ij}\eta_{kl}\eta_{mn} \tag{B.16}$$

where η_{ij}, and so on, are the finite strain values discussed in Eq. (3.75). These constants are usually represented by three-index symbols with the indices running from 1 to 6 in the usual contraction. These constants were first determined by Fumi.[3] Table B.8 shows the three-index symbols, and the relations between various constants for the different symmetry classes shown by the Herman-Mauguin symbols at the top of the table. In Eq. (B.16), i and j, k and l, and m and n can be interchanged. As with energy stored by strains involving second-order elastic constants (the normal constants) i, j can be exchanged with k, l and m, n while k, l can be exchanged with m, n. However, the individual symbols cannot be interchanged. In general, there are 56 third-order moduli. For a cubic crystal of the class $m3m$ (silicon and germanium) there are six third-order moduli, while for the trigonal crystal 32 (class of quartz) there are 14 third-order moduli. Methods for measuring these constants have been discussed by Thurston and McSkimin.[4]

REFERENCES APPENDIX B

1. J. F. Nye, "Physical Properties of Crystals," Chapter XIV. Oxford Univ. Press (Clarendon), London and New York, 1957.
2. W. Voigt, "Lehrbuch der Kristallphysik," pp. 940, 941. B. Tuebner, 1928.
3. F. G. Fumi, *Phys. Rev.* **83**, 1274 (1951); *Ibid.* **86**, 561 (1952).
4. R. N. Thurston *in* "Physical Acoustics" (Warren P. Mason, ed.), Vol. IA, Chapter I. Academic Press, New York, 1964.
5. H. J. McSkimin *in* "Physical Acoustics" (Warren P. Mason, ed.), Vol. IA, Chapter IV. Academic Press, New York, 1964.

[3] See Fumi [3].
[4] See Thurston [4] and McSkimin [5].

Author Index

Numbers in parentheses are reference numbers and indicate that an author's work is referred to although his name is not cited in the text. Numbers in italic show the page on which the complete reference is listed.

A

ABELES, B., 278, *299*
ALERS, G. A., 53(8), *78*
ALTENKIRCH, E., 280, 282, *299*
ANDERSON, L. K., 158, 176, *187*
ANDERSON, O. L., 67, *78*
APLET, L. J., 190, 197, *203*
ARIT, G., 262, *266*
AULD, B. A., 53(9), *78*

B

BALLATO, A. D., 54(16), *78*
BATEMAN, T. B., 59(17), *78*, 223, *240*,
BECKMANN, R., 54(16), *78*
BERGER, S. B., 195(8), *203*
BERLINCOURT, D. A., 96(10), 100(10, 12), *107*
BILLINGS, B. H., 158, 168, *187*
BLATTNER, D., 158(6), 176(6), *187*
BOLT, R. H., 154, *156*
BONNER, W. A., 195(8), *203*
BORN, M., 163, *187*
BOZORTH, R. M., 46, *78*
BRIDGEMAN, P. W., 221, *240*, 284, 292, *299*
BUTTON, M. J., *299*

C

CADY, W. G., 88, 93, *106*, 125, *156*
CALLEN, H. B., 272, 286, *299*
CARPENTER, R. O. B., 158, 169, *187*
CARSON, J. W., 190, 197, *203*
CASIMIR, H. B. G., 208, 209, 210, *220*, 246, *266*

CHANDRASEKHAR, S., 315, 322, *324*
CMOLEK, C., 100(12), *107*
COHEN, R. W., 278, *299*
COKER, E. G., 3, *8*, 158, 180, 183, *187*
COMSTOCK, R. L., 53, *78*
CONDON, E. U., 309, 323, *324*
COTTON, A., 199, *203*
CURRAN, D. R., 96(10), 100(10), *107*

D

DEAN, M., III, 3, *8*
DE GROOT, S. R., 209, 210, *220*, 269, 272, 286, *299*
DEVONSHIRE, A. F., 96, *107*
DIDOMENICO, M., JR., 158, 176, *187*
DOMENICALI, C. A., 272, 282, *299*

F

FARADAY, M., 194, *203*
FELDMAN, W. L., 226(5), *240*
FILON, L. N. G., 3, *8*, 158, 180, 183, *187*
FISCHLER, S., *299*
FOX, A. G., 197(11, 12), 198(11, 12), 199(11, 12), 200(11), *203*
FROCHT, M., 158, 180, 183, *187*
FUMI, F. G., 117, *121*, *340*

G

GEESMAN, L. B., 1(2), *8*
GEUSIC, J. E., 177(15), 178(15), *187*, 190, 197, *203*
GRAHAM, R. A., 1, *8*

341

Subject Index

A

Acceleration, angular, 53
Admittance of free crystal, 139
ADP (ammonium dihydrogen phosphate) $(NH_4H_2PO_4)$, 130
  as electro-optic crystal, 172
  elasto-electric constants of, 132
American Institute of Physics Handbook, 126
Ampere's law, 49, 76
Ampere-turns per meter (definition of units of H), 43
Analyzer (optical), 164
Antimony, 267
Antiresonance frequency, 140
Axes
  crystal, 120
  crystallographic, 38
  fourfold, 113
  fourfold inversion, 113
  handedness of, 24
  onefold, 113
  optic, 188
    directions, 157
    location of, 160
  principal, 19
  rectangular, 11
  sixfold, 113
  sixfold inversion, 113
  threefold, 113
  threefold inversion, 113
  twofold, 113
  twofold inversion, 113

Axes (*cont.*)
  transmission of light
    along general axis, 315
    along optic axis, 312
    at angle to optic axis, 201

B

Bandwidth conditions, 151
Barium titanate, 132
  elasto-electric constants of single crystal, at 25° C, 136
  first ferroelectric phase, 94
  orthorhombic form, 94
  single crystal, 94
  trigonal form, 94
Bequerel's formula for Verdet constant, 323
Birefringence, 2, 158
  along any direction in crystal, 170
  derivation of superposition formula for optical rotation and, 317
  due to difference in principal stresses, 180
  due to electric field, 2
  principle of superposition for optical activity and, 194
Bismuth, 267
Bismuth telluride, 267
Boltzmann constant, 70
  statistics, 224
Bravais-Miller indices, 111, 120
"Brewster," 182
  definition of, 173
Bridgman heat, 284

344